# NEW RESEARCH ON SUPERCONDUCTIVITY AND MAGNETISM

# New Research on Superconductivity and Magnetism

Lannie K. Tran
Editor

**Nova Science Publishers, Inc.**
*New York*

**LIBRARY OF CONGRESS CATALOGING-IN-PUBLICATION DATA**
Available upon request

ISBN 13 978-1-60021-540-7
ISBN 10 1-60021-540-8

*Published by Nova Science Publishers, Inc. ✦ New York*

# CONTENTS

# PREFACE

Superconductivity is the ability of certain materials to conduct electrical current with no resistance and extremely low losses. High temperature superconductors, such as La2-xSrxCuOx (Tc=40K) and YBa2Cu3O7-x (Tc=90K), were discovered in 1987 and have been actively studied since. In spite of an intense, world-wide, research effort during this time, a complete understanding of the copper oxide (cuprate) materials is still lacking. Many fundamental questions are unanswered, particularly the mechanism by which high-Tc superconductivity occurs. More broadly, the cuprates are in a class of solids with strong electron-electron interactions. An understanding of such "strongly correlated" solids is perhaps the major unsolved problem of condensed matter physics with over ten thousand researchers working on this topic. High-Tc superconductors also have significant potential for applications in technologies ranging from electric power generation and transmission to digital electronics. This ability to carry large amounts of current can be applied to electric power devices such as motors and generators, and to electricity transmission in power lines. For example, superconductors can carry as much as 100 times the amount of electricity of ordinary copper or aluminum wires of the same size. Many universities, research institutes and companies are working to develop high-Tc superconductivity applications and considerable progress has been made. This new volume brings together new leading-edge research in the field.

As outlined in Chapter 1, the magneto-superconducting $RuR_{2-x}Ce_xSr_2Cu_2O_{10-\delta}$ (R=Eu and Gd, Ru-1222) materials are the first Cu-O based system in which bulk superconductivity (SC) in the CuO2 planes (below $T_C$ = 32-50 K) and *weak*-ferromagnetism (W-FM) in the Ru sublattice, coexists. The hole doping of the Cu-O planes, which results in metallic behavior and SC, can be optimized with appropriate variation of the R/Ce ratio and/or the oxygen concentration. SC survives because the Ru magnetic moments probably align in the basal planes, which are practically decoupled from the $CuO_2$ planes.

Two magnetic transitions are observed in Ru-1222, at $T_M$~160 K and $T_{2M}$ ~80 K. The higher ordering temperature at $T_M$, does not change with x and one of the disputed question is its *origin*, namely, whether it is intrinsic or not. Below $T_{2M}$ the Ru moments are W-FM ordered and wide ferromagnetic-like hysteresis loops are obtained. They become narrow and disappear at ~ 60-70 K. Above $T_{M2}$ (around 120 K), (i) a small peak is observed in both dc and ac susceptibility curves and (ii) narrow hysteresis loops reappear with a peak in the coercive fields around 120 K. The accumulated results for the higher magnetic phase are interpreted by two scenarios (a) The two observed phenomena are related to intrinsic reduction of pure $Ru^{5+}$ ions (in x=1) to $Ru^{4+}$ caused by the variation of the $Ce^{4+}$ content. This may form $Ru^{4+}$ rich clusters, in which the $Ru^{4+}$-$Ru^{4+}$ exchange interactions are stronger than the $Ru^{5+}$-$Ru^{5+}$ interactions. (b) It is attributed to the presence of nano-particles of a foreign minor extra $Ru^{4+}$ magnetic phase such as Sr-Cu-Ru-$O_3$. The second scenario is supported by Mössbauer spectroscopy of $^{57}Fe$ doped materials.

Chapter 2 discusses analytical expressions of magnetization $M$ and critical current density $J_c$ by an extended critical state model. The formula of *Kimishima model* was one of the universal functions interconnecting the Bean and Anderson-Kim models. The second formula of *Exponential Kimishima model* was derived by the extension of *Kimishima model*, considering the exponential decay of $J_c$ near the irreversible field $H_{irr}$. These theoretical magnetization curves were well fitted to the experimental ones. Estimated magnetic parameters were important to explain the temperature dependences of $J_c$ and the pinning properties of high $T_c$ superconductors.

The past few years have witnessed an increasing interest in studying the static and dynamic phases of heterogeneous ferromagnetic-superconducting systems. Typically, a heterogeneous ferromagnetic-superconducting system consists of a superconductor placed in close proximity with a ferromagnet separated by a layer of insulator oxide. The layer of insulator oxide suppresses proximity effects and spin diffusion hence allowing for the coexistence of two competing order parameters, a feature that is special to these systems. The strong interaction between the ferromagnet and the superconductor gives raise to strong vortex pinning force which enhances the critical current of the superconductor. In this chapter, I will review the important theoretical developments in this area of research. The first part of Chapter 3 will focus on the static properties of these systems; therefore, the general formulation of Londons theory for these systems will be presented. Afterward, I will apply the general theory to study few specific systems with emphasis on the appearance of spontaneous vortex phase and the important role played by the geometry and the magnetization distribution of the ferromagnet on the static and dynamic properties of the superconductor. In the second part, I will concentrate on the dynamic of vortices and antivortices in these systems. Therefore, I will review the anisotropic transport properties of a ferromagnetic-superconducting bilayer then I will study the dynamical properties of a system consisting of a superconducting thin film pierced by a square array of ferromagnetic columnar defects.

Superconducting magnetic sensor SQUID is the most sensitive magnetic sensing element known that has high spatial resolution. Accordingly, scanning SQUID magnetic imaging is currently being developed for applications to NDE (nondestructive evaluation). Laboratory-scale scanning SQUID microscopes (SSM) are useful for academic research as well as for industrial applications that require high-resolution measurements. Chapter 4 reviews current knowledge of (a) temperature dependence of local magnetic distribution in high-Tc superconductors by SSM, and (b) high-resolution measurements using SSM on stress-induced magnetism in austenitic. For practical applications, a cost-effective NDE method that is highly accurate is desired. SQUID-NDE techniques are also reviewed here because they were recently proved capable of detecting damage such as fatigue, deformation, and cracks in materials, especially metals. Considering the demand in industry, conventional stationary SQUID-NDE methods are practically useless for inspecting large structures or material components. To answer this demand, an advanced SQUID-NDE system, called a *traveling SQUID-NDE*, was recently developed. This traveling SQUID-NDE can continuously "travel" over the surface of an object during testing without the need for magnetic shielding. And finally, this system is also reviewed here.

Chapter 5 reviews chemical and macroscopic physical properties of the bilayer sodium cobalt oxyhydrate (BLH) superconductor and its related compounds, focusing on the recent development of research. The exact composition of BLH is $Na_x(H3O)_zCoO2 \cdot y'H_2O$ and it has three adjustable compositional parameters: Na content, $x$, $H_3O$ content, $z$, and $H_2O$ content, $y'$. The former two parameters greatly affect superconductivity and related properties of the system. With adjusting them, a magnetically ordered phase appears in the immediate vicinity of the superconducting phases in a phase diagram, suggesting a certain magnetic origin for superconductivity. Regarding the band structure of BLH, hole pockets near the K points look crucially important relating to the mechanism of superconductivity. Furthermore, there is a possibility that two-gap superconductivity is realized in the present system corresponding to the hole pockets near the K points and a large cylindrical Fermi surface around the Γ point. At the end of the paper, the mother oxide of $\gamma$-$Na_xCoO_2$ is discussed. Its rich physical properties originate from strong electron-electron correlation in the $CoO_2$ layer and seem to relate to the unconventional superconductivity in BLH.

$REBa_2Cu_3O_y$ (RE123: RE=Y, Sm, Gd etc.) superconductive oxides have potential for many applications since they exhibit a high critical current density ($J_c$) under high magnetic fields at 77 K. In this paper, a directional solidification of RE123 (RE=Y and Sm) samples was carried out. The effect of fabrication conditions on the microstructure and transport $J_c$ properties of Y123 fibers were reported, and the continuous growth condition of Sm123 are also investigated. The transport $J_c$ property was measured by the standard d.c. 4 probe method. The continuous growth of the 0.5 mmϕ Y123 fiber was achieved in the samples grown at high pulling rates up to 10 mm/h while the usual Y123 rods of 2 mmϕ could grow continuously up to several mm/h. The size of Y211 particles trapped in Y123 was found to decrease with increasing pulling rate. The Sm123 can continuously grow at higher pulling rates than Y123 for both fibers and rods. The transport $J_c$ properties of Y123 fibers showed the maximum in the sample pulled at 6 mm/h; the best $I_c$ and corresponding $J_c$ values achieved were 134 A and $1\times10^5$ A/cm$^2$ at 77K self-field, respectively. This $J_c$ value is the best transport $J_c$ of the melt grown samples to date and is comparable to the best $J_c$ value of small Y123 bulks. This high $J_c$ value may attributed to the well-controlled microstructure and crystal orientation of the sample fabricated at high pulling rate. That is, the preferential crystal orientation as well as fine Y211 particles is thought to result this high transport $J_c$ of the fiber sample. On the other hand, the enlargement of continuous growth condition for the fiber samples could be fundamentally explained by the difference in effective nucleation rate, which was defined as the product of nucleation frequency and undercooled volume in the liquid. Because of the difference in the cross sectional area, the small nucleation volume of thin fiber leads the reduction of the nucleation rate, and may result in the increase of the continuous growth rate. This difference in nucleation rate is, however, found not to be enough to explain the destruction of continuous growth, and the additional factor such as compositional shift at the interface is needed. To think of a reason causing the compositional shift, a non-steady state growth model is proposed, and the effect of process parameters and growth conditions on the compositional shift is discussed in Chapter 6.

In: New Research on Superconductivity and Magnetism ISBN: 1-60021-540-8
Editor: Lannie K. Tran, pp. 1-33 

*Chapter 1*

# SUPERCONDUCTIVITY AND MAGNETISM IN $RuR_{2-x}Ce_xSr_2Cu_2O_{10-\Delta}$ (R=Eu AND Gd)

***I. Felner***
Racah Institute of Physics, The Hebrew University,
Jerusalem, 91904, Israel

## Abstract

The magneto-superconducting $RuR_{2-x}Ce_xSr_2Cu_2O_{10-\delta}$ (R=Eu and Gd, Ru-1222) materials are the first Cu-O based system in which bulk superconductivity (SC) in the CuO2 planes (below $T_C$ = 32-50 K) and *weak*-ferromagnetism (W-FM) in the Ru sublattice, coexists. The hole doping of the Cu-O planes, which results in metallic behavior and SC, can be optimized with appropriate variation of the R/Ce ratio and/or the oxygen concentration. SC survives because the Ru magnetic moments probably align in the basal planes, which are practically decoupled from the $CuO_2$ planes.

Two magnetic transitions are observed in Ru-1222, at $T_M$~160 K and $T_{2M}$ ~80 K. The higher ordering temperature at $T_M$, does not change with x and one of the disputed question is its *origin*, namely, whether it is intrinsic or not. Below $T_{2M}$ the Ru moments are W-FM ordered and wide ferromagnetic-like hysteresis loops are obtained. They become narrow and disappear at ~ 60-70 K. Above $T_{M2}$ (around 120 K), (i) a small peak is observed in both dc and ac susceptibility curves and (ii) narrow hysteresis loops reappear with a peak in the coercive fields around 120 K. The accumulated results for the higher magnetic phase are interpreted by two scenarios (a) The two observed phenomena are related to intrinsic reduction of pure $Ru^{5+}$ ions (in x=1) to $Ru^{4+}$ caused by the variation of the $Ce^{4+}$ content. This may form $Ru^{4+}$ rich clusters, in which the $Ru^{4+}$-$Ru^{4+}$ exchange interactions are stronger than the $Ru^{5+}$-$Ru^{5+}$ interactions. (b) It is attributed to the presence of nano-particles of a foreign minor extra $Ru^{4+}$ magnetic phase such as Sr-Cu-Ru-$O_3$. The second scenario is supported by Mössbauer spectroscopy of $^{57}Fe$ doped materials.

**Keywords:** Superconducting cuprates, Ferromagnetism, Coexistence

# Introduction

Superconductivity and magnetism are two different ordered states into which substances can condense at low temperatures and in general these states are inimical to one another. The antagonism nature between the two states is one of the fundamental problems of condensed matter physics and has been studied experimentally and theoretically for almost four decades. In conventional superconductors, local magnetic moments break up the spin singlet Cooper pairs and hence strongly suppress SC, an effect known as pair-breaking. Therefore, a level of magnetic impurity of only 1%, can result in a complete loss of SC. In a limited class of inter-metallic systems, SC occurs even though magnetic ions with a local moment occupy all of one specific crystallographic site, which is well isolated and de-coupled from the conduction path. The study of this class of magnetic-superconductors was initiated by the discovery of $RRh_4B_4$ and $RMo_6S_8$ compounds (R=rare-earth), and has been recently revitalized by the discovery of the $RNi_2B_2C$ system. In all three systems, both three dimensional SC and antiferromagnetic (AFM) order states coexist [1-2]. The onset of SC takes place at $T_C \sim 2$-16 K, while AFM order appears at lower temperatures (except for $DyNi_2B_2C$), thus, the ratio $T_N/T_C$ is ~ 0.1-0.5. Many of the high $T_C$ superconducting materials (HTSC) contain magnetic R and are AFM ordered at low temperatures, e.g. in $GdBa_2Cu_3O_7$ ($T_C$ = 92 K), and $T_N$ (Gd)=2.2 K. The R sublattice is electronically isolated from the Cu-O planes, and has no adverse effect upon the superconducting state.

Much attention has been focused on a phase resembling the $YBa_2Cu_3O_7$ materials, having the composition $MR_{2-x}Ce_x Sr_2Cu_2O_{10}$ (M-1222, **M**= Nb, Ru or Ta, R=Eu or Gd) [3]. The M ions reside in the Cu (1) site and the hole doping of the Cu-O planes, which results in metallic behavior and SC, can be optimized with appropriate variation of the R/Ce ratio [4] The Nb-1222 and Ta-1222 materials are SC only, with $T_C$~28-30 K. On the other hand, coexisting of weak-ferromagnetism (W-FM) and SC was discovered first in Ru-1222 materials [5-7], and more recently [8] in $RuSr_2GdCu_2O_8$ (Ru-1212). In both systems, the SC charge carriers originate from the $CuO_2$ planes and the W-FM state which is confined to the Ru layers, coexists and remains unchanged when SC sets in at $T_C$. The Ru-1222 materials display a magnetic transition at $T_M$= 135-170 K and SC occurs ($T_C$ =32-50 K) for Ce contents of x=0.4-0.8 ($T_M > T_c$) depending on oxygen concentration and sample preparation [9]. Specific heat studies show a sizeable typical jump at $T_C$ and the magnitude of the $\Delta C/T$ (0.08 mJ/gK$^2$) indicates clearly the presence of bulk SC [10]. The specific heat anomaly is independent of the applied magnetic field. Scanning tunneling spectroscopy (STM)[5], muon-spin rotation (μSR) [11], Raman [12] and magneto-optic experiments have demonstrated that both states coexist within the same crystalline grain. SC survives because the Ru moments probably align in the basal planes, which are practically de-coupled from the $CuO_2$ planes, so that there is no pair breaking. However, in contrast to the Ru-1212 system in which the AFM nature of the Ru sublattice has been determined by neutron diffraction studies [13-14], the published data up to now have not included any determination of the magnetic structure in Ru-1222.

Generally speaking, the dc magnetic features of the Ru-1222 system exhibit two magnetic transitions, at $T_{2M}$ ~ 80-90 K and at $T_M$ ~130-170 K .(i) At low applied magnetic fields (H<3 kOe), irreversibility in the zero-field-cooled (ZFC) and field-cooled (FC) curves is observed, which disappears at $T_{2M}$. For higher external fields (H>5 kOe), both ZFC and FC curves collapse to a single ferromagnetic-like (FM) behavior. Wide FM hysteresis loops are opened at low temperatures (the coercive field- $H_C$ ~450-500 Oe at 5 K) which become narrower as the temperature increases and practically disappear below $T_{2M}$. This FM-like or the W- FM state, as well as irreversibility phenomena, arise as a result of an antisymmetric exchange coupling of the Dzyaloshinsky-Moriya type [15] between neighboring Ru moments, induced by a local distortion that breaks the tetragonal symmetry of the $RuO_6$ octahedra. (ii) Above $T_{2M}$ a *second* small peak is observed around 120 K in both dc and ac susceptibility curves and narrow hysteresis loops (AFM-like) are observed. The $H_C(T)$ curves show a maximum around 120 K ($H_C$ ~150 Oe) and become zero at $T_M$ [16]. $T_M$ was obtained from the temperature dependence of the saturation moment ($M_{sat}$), and from Mössbauer studies (MS) on $^{57}Fe$ doped materials [5]. One of the disputed question in Ru-1222 materials, is the *origin* of the higher magnetic transition at $T_M$~160 K, whether it is intrinsic or not.

This paper covers the physical properties of the Ru-1222 system only and is organized as follows: In the next section, basic concepts of the crystal structure and of the Ru valence state are described. In sections II and III which are the central content of this paper, an assessment of the most reliable experimental results of the SC and magnetic states are described. Conclusions are summarized in Section IV.

# (I) Basic Concepts

## a Crystal Structure and X-rays Data

The tetragonal Ru-1222 structure (space group I4/mmm) evolves from the $RBa_2Cu_3O_7$ structure by inserting a fluorite type $R_{1.5}Ce_{0.5}O_2$ layer instead of the R layer in $RBa_2Cu_3O_7$, thus shifting alternate perovskite blocks by (a+b)/2 (Fig. 1). The magnetic Ru (as well as M) ions reside in the Cu (1) site. Powder X-ray diffraction (XRD) measurements of $RuEu_{1.5}Ce_{0.5}Sr_2Cu_2O_{10-\delta}$ confirmed the purity of the compounds (~97%) and indicate, within the instrumental accuracy, that all samples have the same lattice parameters a=3.846(1) Å and c=28.72(1) Å. Due to the similarity of the ionic radii of $Eu^{3+}$ (0.94 Å ) and $Ce^{4+}$(0.87Å ), the lattice parameters of Ru-1222 materials are independent of x (and δ) [9]. The diffraction scans of $RuEu_{1.5}Ce_{0.5}Sr_2Cu_2O_{10-\delta}$ at various temperatures indicate clearly that all peaks and their relative intensities remain unaltered down to 10 K (Fig. 2 (a)) [17]. The diffraction scans do not show any anomalies around $T_C$ and $T_M$. Both a and c lattice parameters are linear down to 100 K, and the typical flattening of the thermal expansion behavior is observed at low temperatures (Fig. 2 (b)). The detailed crystal structure and the atomic positions in Ru-1222 were studied by neutron diffraction [18] and XRD diffraction experiments [19-20], which show that the $RuO_6$ octahedra are rotated ~ 14° around the c-axis and that this rotation is essentially the same for x=1 and x=0.6, as well as for Ru-1212. There is no evidence for super-cell peaks in the Ru-1222 samples. Characteristic superlattice reflections corresponding to those of 2a x 2a were found by electron diffraction at 300 K. It is proposed that the supercell structure is a result of an ordering of the $RuO_6$ octahedral around the c axis, which leads to a possible different space group [21].

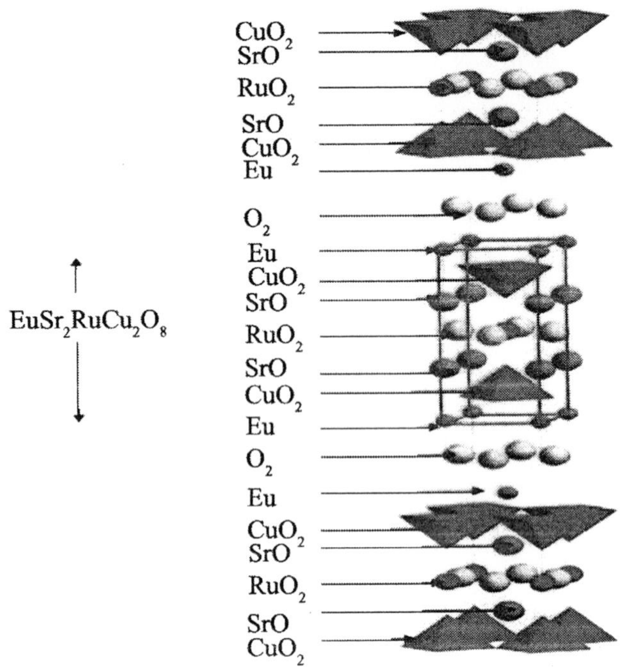


Fig. 1 The crystal structure of $RuEu_{1.5}Ce_{0.5}Sr_2Cu_2\ O_{10-\delta}$ The Ce ions are randomly distributed in the Eu sites. The solid lines mark the Ru-1212 structure.

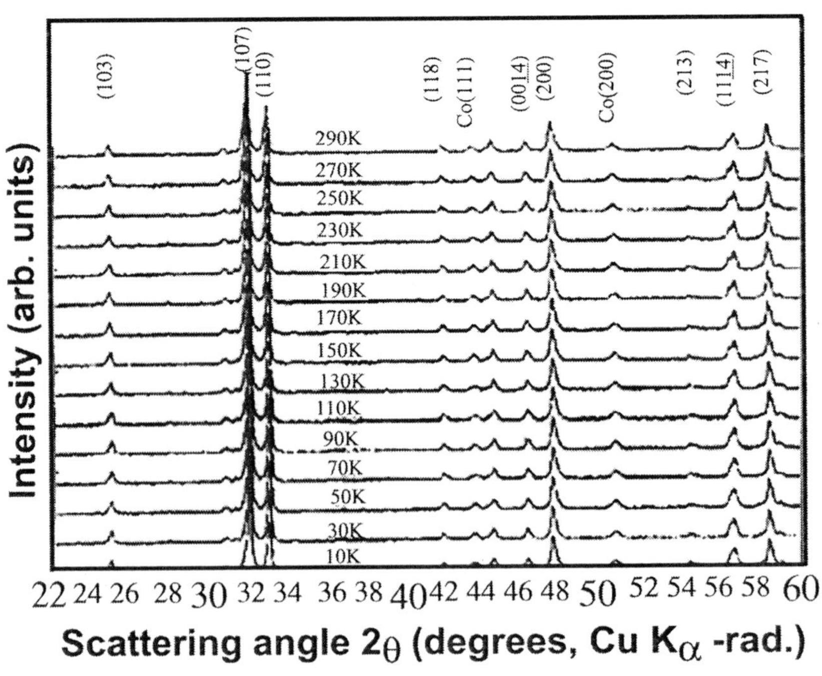


Fig. 2 XRD pattrens of Ru-1222 at various temperatures.

## b The *Pentavalent* State Ru

The valence of Eu, Ce and Ru ions in $RuEu_{1.5}Ce_{0.5}Sr_2Cu_2O_{10-\delta}$ have been studied by MS and x-ray-absorption spectroscopy (XAS) techniques [22]. MS performed on $^{151}Eu$ show a single narrow line with an isomer shift =0.69(2) mm/s and a quadrupole splitting of 1.84 mm/s, indicating that the Eu ions are trivalent with a nonmagnetic J=0 ground state. This is in agreement with XAS taken at $L_{III}$ edges of Eu and Ce which shows that Eu is trivalent and Ce is tetravalent. The local electronic structure in several Ru based compounds was studied by XAS at the K edge of Ru, and the results obtained at room temperature are shown in Fig. 3. Since the valence of $Gd^{3+}$, $Sr^{2+}$ and $O^{2-}$ are conclusive, a straightforward valence counting for $GdSr_2RuO_6$ and $SrRuO_3$ yields Ru, as $Ru^{+5}$ and $Ru^{+4}$ ions respectively. The similarity between the XAS spectra of Ru-1222 and $GdSr_2RuO_6$ indicates clearly, that in Ru-1222 the Ru ions are in a *pentavalent* state. Similar results were obtained by $L_{III}$-edge x-ray-absorption near-edge spectroscopy XANES spectra on $RuGd_{2-x}Ce_xSr_2Cu_2O_{10}$ which show that the average Ru valence is 4.95 irrespective of the Ce concentration [23]. This implies that there is no charge transfer to the RuO layer with increasing Ce concentration. A lower valence value for Ru 4.74 for x=0.5 was obtained in Ref. [24]. This value increase to 4.81 for sample annealed in 100 –atm $O_2$, indicating that the Ru valence is affected by the change of the oxygen content.

Fig. 4 shows a ZFC $^{99,101}Ru$-NMR spectra (at 4.2 K) of both $RuSr_2YCu_2O_8$ and Ru $Eu_{1.4}Ce_{0.6}Sr_2Cu_2O_{10}$ materials. For Ru-1212, two-subspectra are observed which indicates that there exist microscopically two inequivalent Ru sites in the $RuO_2$ plane. The broad low frequency region corresponds to $Ru^{4+}$, whereas the well assigned peaks in the high frequency range are attributed to the two isotopes magnetic internal hyperfine field of $Ru^{5+}$ ($H_{eff}$= 610 Oe) [25]. On the other hand, for the Ru-1222 sample, only the high frequency range is observed ($H_{eff}$ =625 Oe), indicating clearly the presence of $Ru^{5+}$ ions only [26]. This result is consistent with our XAS studies discussed above.

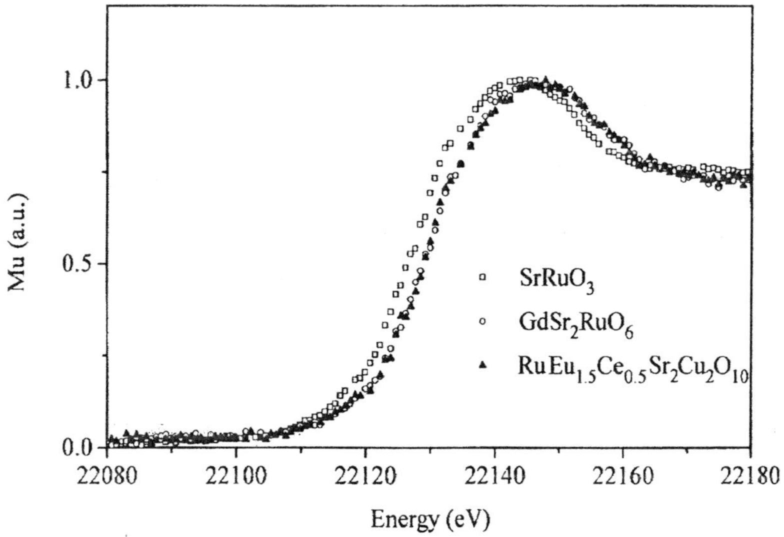


Fig. 3 XAS spectra at the K edge of Ru of Ru-1222 and reference compounds.

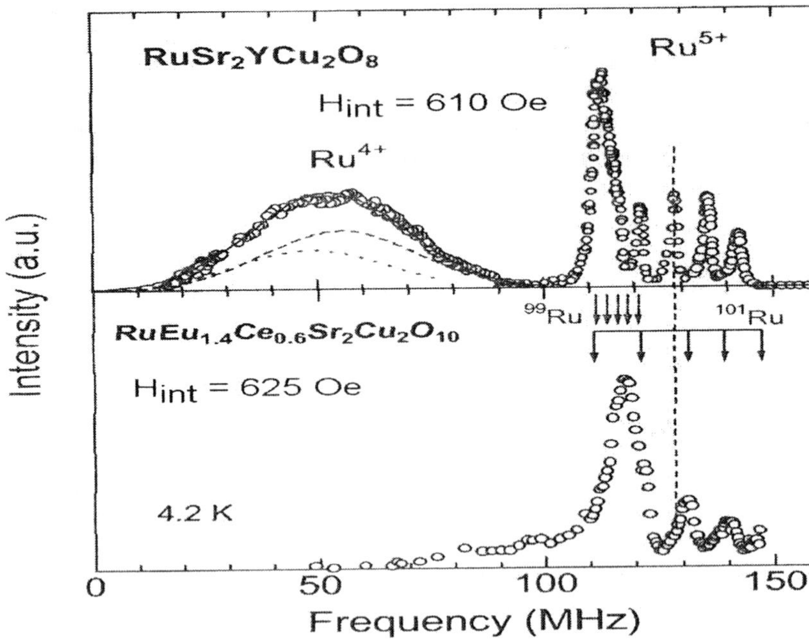


Fig. 4 NMR studies of Ru-1212 and Ru-1222 at 4.2 K. (taken from Ref. 26)

## (II)Superconductivity in $RuR_{2-x}Ce_xSr_2Cu_2O_{10-\delta}$ Results and Model

The existence of SC in Ru-1222 based on resistivity measurements was first discovered by Bauernfeind et. al. [4] and Kaibin et. al.[27]. The nature of Cooper pair formation in the Cu-O planes and SC is most likely to be similar to all other High $T_C$ superconducting (HTSC) materials in particular, $YBa_2Cu_3O_7$ which has a similar crystal structure as discussed above. Using the same terminology it is proposed that the Ru-1222 materials are typical under-doped materials.

The as prepared (asp) Ru-1222 materials were prepared by preheating at 1000° C for 24 hours the mixture of prescribed amounts of the oxides. The products were cooled and sintered at 1050° C for 72 hours in a slightly pressurized oxygen atmosphere and then furnace cooled to room temperature [5]. It should be noted, that the SC as well as the magnetic properties of the compounds, are extremely sensitive to the preparation conditions which are not completely reproducible. Therefore, various asp materials presented here may have different $T_C$ (or magnetic) values. All the series of materials reported below were prepared simultaneously under the same conditions Determination of the absolute oxygen content is difficult because $CeO_2$ is not completely reducible to a stoichiometric oxide when heated to high temperatures. Therefore it is assumed that re-heated the asp material for 24 hours at 800 C under various high oxygen pressures increases the oxygen content.

The absolute thermo-power S(T) of the asp $RuEu_{1.5}Ce_{0.5}Sr_2Cu_2O_{10-\delta}$ annealed under an oxygen pressure of 54 atm., ($T_C$ =45 K) is shown in Fig. 5. S(T) is positive over the entire temperature range indicating that the carriers are holes. This curve shows a break in the slope at $T_C$, followed by a drop to zero at 25 K. The relative broad transition suggests that the oxygen in the sample may be inhomogeneous. Another small peak is evident around the magnetic transition at $T_M$=150 K [28]. S(T) is quite linear with positive slope between $T_C$ and 100 K without any clear feature at $T_{M2}$. The estimated S value for Ru-1222, at 290 is ~ 40 μV/K. Adapting the empirical relation between S and the hole (or carrier) density *p* that was shown to be valid for most HTSC, [29 ] we obtain *p*=0.085 a value which is below the optimal doping value *p*=0.15-0.16. The oxygen concentration (δ) can be modified by annealing the asp sample under high oxygen pressures. As a result a systematic increase of *p* and in $T_C$ is achieved which leads to a reduction in S(T) [7].

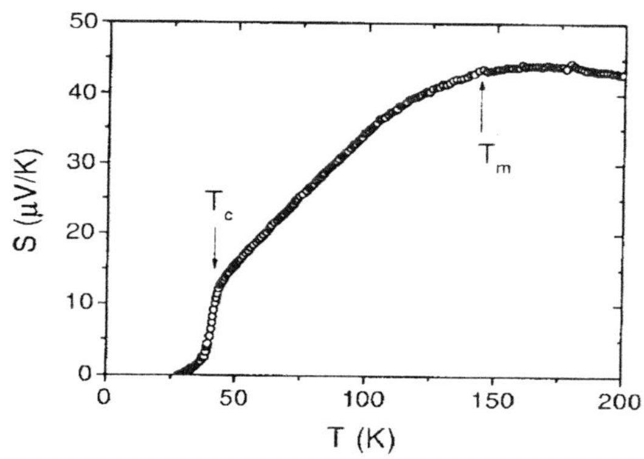


Fig.5 Thermo-power S versus temperature of oxygen annealed $RuEu_{1.5}Ce_{0.5}Sr_2Cu_2O_{10}$

STM [5]and μSR [11] studies have demonstrated that all materials are microscopically uniform with no evidence for spatial phase separation of SC and magnetic regions. Specific heat studies of Ru-1222 show a sizeable typical jump at $T_C$ and the magnitude of the ΔC/T (0.08 $mJ/gK^2$) indicates clearly the presence of *bulk* SC [10]. The specific heat anomaly is independent of the applied magnetic field. SC survives because the Ru moments probably align in the basal planes, which are practically decoupled from the $CuO_2$ planes, so that there is no pair breaking. That is, in Ru-1222 both bulk SC and the magnetic state coexist intrinsically on the microscopic scale.

## a The Effect of Ce and Oxygen Concentrations on $T_C$

The $T_C$ values of $RuR_{2-x}Ce_x$ $Sr_2Cu_2O_{10-\delta}$ system ranges from 32-45 K depending on (i) R (Eu or Gd, (ii) on the Ce concentration (x) and on the oxygen content (δ) [5,30]. Generally speaking, the $T_C$ values of Ru-1222 are (a) higher for R=Gd than for R=Eu (under the same conditions) and (b) are very similar to $T_C$ values obtained in the Ru-1212 system. All these values are much smaller than the optimal doped $RBa_2Cu_3O_7$ compounds, indicating again the typical under-doped nature.

The temperature dependence of the normalized real ac susceptibility susceptibility (χ) curves susceptibility (at $H_{dc}$=0) of $RuSr_2R_{2-x}Ce_xCu_2O_{10-\delta}$ are presented in Fig. 6. It is readily observed that the x=1 and 0.9 samples are not SC and that SC occurs for Ce contents of x=0.8-0.4. Fig 7. exhibits the onset of SC deduced from these ac plots which exhibit a bell shape behavior with a peak at 35 K for Ce=0.6 [9]. A similar bell shape behavior was observed in the non magnetic $NbEu_{2-x}Ce_xSr_2Cu_2O_{10-\delta}$ (x=0.4-1) system. This may suggest that $T_C$ is not being significantly affected by the magnetic state of the Ru-O layers discussed below (e.g. by possible magnetic pair-breaking). Annealing under various oxygen pressures exhibits a monotonic increase of $T_C$. The onset of the asp ($T_C$ = 32 K) sample is shifted to 38 K and up to 49 K for the same material which was annealed under 22 or 150 atm. respectively [5,30]. On the other hand, when hydrogen atoms are loaded, they occupy interstitial sites and suppress SC. This effect is reversible; namely, by depletion of hydrogen, SC is restored [31]. The external pressure dependence of $T_C$ in Ru-1222 is linear up to 2.5 GPa. The pressure coefficient of 4.7 K/GPa obtained [32], is two times larger than that of Ru-1212.

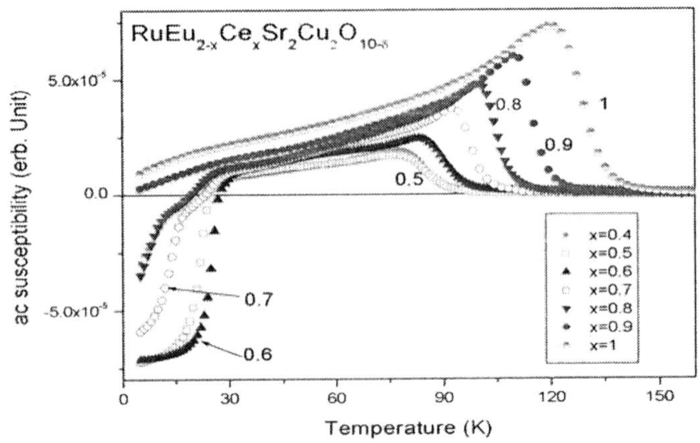


Fig. 6 Normalized ac susceptibility, ($H_{dc}$ =0) of asp Ru-1222 samples.Note, the absence of SC in x=1 and x=0.9 materials.

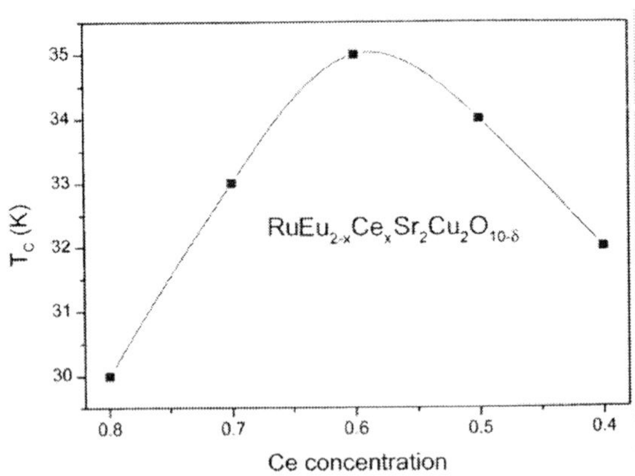


Fig.7 The bell shape SC onset temperature as a function of Ce content

## b A Simple Model for Superconductivity in Ru-1222

Given the variety of crystal structures and the chemical methods used to introduce holes into the $CuO_2$ layers, it is well accepted that a "generic' electronic phase diagram can be sketched for all compounds. The hole density *p* in the $CuO_2$ planes, is a primary parameter which affects $T_C$ in most of the HTSC compounds. In the well-established phase diagram $La_{2-x}Sr_xCuO_4$, the insulating parent $La_2CuO_4$ is AFM ordered and its magnetic interactions are well described by a simple Heisenberg model, with a large exchange interaction value (J= 1500 K). In $La_{2-x}Sr_xCuO_4$ the charge carrier concentration, can be varied by replacing $Sr^{2+}$ for $La^{3+}$. The variation of $T_C$ as a function of *p* exhibits a bell shape behavior, with a peak for the optimally doped material (x=0.15).

We argue that in the Ru-1222 system the $RuEuCeSr_2Cu_2O_{10}$ (x=1) samples serve as the parent stoichiometric insulator compounds (similar to $La_2CuO_4$). Since the valence of $R^{3+}$, $Ce^{4+}$, $M^{5+}$ $Sr^{2+}$ $Cu^{2+}$ and $O^{2-}$ are conclusive, a straightforward valence counting yields a fixed oxygen concentration of 10 (δ=0). Hole doping of the Cu-O planes, which results in metallic behavior and SC, can be optimized with appropriate variation of the $R^{3+}/Ce^{4+}$ ratio (trivalent $R^{3+}$ ions are replaced for $Ce^{4+}$) or the $Ru^{5+}/Cu^{2+}$ ratio in $Ru_{1-y}Sr_2EuCeCu_{2+y}O_{10}$ [33]. SC occurs for Ce contents of 0.4-0.8 (Fig. 7) and the optimally doped sample is obtained for Ce=0.6. This picture is consistent with the positive S(T) curve shown in Fig. 5. Unlike the $La_{2-x}Sr_xCuO_4$ system, this substitution does not appear to significantly alter *p* in the Cu-O planes. This is apparent in Fig. 7 which shows that the change of x from 0.8 to 0.6, results in a small increase in $T_C$. Indeed, if all the carriers were introduced into the Cu-O planes, then for the under-doped (x=0.8) to the optimally doped (x=0.6) samples, p should vary by 0.2 and result in a large shift in $T_C$, as observed in $La_{2-x}Sr_xCuO_4$ and in other HTSC materials. It is thus possible, that in Ru-1222, the extra holes introduced by reducing Ce, are *partially* compensated for by depletion of oxygen [23] and that the oxygen deficiency (δ) increases with $R^{3+}$. This partial hole doping mechanism is also supported by neutron diffraction on $RuGd_{1.3}Ce_{0.7}Sr_2Cu_2O_{10-\delta}$ from which one finds δ =0.22, instead of 0.3 required by charge neutrality [18]. It is also apparent that SC in the M-1222 system exists only for pentavalent M ions such as Nb, Ta and Ru.

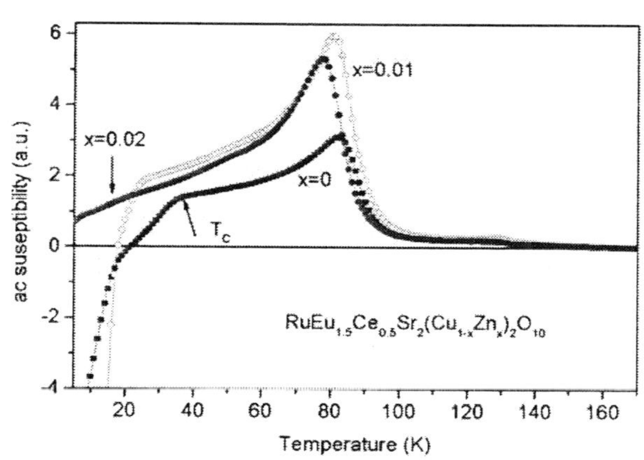


Fig. 8 The normalized real ac susceptibility curves of Zn doped Ru-1222 samples.

## c Substitutions in Ru-1222 System

Among the basic studies characterizing the Ru-1222 materials are many involving substitution of both Cu and Ru elements composing this system. Here, I attempt to review results in only one related area- the effect of substitution on $T_C$, which should offer some insight into the contribution of each element to SC. No attempt is made to survey the literature exhaustively, and to describe the effect of substitution on other physically interesting properties of the SC state. In the next section, the effect of these substitutions on the magnetic parameters will be described as well. As stated above, each series of materials presented here was prepared simultaneously under the same conditions and it is assumed to have the same oxygen concentration.

*Substitution in the Cu site.* The divalent non-magnetic $Zn^{2+}$ and magnetic $Ni^{2+}$ ions substituted for $Cu^{2+}$ were studied[16] . The normalized real ac susceptibility $\chi'(T)$ curves of Zn doped in Ru-1222 are presented in Fig. 8. It is readily observed that $T_C$ is reduced from 38 K for x=0, to 26 K for x=0.01 of Zn. For x=0.025, the sample is not SC down to 4.2 K and is magnetically ordered only. Similar values were obtained by dc magnetic and resistivity measurements.] The magnetic $Ni^{2+}$ ions in the Cu site also reduce $T_C$ but in a more moderate manner, and SC is suppressed totally only at x=0.1. This reminiscent of the typical behavior of Zn and Ni doped HTSC materials.

Both Ni and Zn ions, which are next-door neighbors to Cu in the Periodic Table (ionic radii 0.69 and 0.74 and 0.71Å respectively), induce only small local structural disorders in the Cu-O planes, and slightly affect the magnetic peak position at $T_{M2}$ around 80 K of the Ru sublattice. This indicates that magnetic Ru state is not affected by the presence or absence the SC state (to be discussed later).

*Substitution in the magnetic Ru site.* The $Ru^{5+}$ ions in Ru-1222 can be replaced by the iso- electronic Nb ions to form the $(Ru_{1-x}Nb_x)$-1222 system for all x values. Since pure Nb-1222 is SC at $T_C$= 27 K [3], Nb substitution does not affect much the SC properties. However the presence of Nb up to x=0.5 reduce sharply the magnetic parameters but does not change the typical coexistence of SC and W-FM. [34]. We have studied the effect of non-magnetic $Sn^{4+}$ and $Al^{3+}$ ions substitution for $Ru^{5+}$, on both SC and magnetic states of Ru-1222 [35].

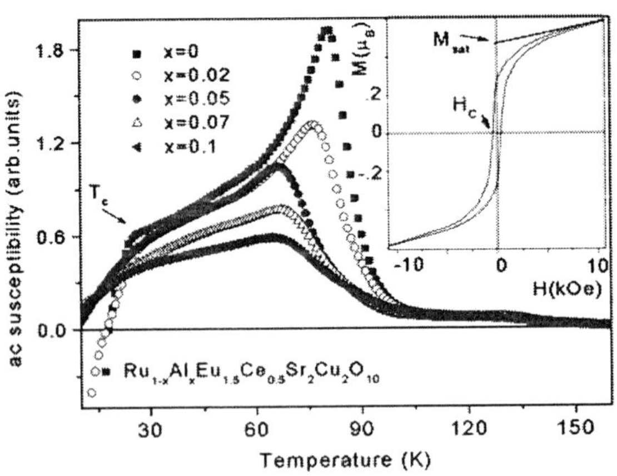


Fig. 9 The normalized real ac susceptibility curves of Al doped Ru-1222 samples. The inset shows the typical FM hysteresis loop at 5K for x=0 .

Doping of these ions in Ru-1222 reminiscence of the typical behavior of Al (and Sn) doped YBCO materials in which both ions substitute for Cu (1). Since the tetragonal Ru-1222 structure evolves from the YBCO one, it is propose that both ions reside mainly in Ru site. The $\chi'(T)$ curves of Al doped samples with x=0, 0.02, 0.05, 0.07 and 0.1 are presented in Fig.

9. The negative susceptibility for samples with x=0, 0.02 and 0.05, indicates that they are SC and that $T_C$ is reduced from 36 K to 30 and 26 K respectively. The x=0.07 and 0.1 samples are not SC and are magnetically ordered only. It appears that the suppression rate of $T_C$ is more moderate than in the Zn doped materials described above. Keeping in mind that such a substitution dilutes the magnetic the $RuO_2$ layers, Fig. 9 shows that Al substitution (up to 10%), slightly affects the magnetic peak position around 80 K and shifts it to lower temperatures. This peak ($T_{M2}$= 82 K for x=0) is shifted to 76, 70, 68 and 67 K for x=0.02, 0.05, 0.07 and 0.1 respectively. These results are consistent with the data reported on the Sb doped for Ru in the Ru-1222 system [36].

The '$\chi$ (T) plots of Sn doped in Ru-1222 are presented in Fig. 10, show that Sn substitution for Ru reduces $T_C$ (39 K for x=0) to 26 (x=0.02) and 15 K (x=0.05) and affects dramatically $T_{M2}$ (91 K for x=0) and shifts it to 76 (a shift of 15 K) and to 51 K for x=0.02 and x=0.05 respectively [35]. The reduction in $T_C$ shown here may also be associated with changing the oxygen concentration since $Sn^{4+}$ or $Al^{3+}$ ions have a lower valence than the host $Ru^{5+}$ ions. It is possible that lattice strain from the mismatch of different ions at the same crystallographic site also contribute to this reduction. In the Sn doped samples, a significant shift is observed $T_{2M}$ (about 40 K) Firm conclusions as to why the substitution of the diamagnetic $Sn^{4+}$ and $Al^{3+}$ for Ru in the Ru-1222 system, behave so differently, cannot be drawn at the present moment. In the case of $Ru_{1-y}Sr_2Eu_{1.5}Ce_{0.5}Cu_{2+y}O_{10-\delta}$ where the extra Cu ions in the Ru site does not affect $T_C$ (up to y=0.4) but reduce the magnetic response and lower the magnetic transition [33]. On the other hand SC is totally suppressed in $(Ru_{0.87}Fe_{0.13})Sr_2Eu_{1.5}Ce_{0.5}Cu_2O_{10-\delta}$ where Fe ions reside in the Ru site [37] and all the magnetic parameters are diminished.

As a final point of interest, R=Eu or Gd can be replaced completely by Dy, Y and Ho when synthesized under high pressure (6Gpa at 1200 C). These compounds exhibit similar magnetic behavior to the R=Eu, Gd materials, but none of them is SC down to 4 K [38].

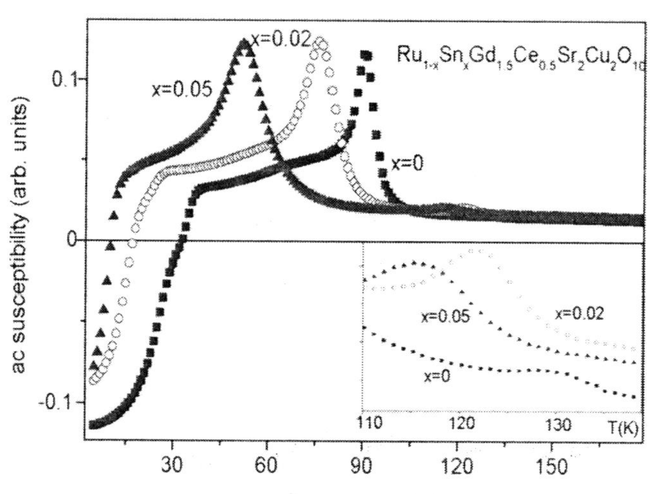


Fig. 10 The normalized real ac susceptibility curves of Sn doped Ru-1222 samples. The susceptibility above $T_{2M}$ in an extended scale is shown in the inset

## d Physical Properties of the SC State

Up to now, the determination of the most important SC parameters in Ru-1222 is unexplored. The strong magnetic contribution to the magnetization from the Ru-planes makes impracticable the use of isothermal magnetization curve to determine relevant intrinsic properties such as the lower critical field ($H_{C1}$) and the critical current density ($J_C$). However a rough estimation of $H_{C1}$=25 Oe was reported on the basis of the diamagnetic minimum in the virgin M(H) curve [30]. A much higher value was determined by detailed processing of magneto-resistance loops R(H) [39]. However, our previous magnetization data reveal that the Ru sublattice generates an internal field of a few hundreds Oe. in Cu-O planes below $T_{2M}$. Such an internal field which is higher than $H_{C1}$ is suggested to be responsible for the appearance of the spontaneous vortex phase (SVP) in the system which obscures $H_{C1}$[6]. Therefore, it is suggested that $H_{C1}$ parameter in the Ru-1222 system is a meaningless term.

A careful examination of the R(H) curves in the high field regime of Ru-1222 sample (prepared in the sol-gel technique and annealed under 95 atm. pressure of oxygen), yields an upper critical field (along the $CuO_2$ plane) of $H_{C2}(0)$ = 39 T, a value which is line with values for other HTSC materials [40]. An out of plane SC coherence length $\xi_0$ =28 Å was deduced by using the relation $H_{C2}(0)= \Phi_0/2\pi\, \xi_0^2$ where $\Phi$ is the quantum of magnetic flux. Assuming an anisotropy factor of 5 a value of $\xi_{ab}$ = 140 Å is obtained which indicates that SC is limited spatially to the submicron domain lengths.

The synthesis process for Ru-1222 described above, leads to the formation of ceramic granular materials. Due to the weakly coupled granular superconductivity of the ceramic samples, the intrinsic properties such as the transport critical current density ($J_C$) of the SC state are further complicated.

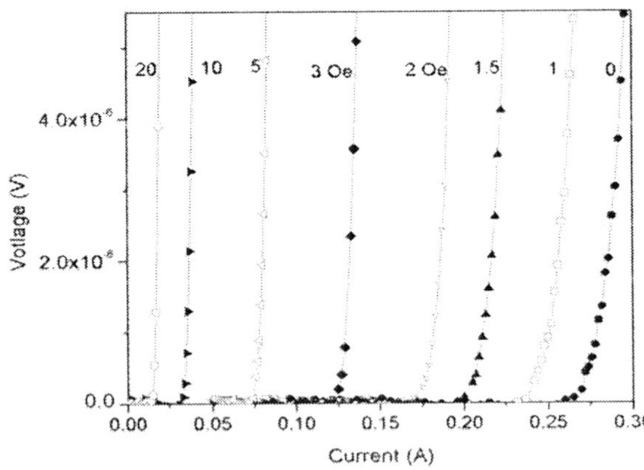


Fig. 11 Current Voltage curves at 5 K for annealed $RuSr_2Gd_{1.4}Ce_{0.6}Cu_2O_{10}$ at various applied fields in Oe (marked in the figure).

The low intragrain $J_C$ at 5 K, of high quality annealed $RuGd_{1.4}Ce_{0.6}Sr_2Cu_2O_{10-\delta}$ , was measured by direct I-V curves at various applied fields (Fig. 11). The onset of the resistivity changes dramatically to lower currents with the applied field below 10 Oe, but is almost unchanged for $H_{ext}$ higher than 25 Oe [41]. The deduced field dependence of $J_C$ is exhibited in Fig. 12, and the solid line is the fit of curve to: $J_C \propto H_{ext}^{-0.45}$. The fit yields an intragrain $J_C$

value (at H=0) of 22 A/cm$^2$ obtained. An higher intergranular $J_C$= 110 A/cm$^2$ was achieved by tracing the frequency shift in the peak of the imaginary part of the ac susceptibility [39]. The intergrain $J_C$ of Ru-1222 at 5 K can be evaluated from the remnant moment, $M_{rem}$=17.5 Oe. of the non magnetic **Nb**$Sr_2Gd_{1.4}Ce_{0.6}Cu_2O_{10-\delta}$, assuming similar SC properties of the two iso-structural systems. Using the Bean model $J_C$=30*□M/d (where = 0.3 cm is the largest dimension in the sample) $J_C$~ 1575 A/cm$^2$ was obtained, a value which is a few orders of magnitude lower than the $J_C$ for Ru-1222 reported in ref [41].

## e The Granular Behavior of the SC State

The microstructure of Ru-2122 (x=0.5) exhibits well-defined grains with a typical size of a few $\mu$m and pronounced grain boundaries [42]. The transport transition to the SC state occurs via two intermediate stages[41,43-45]. Figure 13. shows the temperature dependence of resistance of asp $RuEu_{1.4}Ce_{0.6}Sr_2Cu_2O_{10-\delta}$ at a fixed current (0.1 mA) at various $H_{ext}$. The intra-grain sharp drop of the resistivity at $T_C$ =43 K occurs when the grains become superconducting. At this point the weak-links (WLNs) between the grains contribute non-zero resistance across the sample. At lower temperature ($T_P$ =38 K) the WLNs become superconducting. This transition which is due to weak-Josephson inter-grain coupling, is much more sensitive to $H_{ext}$. The broadening of the resistive transition was already observed in oxygen annealed Ru-2122 materials [5], and was related to inhomogeneity in the oxygen concentration which causes a distribution in WLNs and in the $T_C$ values. Below $T_P$ the resistivity (as well as the ac susceptibility) is governed mainly by the WLN properties and it is sample dependent. Various parameters affect the WLN properties such as: heat treatment, annealing conditions and the way the samples have been stored prior to measuring.

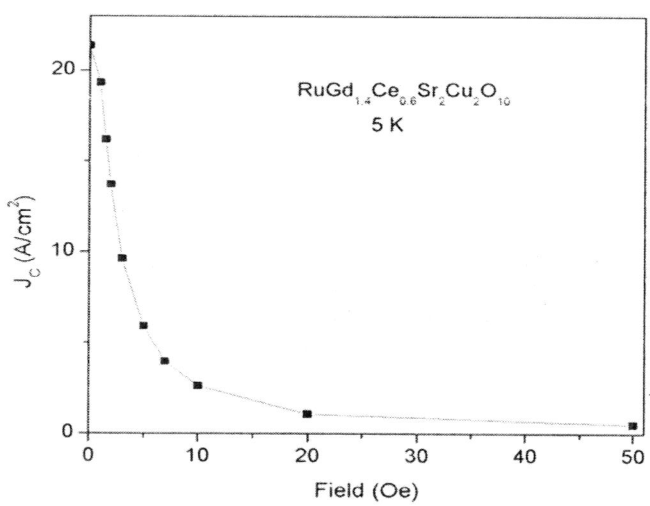


Fig. 12 The field dependence of the critical current density at 5 K.

The isothermal magneto resistance R(H) curves for temperatures below $T_P$ are shown in Fig. 14. Surprisingly, as the applied field is decreased, an anomalous magnetic hysteresis is observed in the R(H) curves. (i) The resistance in the decreasing $H_{ext}$ curve is lower than that in the increasing $H_{ext}$ curve above a crossover field. (ii) Below the crossover field the trend is reversed and at $H_{ex}$ =0 the resistance is finite after the field-cycle. The loop's width decreases with temperature and no hysteresis is observed above $T_P$; thus the hysteresis is the result of the granularity of the material. The general trend is that the temperature in which the hysteresis appears as well as the width of loops depend on the WLN properties namely, the heat treatments and annealing procedures of materials.

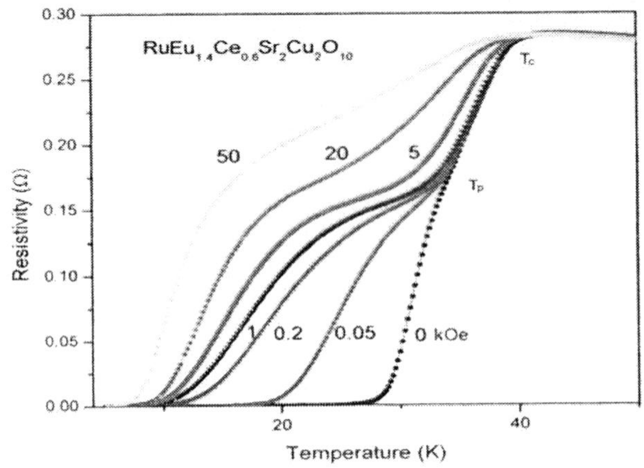


Fig 13. The temperature dependence of resistivity in magnetic fields up to 50 kOe

Similar hysteresis loops were obtained for the non-magnetic Nb-1222 sample (Fig. 15) indicating clearly, that this hysteresis is not connected to the magnetic state of Ru-1222 and to its coexistence with SC below $T_C$ [41]. Qualitatively speaking the hysteresis in the R(M) curves exhibited in Figs (14,15) closely resemble the behavior predicted by the two-level critical-state model of Ji et. al [46], which explains qualitatively the surface resistance of granular superconductors.

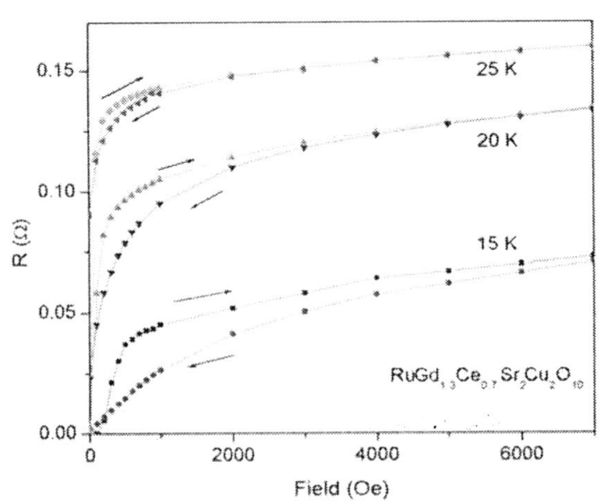


Fig. 14 Hysteresis loops of the magneto-resistance of Ru-1222. The arrows show the direction of the field change during the measurements.

According to this model, two distinct different current densities are taking place for granular systems. (i) A large $J_C$ inside the grains, and (ii) a much weaker one reflecting the inter-granular Josephson coupling. In other words, two types of fluxons according to their dc pinning characteristics are defined. (a) the strong grain-pinned fluxons and (b) the free grain-boundary fluxons which are those that never move through grains. The gradient of the flux density inside the grains is much larger than that in the boundaries between the grains. On a macroscopic scale, the flux density averaged over many grains should have a gradient determined by the weak inter-grain properties. The resistivity is dominated by the flux density in the grain boundaries ($n_j$) due to its smaller pinning strength. The $n_j$ is further affected by the combination of the external field $H_{ext}$ and the field of the local intragrain supercurrents (or the magnetic domains) and experience an *effective* field of $H_{ext}$-$\alpha 4\pi M$, where $\alpha$ is a numerical factor. Theoretical calculations of $n_j$ as a function of $H_{ext}$ show that, contrary to homogeneous samples, for the same value of $H_{ext}$, $n_j$ in the decreasing branch of $H_{ext}$ is *lower* than that in the increasing $H_{ext}$ branch, except near the remnant state. The explanation for this reduction is as follows. When the field is reduced, some of trapped fluxons are no longer supported by the external field and form closed loops through the grain boundaries. The opposite direction of the flux lines, effectively drives a *reverse* flux component (-$\alpha 4\pi M$) into the sample through the weakest junctions. Because of magnetic field superposition, the average field within the junctions of the weak-links thus becomes ($H_{ext}$-$\alpha 4\pi M$) less than the applied field. As a result, $J_C$ in descending field will be correspondingly increased over the initial $J_C$ curve. Thus, the smaller effective field in the grain boundaries leads to the hysteresis observed in Figs. (14-15), and this behavior is typical of any granular material.

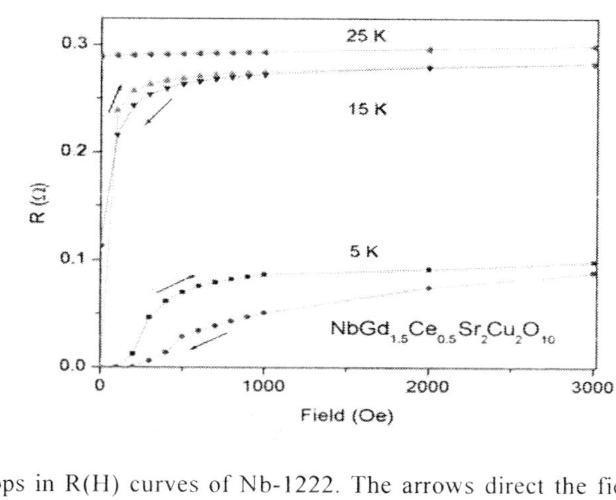


Fig. 15.Hysteresis loops in R(H) curves of Nb-1222. The arrows direct the field sequence during the measurements.

## (III) The Magnetic State in $RuSr_2R_{2-x}Ce_xCu_2O_{10-\delta}$ Results and Model

One of the most disputed questions is the magnetic structure in Ru-1222, and the detailed magnetic features of the Ru-1222 system are lacking. The complexity of the magnetic features is presented in Fig. 16. The ZFC and FC curves for oxygen annealed

$RuGd_{1.4}Ce_{0.6}Sr_2Cu_2O_{10-\delta}$ sample measured at very low external field, ($H_{ext}$= 80 mOe.) are shown in Fig 16. The typical expulsion of magnetic flux lines at $T_C$ =43 K in both ZFC and FC curves, as well as the flatness below $T_C$, are readily observed. The $\Delta M_{FC}/H_{ext}$ at $T_C$ is ~70% of 1/4$\pi$ value indicating bulk SC [41]. At higher temperatures, two magnetic transitions are observed, at $T_{2M}$ (around 85 K) and at $T_M$ (around 150 K). At $H_{ext}$<2-3 kOe, irreversibility in the ZFC and FC curves is observed, which disappears at $T_{2M}$. The dc ZFC curves (as well as the ac susceptibility plots Figs 6,8) show a well-distinguished peaks at temperatures, which are a bit lower than $T_{2M}$. For $H_{ext}$ >5 kOe, both ZFC and FC curves collapse to a single ferromagnetic-like behavior.

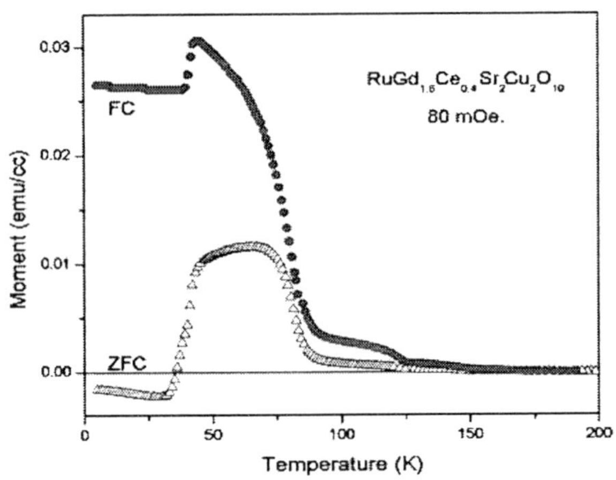


Fig. 16 ZFC and FC susceptibility curves for $RuSr_2Gd_{1.4}Ce_{0.6}Cu_2O_{10}$ annealed under 300 oxygen atm. at 600 C.

Since all the Ru-1222 materials have basically the same complicated magnetic behavior, we show here a detailed magnetization study of the non-SC $RuSr_2Y_{1.5}Ce_{0.5}Cu_2O_{10}$ (Ru-1222Y) which has been synthesized at 6GPa at 1200° C [47]. Here, Y ions which replace Eu or Gd, are non-magnetic and permit an easier direct interpretation of the intrinsic Ru magnetism.

## a The Magnetic Properties of $RuY_{1.5}Ce_{0.5}Sr_2Cu_2O_{10}$ (Ru -1222Y)

The temperature dependence of the ac susceptibility for the non-SC Ru -1222Y, is presented in Fig. 17. Two peaks at 105(1) K and at 127(1) K are observed. Similar peaks were observed in the ZFC curve at measured 15 Oe. All isothermal M(H) curves below $T_M$ depend strongly on H (Fig. 18). The remarkable feature is the apparent tendency toward saturation but not reaching full saturation even at 5 K and 50 kOe. The Ru moment at 5 K and 50 kOe, is M=0.71$\mu_B$/Ru. By plotting $M^2 \alpha 1/H$ (for high $H_{ext}$ values) and extrapolating to 1/H=0, $M_{max}$= 0.75$\mu_B$/Ru is obtained. The linear part of M(H) in Fig. 18 can be described as: M(H)= $M_s$+ $\chi$ H, where $M_s$ =0.63(1)$\mu_B$ is the value obtained by the extrapolation to H=0 and $\chi$=dM/dH is the slope (8.5*$10^{-3}$ emu/mol*Oe =1.53*$10^{-2}\mu_B$/T). Similar M(H) curves have been measured at various temperatures and the temperature dependence of $M_s$ yields $T_M$(Ru)=152(2) K.

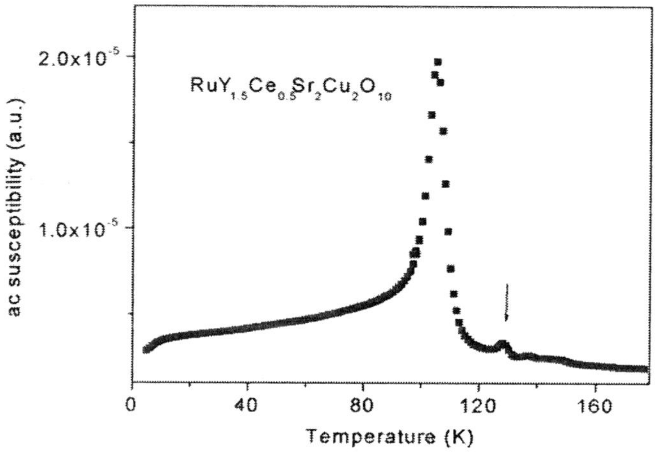


Fig. 17 ac susceptibility of Ru-1222Y. Note the small peak at 125 K

At low temperatures, at low applied fields the M(H) curves exhibit a typical ferromagnetic-like hysteresis loop (Fig. 18 inset) and values deduced at 5 K are: $M_{rem}$= 0.31$\mu_B$/Ru and $H_C$ = 410 Oe. Both $M_{rem}$(T) and $H_C$(T) decrease with temperature and become zero around $T_{2M}$ =100 K (Fig. 19). Above $T_{2M}$ reappearance of the hysteresis loops is observed with a peak of $M_{rem}$(T) and $H_C$(T) around 125 K (Fig. 19 inset) close to the minor peak in Fig. 17. In contrast to the FM-like hystersis loop obtained at low temperatures, the loops above $T_{2M}$ exhibit an AFM like feature (Fig. 20) [47]. Note the difference in moment scale between Figs. 18 (inset) and 20. In the paramagnetic range, the $\chi$(T) curve follows the Curie-Weiss (CW) and the net paramagnetic effective moment and Curie-Weiss temperature extracted are: $P_{eff}$ =2.05$\mu_B$/Ru and $\theta$= 136(1) K. This positive $\theta$ value is in fair agreement with $T_M$(Ru). The $P_{eff}$ is somewhat greater than 1.73 $\mu_B$, the calculated value of $Ru^{5+}$ in the low-spin state and supports the existence of a minor $Ru^{4+}$ fraction of ($P_{eff}$ =2.83 $\mu_B$, in the low-spin state) to be discussed below.

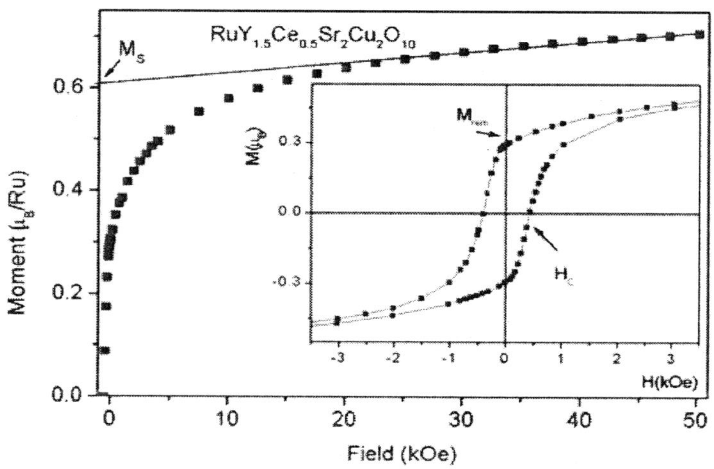


Fig. 18 The isothermal magnetization and the FM hysteresis loop (inset) at 5 K

## b The Qualitative Magnetic Structure of (Ru -1222Y).

While our data described here do not include any determination of the magnetic structure of the Ru sublattice, the results are compatible with a simple model, which is, however, of use in understanding the qualitative features at low applied fields. Starting from low to high temperatures, the magnetic behavior is basically divided into two regions.

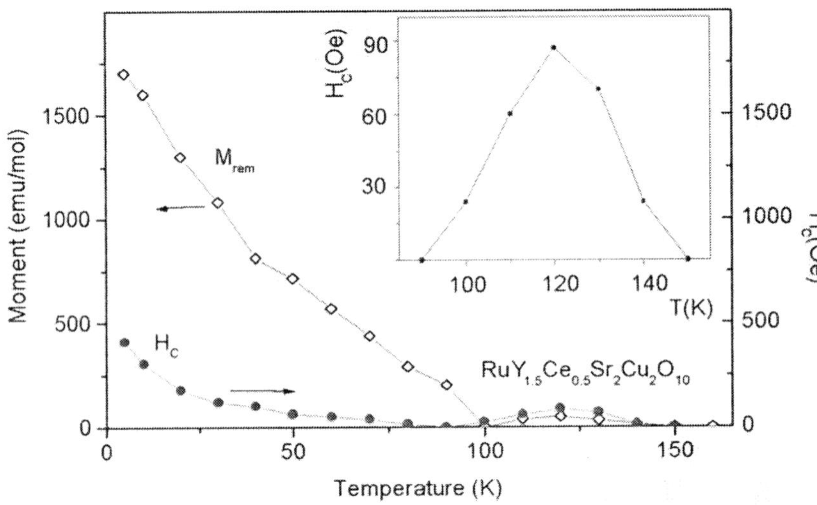


Fig. 19 The temperature dependence of the coercive field ($H_C$) and remnant moment for Ru-1222Y. The inset shows $H_C$ in an extended scale.

(i) Wide FM hysteresis loops are opened at low temperatures (the coercive field- $H_C$ ~450 Oe at 5 K) which become narrower as the temperature increases and practically disappear below $T_{2M}$. The extrapolated moment value is $M_{max}$=0.75$\mu_B$/Ru, a value which is smaller than 1$\mu_B$, the expected moment for $Ru^{5+}$ in the low-spin state (g=2 and S=0.5). For R= Eu and Gd, SC is induced at $T_C$. Below $T_C$, both SC and weak-ferromagnetic states coexist and the two states are practically decoupled as described below.

(ii) For $T_{2M}<T<T_M$, the Ru sub-lattice becomes basically AFM ordered at low applied fields. This interpretation is supported, by the small peaks observed in both the ac susceptibility (Fig.17) and $H_C$ (Fig.19). At higher applied fields a metamagnetic transition is induced. The AFM Ru moments are realigned through a spin-flip process by $H_{ext}$, to form the AFM-like shape hysteresis loops observed in Figs 20 (the coercive field- $H_C$ ~50-80 Oe. In this state, both the anisotropy and AFM exchange energies are temperature dependent, and the competition between them may lead to this unusual maximum and to the disappearance of $H_C$.

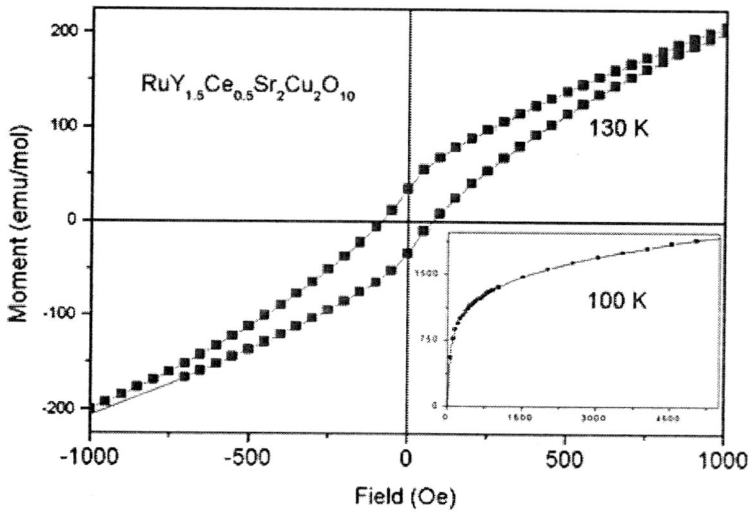


Fig. 20 The hysteresis loop opens at 130 K and the magnetization at 100 K(inset)

## c Decoupling of the two SC and Magnetic States

Fig. 8 shows three Zn doped samples of which two are SC ($T_C$ 38 K for x=0, and 26 K for 1% at. Zn) and a third material (x=0.025), in which SC is totally suppressed. On the other hand, the presence of Zn in the Cu-O layers, slightly affects the peak position at 80 K. Fig. 21 shows the temperature dependence of $M_s$ for the three materials studied, which becomes zero at $T_M$(Ru)=165(2) K, regardless of their Zn concentration. Note the FM-like shape of the curves. Similar to the Ru-1222Y compound described above, relatively wide FM hysteresis loops opened at low temperatures ($H_C$ ~400-500 Oe at 5 K). They become narrow as the temperature increases and practically disappear below $T_{2M}$. Above $T_{2M}$, small AFM hysteresis loops are reopened, with a maximum of $H_C$ (~130 Oe.) and $M_{rem}$ at 120 K for all three samples (Fig. 22). Similar features were observed for the Al doped materials presented in Fig. 9. [35].

Therefore we may suggest that (i) The SC and the non-SC materials have basically the same $T_M$ and a similar $H_C$ trend which are not affected by the presence of non-magnetic ions in the Cu or the Ru sites. This means that the SC and magnetic states are decoupled from each other. (ii) The materials are W-FM at T< $T_{M2}$ and canted AFM ordered at $T_{M2}$ <T < $T_M$.

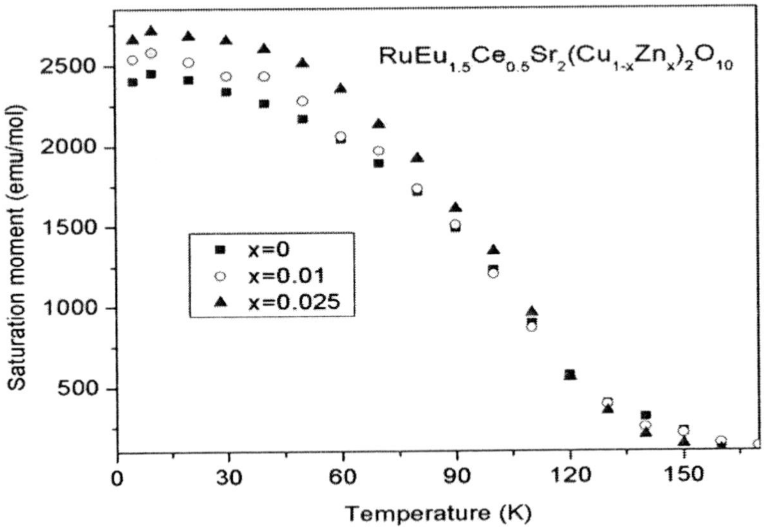


Fig.21 The temperature dependence of saturation moment for the Zn doped materials

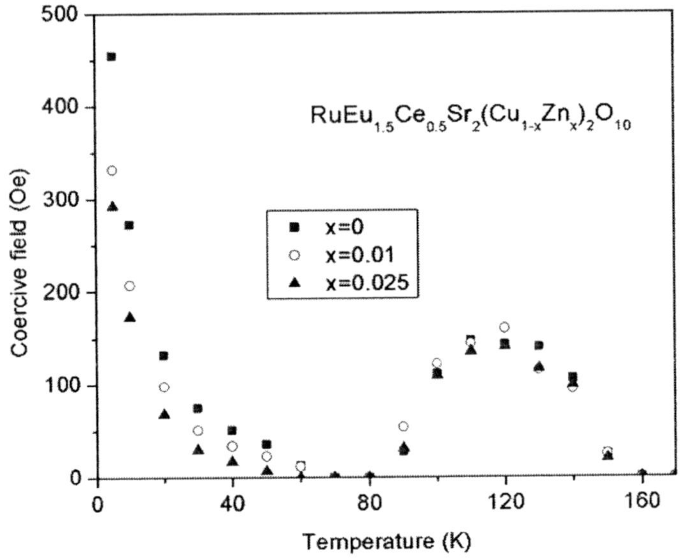


Fig. 22 The temperature dependence of the coercive fields for the three samples

## d The Magnetic Properties of $RuEu_{2-x}Ce_xSr_2Cu_2O_{10-\delta}$ System

We suggest the general picture described for Ru-1222Y to describe qualitative magnetic properties of the entire Ru-1222 system, in which the Gd or Eu ions paramagnetic moments obscure the net Ru contribution. The phase diagram of the $RuEu_{2-x}Ce_xSr_2Cu_2O_{10-\delta}$ system was discussed in length in a previous publication [9] and for the sake brevity we display here only the data which supports our seggested models discussed below.

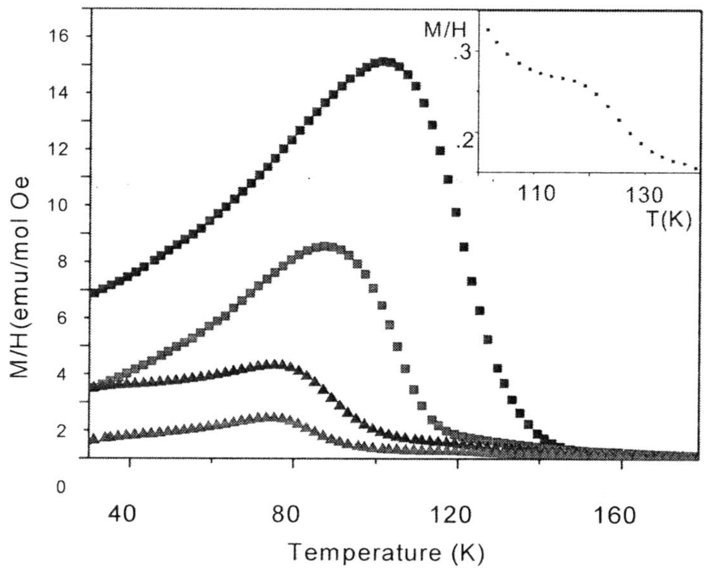


Fig. 23 ZFC dc curves measured at 4 Oe for asp Ru-1222 materials. The inset shows the expanded high temperature range for x=0.4.

Generally speaking, due to the similarity of the ionic radii of $Eu^{3+}$(0.94 Å) and $Ce^{4+}$(0.87Å) the lattice parameters of all materials studied are basically unchanged [9]. $RuEuCeSr_2Cu_2O_{10}$ (x=1) is magnetically ordered only, and SC appears for samples with x=0.8-0.4 (Fig. 6). The ZFC plots (measured at 4 Oe) of the asp $RuEu_{2-x}Ce_xSr_2Cu_2O_{10-\delta}$ (x=1-0.4) materials are shown in Fig. 23. The major peak for x=1 at 108 K is shifted to 75 K for x=0.4. This picture agrees well with the ac curves shown in (Fig. 6). Fig. 24 shows the the ZFC and FC curves for both x=1, and x=0.5 samples. It appears, that for x=1, the magnetic properties due to the Ru are all *enhanced*, as compared to the x=0.5. The latter sample is SC and its ZFC branch starts from negative values. The diamagnetic signal (at 50 Oe.) due to the high shielding fraction of the SC state dominates, and the net moment at low temperatures (up to 18 K) is negative. At higher external fields or when the ZFC process was not performed properly, the negative signal is not visible. The inflection in both ZFC and FC branches at 38 K agrees well with Tc determined from the ac curve (Fig. 6).

For x=1neither the peak at 108 K nor $T_M$ =160 K are altered by further annealing or quenching of the sample. On the other hand, reheating the x=0.8-0.4 asp materials, shifts the peak to higher temperatures (from 75 K to 85 K for x=0.4). The second peak (around 120 K) appears clearly in the SC materials (x=0.8-0.4), in both ZFC and ac curves (Fig. 23 inset), whereas *no second peak* is observed whatsoever for the x=1 materials. Fig. 25 shows the $M_s(T)$ for the various samples. The deduced information is: (i) All $M_s$ curves have a typical FM like behavior and merge at $T_M$ =160 K. (ii) The $M_s$ values gradually decrease with x, from $0.88\mu_B$ for x=1 to $0.39\mu_B$ for x=0.4 (at 5 K).

Wide FM hysteresis loops are opened at low temperatures, and the $H_C$ values depend also on x and vary (at 5 K), from 460 Oe for x=0.4 to ~320 Oe for x=0.8 and 1 (Fig. 26). The hysteresis loops become narrower and close themselves at ~60 K ($H_C$=0), thus essentially no discernible hysteresis exists above 60 K (Fig. 22). At T>80 K for x=0.4-0.8, AFM like hysteresis loops reappear and the temperature dependence of $H_C$ exhibits a bell shape behavior with a peak around 110-120 K (close to the second peaks temperatures). For x=1, the $H_C$ values for both the asp and quenched samples are the same (Fig. 26 inset), and no reappearance of $H_C$ at higher temperatures is observed.

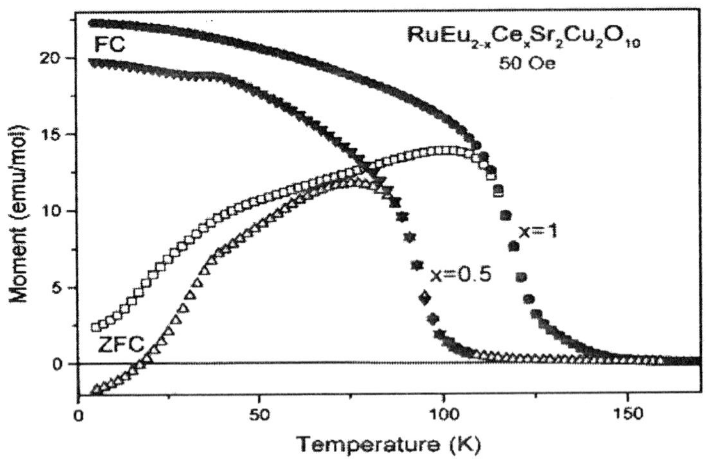


Fig. 24 ZFC and FC curves for SC and non-SC Ru-1222 materials

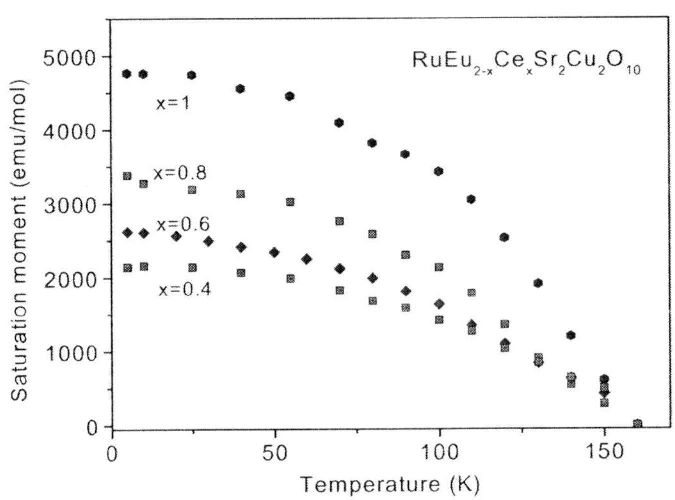


Fig. 25 Temperature dependence of the saturation moment of $RuEu_{2-x}Ce_xSr_2Cu_2O_{10}$

Figs. 24-26 lead to the following conclusions: (i) The susceptibility second peak is connected with the bell shape of $H_C$. For x=1, neither a reappearance of $H_C$ nor a second peak are observed (ii) The general trend is that the maximum in $H_C$ (Fig. 26) and the second peak, both increase with the decrease of x. The intriguing question arises as to why both observations do not appear in the x=1 material. As stated above, the straightforward valence counting for x=1, yields a fixed oxygen concentration of 10, which does not change either by further annealing or by quenching the sample. As a result the magnetic parameters such as: $T_M$, $M_s$ and $H_C$ are the same for all three (asp, reheated and quenched) materials.

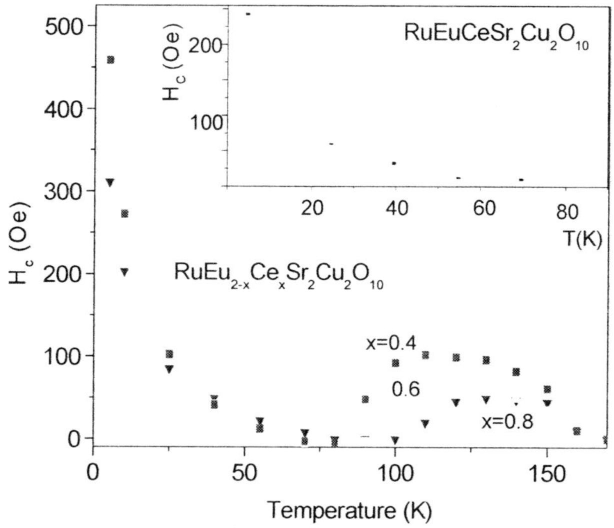


Fig. 26 The coercive filds of various Ru-1222 and the x=1 (inset) materials

In the paramagnetic range, the $\chi(T)$ curve measured at 10 kOe for x=1, follows the CW law. The net paramagnetic Ru contribution to $\chi(T)$, was obtained by subtracting $\chi(T)$ of **Nb**$Sr_2EuCeCu_2O_{10}$ (the reference material) from the measured data. This procedure yields: $\chi_0$ =0.0063, $P_{eff}$ =2.13 $\mu_B$ and $\theta$= 146(1) K. (for x=0.5, $P_{eff}$=2.15 $\mu_B$ and $\theta$= 134(1) K). These values are in fair agreement with the parameters deduced for Ru-1222Y, indicating a similar net paramagnetic Ru contribution in all Ru-1222 compounds.

## e The Major W-FM Magnetic Region

The exact FM-like structure of the Ru spin ordering below $T_{2M}$ is still debated and no conclusions have been reached as yet. Neutron diffraction measurements are required to precisely determine the nature of the magnetic order in this region. Meanwhile three scenarios may be suggested.

(i) This canting of the Ru moments, as well as irreversibility phenomena at low external fields, arise as a result of an antisymmetric exchange coupling of the Dzyaloshinsky-Moriya type [15] between neighboring Ru moments, induced by a local distortion that breaks the tetragonal symmetry of the $RuO_6$ octahedra [18]. The saturation moment of $Ru^{5+}$ in the low spin state is 1 $\mu_B$/Ru (3 $\mu_B$/Ru for $Ru^{5+}$ in the high spin state). The $M_s$ =0.88- 0.40 $\mu_B$/Ru obtained (at 5 K well below $T_M$), are a large fraction of 1 $\mu_B$, implying a large canting angle of the AFM Ru moments. This angle changes with x and depends on the oxygen content. The large $M_{rem}$ /$M_{max}$ (0.41 for Ru-1222Y) and the high positive $\theta$ values are also consistent with this W-FM interpretation.

(ii) Alternatively, a frequency dependent cusp was observed in the ac susceptibility of Ru-1222, (Fig. 27) which may be interpreted as a spin glass (SG) behavior [48]. A SG state is a collection of frozen and disordered magnetic moments. To produce such a state, two ingredients are necessary:

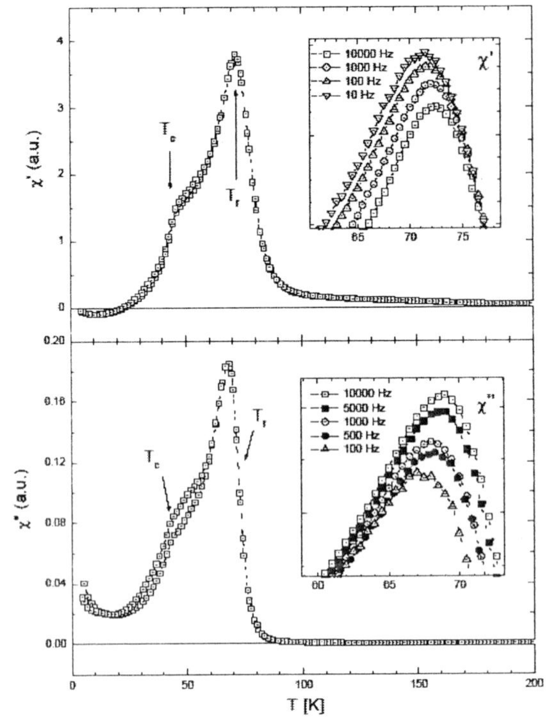


Fig. 27 Real and imaginary ac susceptibility curves and the frequency dependence (inset) of Ru-1222 (taken from Ref.48)

frustration and partial randomness of interaction between the magnetic moments. Although the $Ru^{5+}$ ions in Ru-1222, are arranged in a strictly periodic order, it is possible that the incomplete oxygen stoichiometry and/or the small distortion of the oxygen octahedra are involved in inducing frustration, which, in turn leads to glassy behavior. The peaks in the ac '$\chi$(T) and "$\chi$ plots represents the freezing temperature to the SG state ($T_f$ in the major panels of Fig. 27) and its frequency dependence is a typical feature of the dynamics of this SG states.

(iii) The unsaturated M(H) curve (Fig. 18) at low temperatures may suggest (similar to $SrRuO_3$) itinerant-electron magnetism in Ru-1222. Consequently, the conventional so-called Arrott-plots to determine the magnetic transition was applied [47], and the extracted magnetic transition is 125 (3) K, midway between $T_{2M}$ and $T_M$. Supporting evidence to this scenario are: (a) the $P_{eff}$ values ($P_{eff}$ =2.05-2.15 $\mu_B$) which exceed the expected value for a localized $Ru^{5+}$ low spin state and (b) the relatively high $P_{eff}/M_s$~2.3-5.4 ratio.

## f Mössbauer Effect and μSR Studies

Mössbauer effect studies (MS) on $^{57}$Fe doped samples have been proved to be a powerful tool in the determination of the magnetic nature of the Fe site location. When the Ru ions become magnetically ordered, they produce an exchange field on the Fe ions residing in thier site. The Fe nuclei experience a magnetic hyperfine field leading to a sextet in the observed MS spectra. As the temperature is raised, the magnetic splitting decreases and disappears at $T_M$.

Some of the MS spectra of $^{57}Fe$ doped (1 at.%) in Ru-1222 are shown in Fig. 28. The experimental spectra are divided into three groups. (i) The 300 K spectrum (well above $T_{2M}$) indicates that all $^{57}Fe^{3+}$ ions which replace $Ru^{5+}$ ions, have distorted environments, leading to a superposition of three doublets corresponding to different local environments [35].

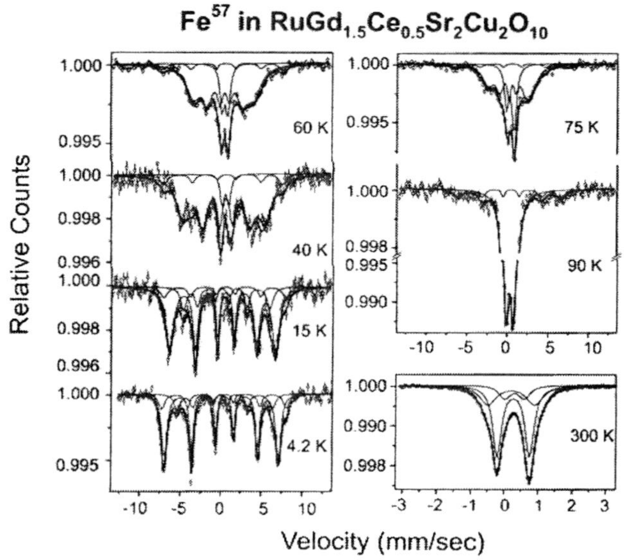


Fig. 28 Mossbauer spectra of Ru-1222. Note the smaller scale at 300 K

(ii) The spectrum at 90 K ($T_{2M}<T<T_M$) exhibits a superposition of two sub-spectra: a *major* paramagnetic doublet (similar to the 300 K spectrum) and a *minor* one which is a magnetically split sextet. The magnetic hyperfine field ($H_{eff}$) of this *minor* fraction (11-20%) increases when lowering the temperature and reaches almost saturation around $T_{2M}$ (Fig. 29 upper panel). (iii) The spectra below $T_{2M}$, *are much different* than that at above $T_{2M}$. The rest of the $^{57}Fe^{3+}$probe ions becomes magnetically split, with a distribution of hyperfine interaction parameters, according to the inequivalent iron probe locations. The spectra were analyzed in terms of three subspectra, one corresponding to the *minority* fraction ordered at $T_M$, the others to the majority fraction ordered at $T_{2M}$. The temperature dependences of $H_{eff}$ for the subspectra and the total magnetic fractions below and above $T_{2M}$ are shown in Fig. 29. It appears that the minor fraction $H_{eff}$ values are higher than these belong to the majority fraction. Similar results were obtained by MS of $^{119}Sn$ (2 at%) doped for Ru in the same material [49].

ZF-μSR experiments [11] which provide evidence of magnetically ordered states in the SC Ru-1222, confirm also the presence of a minor (15% of the sample volume) phase which orders magnetically at $T_M$. Below $T_{2M}$ there is an additional sharp magnetic order accounts for the rest of the sample volume. Fig. 30 exhibits the temperature dependence of the different internal magnetic fields and of the magnetic volume fractions. Similar to Fig. 29, two oscillating signals (high (HF2) and low (LF)) appear at $T_{2M}$ and a third one (HF1) corresponds to a large internal magnetic field arises at $T_M$. It is also evident that the magnetic order persists unchanged in the SC state. From the amplitude of the oscillations signals it is deduced that below $T_{2M}$ all the material is magnetically ordered and no non-magnetic phase exists [11].

Both different MS and μSR techniques indicate clearly the presence of a *minor* magnetic phase at $T_{2M}<T<T_M$, its origin is discussed below. Below $T_{2M}$ two magnetic subspectra exist which accounts to the major fraction. In addition, (unexpectedly) the magnetic field (HF1 or $H_{eff}$) of the *minor* phase is higher than these belong to the major fraction.

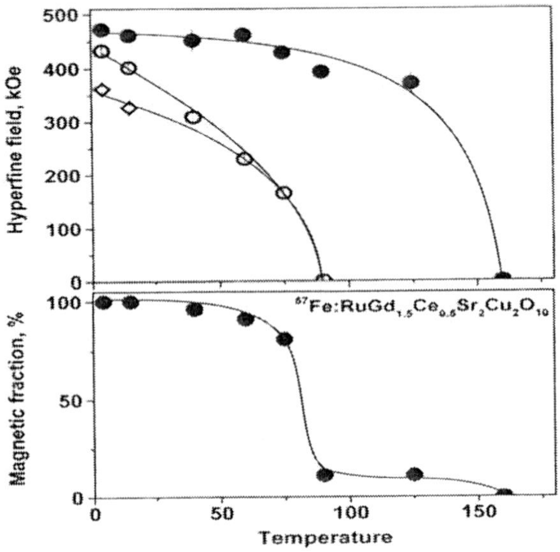


Fig. 29 Temperature dependence of the MS hyperfine fields and total magnetic fraction for 1% $^{57}Fe$ doped in $RuGd_{1.5}Ce_{0.5}Sr_2Cu_2O_{10}$

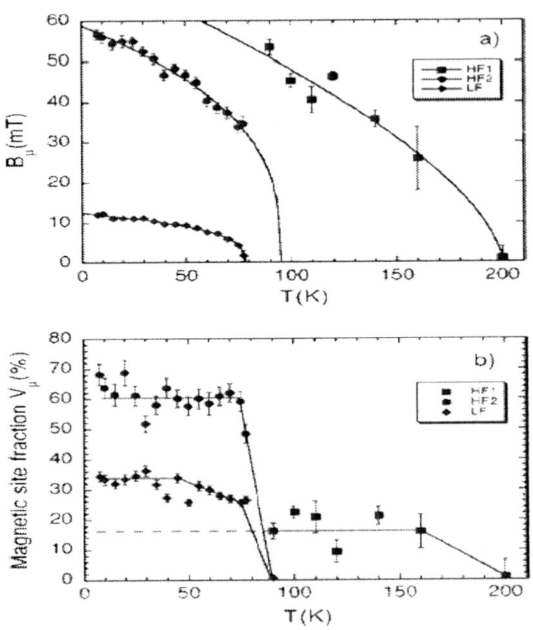


Fig. 30 Temperature dependence of different μSR internal fields (a) and the volume fraction in $RuEu_{1.4}Ce_{0.6}Sr_2Cu_2O_{10}$

## (f) The Minor AFM Magnetic Region

The main interest now is the origin of this minor AFM state in Ru-1222 which exhibits two connected magnetic phenomena namely: the second peak in $\chi(T)$ and the peculiar bell shape behavior of $H_C$. The two intriguing questions arise are: (i) why both observations do not appear in $RuEuCeSr_2Cu_2O_{10}$ (x=1) and (ii) do the two W-FM (major) and AFM (minor) correspond to two *intrinsic* magnetic regions in the *same* material or to different magnetic phases.

In the absence of microscopic information on the magnetic structure, an interpretation of these anomalies is not straightforward. The key model to this puzzle is the absence of the minor AFM state in the x=1 material. As stated above, the hole doping of the Cu-O planes, which results in metallic behavior and SC in for the x<1 materials, is obtained when $Ce^{4+}$ ions are replaced by trivalent $Eu^{3+}$. Reducing the $Ce^{4+}$ content is partially compensated by depletion of oxygen; thus the oxygen deficiency increases with $Eu^{3+}$. It is possible that this depletion is not homogeneous throughout the whole material and one cannot exclude tiny differences in the oxygen content in various clusters. There are clusters with high oxygen deficiency, and in order to maintain neutrality, those $Ru^{5+}$ ions (say about 5-10%) that are surrounded by less oxygen as first neighbors are reduced to $Ru^{4+}$. This is consistent with the average Ru valence of 4.95(5) reported in Ref. 23. The $Ru^{4+}$- $Ru^{4+}$ exchange interactions (S=1) are stronger than the $Ru^{5+}$-$Ru^{5+}$ones (S=1/2) and cause these clusters to order magnetically at a higher temperature. According to this scenario, the two magnetic transitions in Ru-1222, are due to different *intrinsic* regions, which differ in their oxygen distribution. The major W-FM fraction is the one in which $Ru^{5+}$ orders at $T_{2M}$ (80-90 K), whereas the *minor* fraction which is rich with $Ru^{4+}$ ions orders *ferrimagnetically* at $T_M$ (around 160 K). It is reminiscent the general trend in the ruthenates, where the $Ru^{4}$ ions (e.g. SrRuO3 $T_M$=165 K) order at much higher temperatures than materials with $Ru^{5+}$ ions (e.g. $EuSr_2RuO_6$ $T_M$=30 K). The maximum in $\chi(T)$ stems from different temperature dependencies of the opposing $Ru^{4+}$ and $Ru^{5+}$ sublattice magnetic moments. In this AFM state, both the anisotropy and AFM exchange energies are temperature dependent. The competition between them may lead to this unusual maximum and to the disappearance of $H_C$. Thus, the two magnetic transitions in Ru-1222 may correspond to two intrinsic magnetic regions in the same material.

## (g) The Sr-Ru-Cu-$O_3$ system

In view of the results presented so far, it appears tempting to argue that the *minor* magnetic fraction in Ru-1222 and the anomalies discussed above, are due to the itinerant ferromagnetic SrRuO3 ($T_M$ =165 K), which may exist as an extra impurity phase. However, the comparison between the MS of Fe doped Ru-1222 and SrRuO3 samples exclude this assumption [5,50]. In Ru-1222, the dilute Fe probe, successfully follows the Ru magnetization and well-defined magnetic hyperfine fields are visible in the MS spectra, up to $T_M$ Fig. 28). On the other hand, the MS spectra of SrRuO3 at T>90 K, do not show any sign whatsoever of magnetic order, indicating a weak Fe–Ru coupling at elevated temperatures. It was shown that all the FC magnetic curves of the Sr-Cu-$RuO_3$ materials, in which Cu substitutes Ru and/or Sr, have a typical FM shape and resemble the FM features of pure $SrRuO_3$ [49]. All isothermal M(H) curves below $T_M$, are strongly dependent on the field (up to 7-10 kOe), until a common slope is reached. Fig. 31 shows that at low applied fields, the M(H) curve exhibits a typical FM-like hysteresis loop. The temperature dependence of $H_C$ for $SrCu_{0.1}Ru_{0.9}O_3$ decreases rapidly with T up to 70 K, then increases with a peak at 100 and becomes zero at $T_M$ (Fig. 31). For higher Cu content, the peaks shifts to 110 K. The peaks in $H_C$ appear at the *same temperature range* as the peaks in $H_C$ in Ru-1222 presented in Figs.19,22 and 26. In fact the 1% $^{57}Fe$ MS spectrum of $SrCu_{0.1}Ru_{0.9}O_3$ shows one singlet, (similar to that of $SrRuO_3$) but the material is FM up to 155 K [49]. The similarity between the bell shape of $H_C(T)$ in $SrCu_{0.1}Ru_{0.9}O_3$ and Ru-1222, may lead to the suspicion that the minor magnetic phase in Ru-1222 may be caused by a small *extra* magnetic phase of Sr-Cu-Ru-$O_3$ phase, possible as small nano-particles, undetected by XRD. Its exact composition and fraction depend on Ce/O concentrations as well as on the preparation procedures.

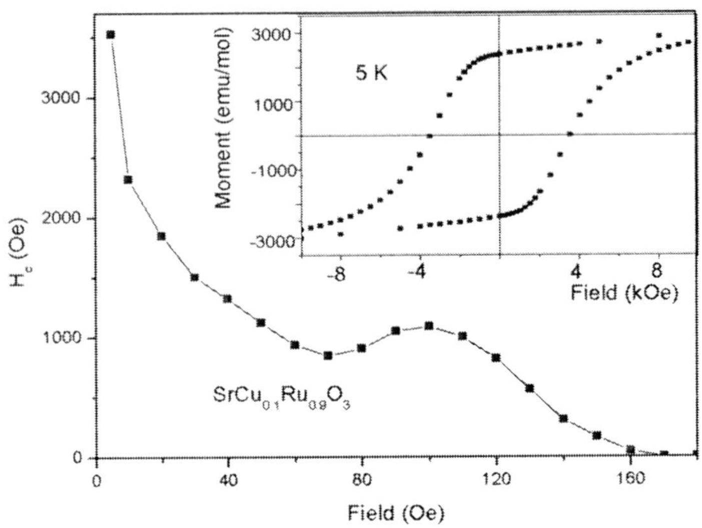


Fig.31 The temperature dependence of the coercive fields and the hysteresis loop at 5 K for $SrCu_{0.1}Ru_{0.9}O_3$

## (h) Anisotropic Magnetization of $RuEu_{1.5}Ce_{0.5}Sr_2Cu_2O_{10}$ Thin Films

Since single crystals of Ru-1222 are unavailable at the present time, all magnetic investigations reviewed here were so far limited to ceramic polycrystalline materials. Hence, the experimental magnetic results represent average values only. The first report on highly c-oriented Ru-1222 thin films by W.Q. Lu et. al.[51] exhibits SC at $T_C$=32 K, and magnetic ordering around 130 K. Following the same experimental details, we managed to grow highly c-axis oriented epitaxial non-SC $RuEu_{1.5}Ce_{0.5}Sr_2Cu_2O_{10}$ thin films on (100) $SrTiO_3$ and to measure their magnetic properties [52].

The ZFC and FC curves measured at applied field parallel ($H_{PAR}$) and perpendicular ($H_{PERP}$) of 90 Oe, are presented in Fig. 32. The two branches merge in both orientations at $T_M$ around 180 K which is somewhat higher than $T_M$ obtained for non-SC ceramic Ru-1222 bulk material (Fig. 24). Isothermal M(H) measurements (at 5 K) in both directions are exhibited in Fig. 33. The saturation moments obtained at 5 K are $M_{sat}$ = 3.0(2) and 0.40(3) $\mu_B$ for $H_{PAR}$

and $H_{PERP}$, respectively. Similar M(H) curves have been measured at various temperatures. At low applied fields, the M(H) curve exhibits a typical ferromagnetic-like hysteresis loop (Fig. 33 inset) for both $H_{PAR}$ and $H_{PERP}$. The coercive field ($H_C$ = 500(40) Oe at 5 K) is similar to that of Ru-1222 ceramic materials.

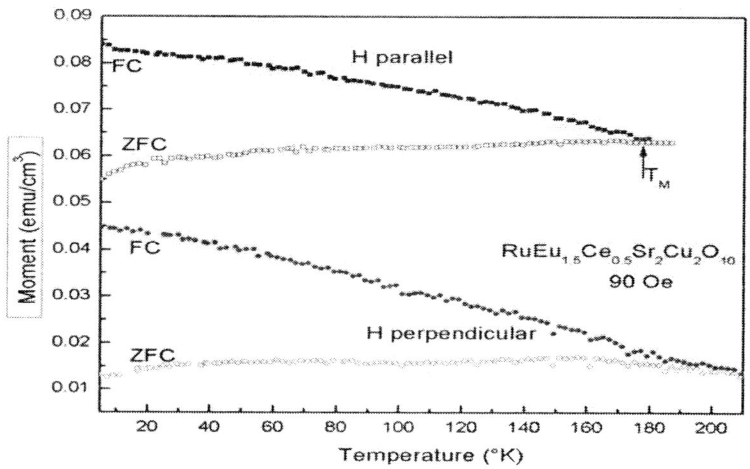


Fig.32 ZFC and FC moments versus temperature for the 200 nm-thick $EuEu_{1.5}Ce_{0.5}Sr_2Cu_2O_{10}$ thin film.

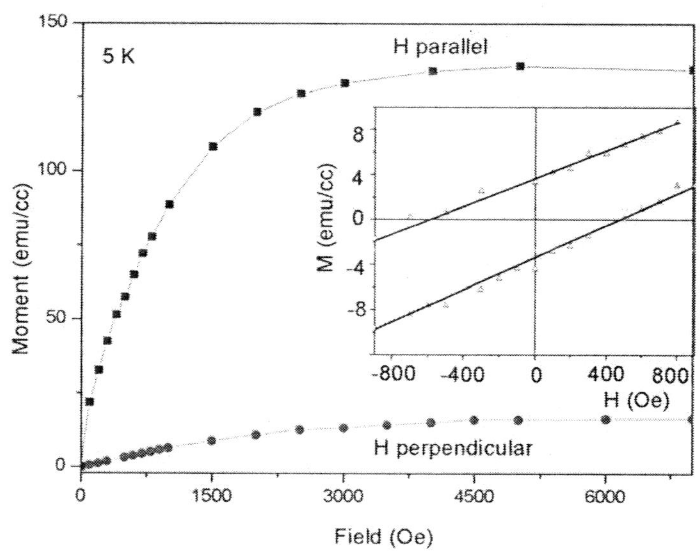


Fig 33 Magnetization curves at 5 K for the 200 nm-thick $RuEu_{1.5}Ce_{0.5}Sr_2Cu_2O_{10}$ film measured in both directions. The inset shows the low field hysteresis loop obtained at 5 K for H parallel

Three significant conclusions can be made. (i) The higher magnetic moment value obtained for $H_{PAR}$ indicates that the easy axis of the magnetization is confined to the basal plane. This is consistent with the easy axis determination based on MS studies on $^{57}Fe$ doped Ru-1222 ceramic materials [5]. Further, as the internal magnetic field is confined to the Ru-O plane, while SC resides mostly in the Cu-O layers. Therefore, magnetism and SC are mostly decoupled and can coexist. (ii) The measured $M_{sat}$= 3 $\mu_B$ value, is identical to the fully saturated 3$\mu_B$ moment expected for $Ru^{5+}$ in the high-spin state, i.e. $g\mu_B S$ for g=2 and S=1.5. This value is higher than the FM $M_{sat}$ ~0.9-08$\mu_B$ and the average $P_{eff}$=2.05-2.15$\mu_B$ obtained for ceramic materials. (iii) The anisotropy factor of Ru-1222 for $H_{PAR}$ and $H_{PERP}$ is 7.5.

## (IV) Summary and Conclusions

### Proposed Models for Magnetic States of Ceramics $RuEu_{2-x}Ce_x Sr_2Cu_2O_{10-\delta}$

Without a detailed magnetic structure from neutron diffraction studies, it is difficult to determine the exact nature of the magnetic structure in Ru-1222 system. The summary of the magnetic features of ceramic materials is as follows. The dc and ac magnetic features exhibit two magnetic transitions. A bulk magnetic transition at $T_{2M}$ =80-90 K and at $T_M$=150-160 K. Below $T_{2M}$, at low applied magnetic fields (H<3-4 kOe), irreversibility in the ZFC and FC curves is observed, which disappears at $T_{2M}$. For higher $H_{ext}$, both curves collapse to a single FM-like behavior. The dc ZFC and ac '$\chi$(T) curves show a well-distinguished peak at temperatures which are a bit lower than $T_{2M}$. Wide FM hysteresis loops are opened at low temperatures ($H_C$ ~450-500 Oe at 5 K) which become narrower as the temperature increases and practically disappears around 60-70 K. This major W-FM magnetic state is due to canting of the Ru moments, arises as a result of an antisymmetric exchange coupling of the Dzyaloshinsky-Moriya (DM) type [15] between neighboring Ru moments, induced by a local distortion that breaks the tetragonal symmetry of the $RuO_6$ octahedra [18]. The $M_s$ of 0.89-0.39 $\mu_B$/Ru is only a large fraction of 1 $\mu_B$, the calculated low spin state of $Ru^{5+}$, implying a very large canting angle of the AFM Ru moments. The large $M_{rem}$ /$M_s$ ratio and the high positive $\theta$ values are also consistent with this W-FM interpretation. This state exhibits some spin-glass features or alternatively may be considered as an itinerant W-FM, as discussed above. The high $M_s$ =3 $\mu_B$/Ru obtained (corresponds to $Ru^{5+}$ in the high spin state) for Ru-1222 thin film ($H_{PAR}$) indicates that the easy axis of the magnetization confined to the basal plane. Above $T_{2M}$, a *second* small peak around 120 K is observed in both dc and ac susceptibility curves. Narrow hysteresis loops (AFM-like) are observed and the $H_C$(T) curves show a maximum around 120 K.

The accumulated results were interpreted by three scenarios. Scenario (A) also supported by ESR studies [53], assumes that the materials are *uniform as a whole* and that *all* the Ru ions behave in the same manner: The Ru sublattice orders AFM at $T_M$ and W-FM at $T_{M2}$ and then becomes SC at $T_C$. (B) Detailed analysis of the magnetization under various thermal-magnetic conditions suggests a phase separation of Ru-1222 into FM and AFM [7,54] nano-domain species inside the crystal grains. A *minor* part of the material becomes FM (at $T_M$), whereas the *major* part orders AFM and then becomes SC at $T_C$. In this scenario, the unusual SC state is well understood. However, the reopening of the hysteresis loops above $T_{M2}$, and the peak observed in $H_C$ cannot be reconciled, because the hysteresis loops opened at $T_M$ would remain all the way down to low temperatures and $H_C$ would increase continuously, or at least remain constant. The preferred scenario (C), which is a combination of the two (A) and (B) models, also assumes phase separation but in a different manner. The detailed MS [49] analysis and $\mu$SR studies [11] indicate with little doubt the occurrence of two magnetic phases, which order magnetically at two different temperatures. Due to inhomogeneity of the oxygen deficiency across the samples, a *minority* fraction (~10%) rich with $Ru^{4+}$ orders AFM at $T_M$, displaying hysteresis loops. When the temperature is lowered, the AFM alignment of the spins strengthens. $H_C$ goes through a maximum, and then decreases and disappears. Alternatively, this minor fraction is an impurity phase of scattered islands of Sr-Cu-Ru-$O_3$ nano-particles exhibiting typical FM single domain hysteresis loops [55]. The *majority* fraction becomes W-FM ordered at $T_{M2}$ and displays the typical FM-like hysteresis loops.

The SC and magnetic properties of this new Ru-1222 system have been extensively investigated. We attribute the magnetic order to the Ru sublattice, whereas SC is confined to the $CuO_2$ planes. Both sites are practically decoupled from each other. The SC state is well established and understood in the frame of granular superconductors. Hole doping of the Cu-O planes, which results in SC, can be optimized with either appropriate variation of the $Ce^{4+}$ and oxygen content. The non-SC magnetic $RuY_{1.5}Ce_{0.5}Sr_2Cu_2O_{10}$, is used to describe the magnetic behavior of the Ru-1222 system. The magnetic structure of all $RuR_{2-x}Ce_x Sr_2Cu_2O_{10-\delta}$ materials studied is practically the same. The two magnetic transitions found are very complex and their interpretation is not straightforward. The high Meissner fraction ( ~70%), the specific heat and STM measurements demonstrate bulk SC. Taking into account the μSR and Mossbauer studies which indicate clearly that below $T_{2M}$ ((at least) the whole materials are magnetically ordered, it can be concluded with high confidence, that in Ru-1222, both SC and weak-ferromagnetic states coexist below $T_C$. The two states are practically decoupled. SC survives because the Ru magnetic moments align in the basal planes. The exact nature of the Ru spins magnetic ordering is still debated and further studies such as neutron diffraction and $^{99}$Ru Mössbauer spectroscopy studies are warranted to investigate the magnetic structure of the Ru-2122 system.

## Acknowledgments

This research is supported in part by the Israel Science Foundation (ISF, 2004 grant number: 618/04), and by the Klachky Foundation for Superconductivity.

## References

[1] R. Nagarajan, C. Mazumdar, Z. Hossain, S,K, Dhar, K.V. Gopalakrishnan, L.C. Gupta, C, Godar, B,D, Padalia and R. Vijayaraghavan, *Phys. Rev. Lett.* **72**, 274 (1994).

[2] R .J. Cava , H. takagi,B. Batlogg, H,W, Zandbergen, J.J. Krajewski, W.F. Peck Jr, R.B.Van Dover, R,J, Felder, T. Siegrist, K. Mizuhashi, J.O. Lee, E. Eiski, S.A. Carter and S. Usida, *Nature* **367**, 146 (1994); ibid, 367, 252 (1994).

[3] R.J. Cava, J.J. Krajewski, H. Takagi, H. Zandbergen, R.W. R.B.Van Dover, W.F. Peck Jr, and B. Hessen, *Physica C* **191**, 237 (1992). W. Shiwei et. al. *Physica C* **210** 463 (1993).

[4] L.Bauernfeind, W. Widder and H.F. Braun, *Physica C* **254**, 151 (1995).

[5] Felner, U.Asaf, Y. Levi, and O. Millo, *Phys. Rev. B* **55**, R3374 (1997); *ibid, Physica C* **334**, 141 (2000).

[6] E.B. Sonin and I. Felner, *Phys. Rev. B.* **57**, R14000(1998).

[7] Y.Y. Xue, B. Lorenz, A. Baikalov, D.H. Cao, Z.G. Li and C.W. Chu, *Phys. Rev. B* **66**, 014503 (2002).

[8] C. Bernhard, J.L.Tallon, Ch. Niedermayer, Th. Blasius, A. Golnik, B. Btucher, R.K. Kremer, D.R. Noakes, C.E. Stronach and E.J. Ansaldo, *Phys. Rev. B* **59**, 14099 (1999).

[9] C. Bernard, J. L. Tallon, E. Brucher and R. K. Kremer, *Phys. Rev. B* **61**, R14960 (2000).

[10] I Felner, I. Asaf, U and E. Galstyan, *Phys. Rev. B* **66**, 024503 (2002).

[11] X.H. Chen, Z. Sun, K. Q. Wang, S.Y.Li, Y.M. Xiong, M. Yu and I.Z. Cao, *Phys. Rev. B* **63**, 064506 (2001).

[12] Shengelaya. R. Khasanov, D.G.Eshchenko, I.Felner, U. Asaf, H. Keller, K.H. Muller, *Phys. Rev. B* **69**, 024517 (2004).

[13] G.V.M. Williams and M. Ryan, *Phys. Rev. B* **64**, 094515 (2001).

[14] J.W. Lynn, B. Keimer, C. Ulrich, C. Bernhard and J.L. Tallon., *Phys. Rev. B* **61**, 964 (2000).

[15] J.D. Jorgensen, O. Chmaissem, H. Shaked, S. Short, P.W. Klamut, B, Dabrowski and J.I. Tallon, *Phys. Rev. B* **63**, 05444 (2001).

[16] J. Dzyaloshinsky, *J. Phys. Chem. Solids* **4**, 241 (1958).

[17] I, Felner, E. Galstyan, V.P.S. Awana and E. Takayama-Moromachi, *Physica C* **408**-410, 161 (2004).

[18] I.Felner, U. Asaf, F. Ritter, P.W. Klamut and B. Dabrowski, *Physica* **C 364**-368, 368 (2001).

[19] C.S. Knee, B.D. Rainford and M.T. Weller *J. Mater. Chem.* **10**, 2445 (2000).

[20] G.V.M. Williams, A.C. McLaughlin, J.P. Attfield, S. Kramer and H.K. Lee, *Cond-Mat* **0108521** (2001).

[21] G.M. Kuzmicheva, V.V. Luparev, E.P. Khlybov, I.E. Kostyleva, A.S. Andreenko and K.N. Gavrilov, *Physica C* **350**, 105 (2001).

[22] T. Yokosawa, V.P.S. Awana, K.Kimoto, E. Takayama-Muromachi, M. Karppinen, H.Yamauchi and Y. Matsui, *Ultramicroscopy* **98**, 283 (2004).

[23] Felner, U. Asaf, C. Godart and E. Alleno, Physica B **259**-261, 703 (1999).

[24] G.V.M. Williams and L.Y. Yang and R.S. Liu, *Phys. Rev. B* **65**, 064508 (2002).

[25] V.P.S. Awana , M. Karppinen, H. Yamauchi , M. Matvejeff, R.S. Liu and L-Y. Yang, *J. Low Temp. Phys.* **131**, 1211, (2003)

[26] Y. Tokunaga, H. Kotegawa, K. Ishida,Y. Kitaoka, H. Takagiwa and J. Akimits *Phys. Rev. Lett.* **86**, 5765 (2001).

[27] Ishida (privete communication)

[28] T. Kaibin, Q. Yitai, Z. Yadun, Y. Li, C. Zuyao and Z. Yuheng, *Physica C*, **259**, 168 (1966).

[29] B.D. Hennings, K.D.D. Rathnayaka, D.G. Naugle and I. Felner, *Physica C*, **370**, 253 (2002).

[30] S.D. Obertelli, J.R. Cooper and J.L. Tallon, *Phys. Rev. B* **46** 14928 (1992).

[31] V.P.S. Awana , E. Takayama- Muromachi, M. Karppinen, H. Yamauchi, *Physica C* **390**, 233 (2003).

[32] Felner, I. Asaf U. Goren, S. and C. Korn., *Phys. Rev. B.* **57**, 550 (1998).

[33] G. Oomi, F. Honda, M. Ohashi, T. Eto, D.P. Hai, S. Kamisawa, M. Watanabe and K. Kadowaki, *Physica B* **312**, 88 (2002).

[34] P.W. Klamut, B. Dabrowski, S.M. Mini, M. Maxwell, J. Mais, I. Felner, U. Asaf, F. Ritter, A. Shengelaya, R. Khasanov, I.M. Savic, H. Keller, A. Wisnieewski, R. Puzniak, I.M. Fita, C. Sulkowski and M.Matusiak, *Physica C* **387**, 33 (2003).
[35] Felner, E.B. Sonin, T. Machi and N. Koshizuka, *Physica C* **341**-348 715 (2000).
[36] Felner, E. Galstyan, R. Herber and I. Nowik, *Phys. Rev. B* **70**, 094504 (2004).
[37] L.Shi, G. Li, X.D. Zhang, S.J. Feng and X.G. Li, *Physica C* **383,** 450 (2003).
[38] Felner and U. Asaf, *Physica C* **292,** 97 (1997).
[39] V.P.S. Awana and E. Takayama-Muromachi, *Physica C* **390**, 101 (2003).
[40] M.G. das Virgens, S. Garcia, M.a. Continentino and L. Ghivelder, *Phys. Rev. B* **71** (2005) in press.
[41] M.T. Escote,V.A. Meza, R.F. Jardin, L.Ben-Dor, M.S. Torikachvili and A.H. Lacerda, *Phys. Rev. B* **66**, 144503 (2002).
[42] Felner, E. Glastyan, B. Lorenz, D. Cao, Y.s. Wang, Y.Y. Xue and C.W. Chu, *Phys. Rev. B* **67**, 134506 (2003).
[43] Zivkovic, Y. Hirai, B.H. Frazer, M. Prester, D. Drobac, D. Atiosa, H. Berger, D. Pavuna, G. Margaritondo, I. Felner and M. Onellion, *Phys. Rev. B* **65**, 144420 (2002).
[44] Y.Y. Xue, B. Lorenz R.L. Meng, A. Baikalov and C.W. Chu, *Physica C* **364-365**, 251 (2001).
[45] R.E. Freitas, J.E. Musa, L. Ghivelder, E, Baggio-Saitovitch, and S. Gracia, *Physica C* **408**, 191 (2004).
[46] L. Shi, S.J. Feng, G. Li, X.F. Sun and X.G. li *Physica C* 386, **65** (2003).
[47] L. Ji, M.S. Rzchowski, N. Anand and M. Tinkham, *Phys. Rev. B* **47**, 470 (1993).
[48] Felner, V.P.S Awana and E. Takayama-Muromachi, *Phys. Rev. B* **68**, 094508 (2003).
[49] C. A. Cardoso, A. J. C. Lanfredi, A. J. Chiquito, V. P. S. Awana, H. Kishan, R. L. de Almeida, O. F. de Lima, F. M. Araujo-Moreira, *Phys. Rev. B* **67**, 020407 (2003); ibid *Physica C* **405**,212 (2004).
[50] Felner, E. Galstyan and I. Nowik, *Phys. Rev. B*,**71** 064510 (2005), ibid 70, 094504 (2004).
[51] Nowik and I. Felner, *Hyperfine Inter.* **141-142**, 213 (2002).
[52] W.Q. Lu, Y. Yamamoto, I. Ohkubo, V.V. Petrykin, M. Kakihana, Y. Matsumoto and H. Koinuma, *Physica C* **405,** 21 (2004).
[53] Felner, G. Kopanov and G. Koren *Physica C* (2005) submitted.
[54] K. Yoshuda, H. Kojima and H. Shimiza, *J. Phys. Soc. Japan*, **72**, 3254 (2003).
[55] Y.Y. Xue, D.H. Cao, B. Lorenz and C.W. Chu, *Phys. Rev. B* **65**, 020511(R) (2002).
[56] S.I. Ikeda and N. Shirakawa, *Physica C* **341**-348, 785 (2000).

In: New Research on Superconductivity and Magnetism ISBN: 1-60021-540-8
Editor: Lannie K. Tran, pp. 35-81 

*Chapter 2*

# APPLICATIONS OF EXTENDED CRITICAL STATE MODEL TO HIGH $T_C$ SUPERCONDUCTORS

***Yoshihide Kimishima***
Department of Physics, Division of Physics, Electrical and Computer Engineering, Graduate School of Engineering, Yokohama National University
Tokiwadai 19-5, Hodogaya-ku, Yokohama 240-8501, JAPAN

## Abstract

Analytical expressions of magnetization $M$ and critical current density $J_c$ by an extended critical state model. The formula of *Kimishima model* was one of the universal functions interconnecting the Bean and Anderson-Kim models. The second formula of *Exponential Kimishima model* was derived by the extension of *Kimishima model*, considering the exponential decay of $J_c$ near the irreversible field $H_{irr}$. These theoretical magnetization curves were well fitted to the experimental ones. Estimated magnetic parameters were important to explain the temperature dependences of $J_c$ and the pinning properties of high $T_c$ superconductors.

## Introduction

Quantitative studies on the pinning properties of high $T_c$ superconductors are very important for the superconductor physics and applications. In this chapter, the analytical expressions of magnetization are given by the extended critical state model. In earlier papers [1,2], the author presented the analytical expressions of AC magnetic susceptibility $\chi_{ac}$ by some critical state models for type-II superconductors. He indicated the importance of a lower critical field $H_{c1}$ and surface equilibrium (reversible) magnetization $M_{eq}$ to analyze the magnetization curve and magnetic susceptibility.

In Fig. 1, semi-logarithmic plots of ZFC $M$-$H$ loop between 0 T and 5 T were shown for the sintered high $T_c$ type-II superconductor of $Bi_{1.8}Pb_{0.2}Sr_2Ca_2Cu_3O_y$ (Bi2223) at 50 K [3]. Since the present sample was composed of thin flakes of Bi2223, aligning nearly parallel to the pressed plane and the external field $H$, the demagnetization effect was ignored. The broken line shows the $M_{eq}$ estimated by the relation of $M_{eq}=(M_U+M_L)/2$ [4,5], where $M_U$ is shown by the upper hysteretic magnetization curve on the decreasing field process after the initial magnetized process, and $M_L$ is shown by the lower hysteretic magnetization curve on the increasing field process. From this figure, following arguments are possible on the basis of theoretical results for the critical state models [1,2]. One argument is that the *lower critical field* $H_{c1}$ is much smaller than $H_p$ at which the $M_{vir}$ and $M_L$ coincide with each other, where $H_p$ is called as the *external penetration field* above which the *internal flux density* $B_{int}$ has a non-zero value at the center of the sample. Theoretically, the initial magnetization $M_{vir}$ must equal $M_{eq}$ below $H_{c1}$, but the $M_{vir}$ in Fig. 1 does not agree with $M_{eq}$ down to a few mT. The $M_{eq}$ in this figure seems to have a minimum value near $\mu_0 H_p$ much larger than $\mu_0 H_{c1}$, where $\mu_0$ is the vacuum permeability. From these characteristic behaviors of magnetization, the present author constructed the *Kimishima* (*Kimi.*) *model* as an extended critical state model for the Bi2223 system [3].

In Ref. [2], the $M_{eq}$ was assumed to have a discontinuity at $H_{c1}$ and approximated to be constant above $H_{c1}$, which has been called as the "triangular approximation" [6,7]. While Ginzburg-Landau theory [8] gave the $M_{eq}$ showing a discontinuity at $H_{c1}$ and logarithmic dependence on the *external field H* above $H_{c1}$. In the present *Kimi. model*, the $M_{eq}$ is continuous at $H_{c1}$, and the derivative of $dM_{eq}/dH$ becomes larger than -1 above $H_{c1}$. It also means that the *equilibrium flux density at the inner surface* $B_{eq}$ is continuously 0 at $H_{c1}$ where $B_{eq}$ is given by $\mu_0(H + M_{eq})$.

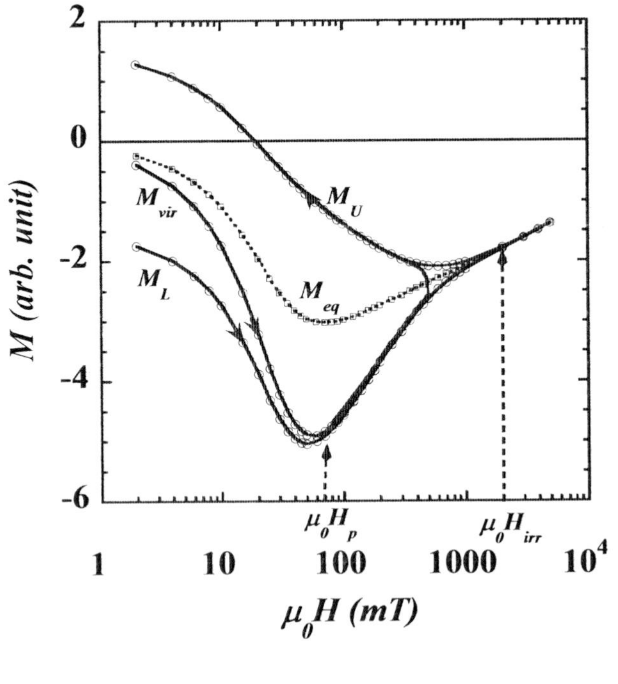


Figure 1. Magnetization curve of sintered Bi2223 high $T_c$ superconductor at 50K under the field up to 5T. $M_{vir}$ is the initial magnetization, and $M_{eq}$ is the reversible (equilibrium) magnetization. $M_U$ and $M_L$ are hysteretic upper and lower curve. *M-H* curve with field amplitude of 500mT is also shown by solid line.

# Extended Critical State Model with $H_{c1}$ and $M_{eq}$ (*Kimishima Model*)

All of following formulae of the *Kimishima (Kimi.) model* are applicable to the $2a\times 2b$ orthorhombic long column. The origin of $x$-axis is located at the center of the column. If $a<<b$, they represent the case of slab geometry with a thickness of $2a$ along the $x$-direction perpendicular to the slab surface. On the basis of Kim's idea [9], the author assumed the internal current density equals the *critical current density* of

$$\mu_0 |J_c^{int}(x,H)| = \frac{B_{eq}^*(B_{eq}^* + 2B_0)}{2a[|B_{int}(x,H)| + B_0]} \quad (1).$$

Here $B_{int}(x,H)$ is the *internal flux density* depending on $x$ and $H$, and $B_{eq}^*$ is defined as $B_{eq}(H_p)$ under the external field of $H_p$, namely $B_{eq}^* = B_{eq}(H_p) = B_{int}(\pm a, H_p) = \mu_0 [H_p + M_{eq}(H_p)]$. When the external field $H$ becomes larger than $H_p$, the $B_{int}(0,H)$ at the center of the sample has a non-zero value. While $B_0$ in Eq. (1) is the specific constant in the present *Kimi. model*. If $(B_{int}/B_0)>>1$ holds, the relation of $J_c^{int} \propto 1/B_{int}$ is obtained from this equation. Then the *pinning force density* $f_p = J_c^{int} B_{int}$ to each quantum flux is independent of $B_{int}$, which is the case of Anderson-Kim's (or simplified Kim's) model [1,10]. On the other hand, if $(B_{eq}^*/B_0) <<1$ and $(B_{int}/B_0)<<1$ hold, $J_c^{int}$ is constant and $f_p$ shall be linear with $B_{int}$ as in the case of Bean's model [1,11]. Therefore, Eq. (1) is one of the *universal* expressions of the critical state model. Since the above model was based on the lattice formation of quantum flux, the flux glass and flux liquid phase transitions are not problems in the present model.

When the external field $H$ approaches the *irreversible field* $H_{irr}$, above which the magnetization becomes reversible and coincide with $M_{eq}(H)$ as shown in Fig. 1, both of the $J_c^{int}$ and $f_p$ should be simultaneously reduced to 0. Hereafter, I assume the inequality of $0<H_{c1}<H_p<<H_{irr}$. The principal characteristics of the present model is to adopt the semi-empirical $M_{eq}$ into the expressions of $M$. Discussion about the theoretical form of $M_{eq}$ on $H$ will be given in a later section.

# Magnetic Flux Density Distribution with Surface Shielding by $M_{eq}$

When the *external field amplitude* is $H_m$, the profiles of *internal flux density* $B_{int}(x,H)$ are given in Figs. 2 and 3. Figs. 2(a) and (b) are illustrated for two kinds of initially magnetized processes. Meanwhile, three kinds of decreasing field processes –after the initial process– are given in Fig. 3(a), (b) and (c). The definition of $K_m = B_0 + |B_{eqm}|$ were used in these figures, where $B_{eqm} = B_{eq}(H_m)$. $B_{int}(x,H)$ was derived from Maxwell's equation of

$$\mu_0 J_c^{int}(x,H) = -\partial B_{int}(x,H)/\partial x \quad (2),$$

and Eq. (1). Here I will give the expressions of $B_{int}(x,H)$ for the initially magnetized process and field decreasing process for $0\leq x\leq a$. For $-a\leq x\leq 0$, we can obtain the internal magnetic flux density by the transformation of $B_{\rm int}(-x,H)$ for the following $B_{\rm int}(x,H)$-expressions.

## Initially Magnetized Process

In the initially magnetized process, $B_{\rm int}(x,H)$ in $0 \leq H \leq H_{\rm c1}$ is kept zero by the perfect Meissner effect. When $H_{\rm c1} \leq H \leq H_{\rm m}$,

$$B_{\rm int}(x,H) = -B_0 + \sqrt{\frac{A}{a}\{x - x_0(H)\} + B_0^{\,2}} = -B_0 + \sqrt{K(H)^2 - A\left(1 - \frac{x}{a}\right)} = B_{\rm vir}(x,H) \qquad (3),$$

where $A = B_{\rm eq}^{\ *}(B_{\rm eq}^{\ *} + 2B_0)$, $K(H) = |B_{\rm eq}(H)| + B_0$ and $x_0(H) = a\left\{1 - \frac{K(H)^2 - B_0^{\,2}}{A}\right\}$.

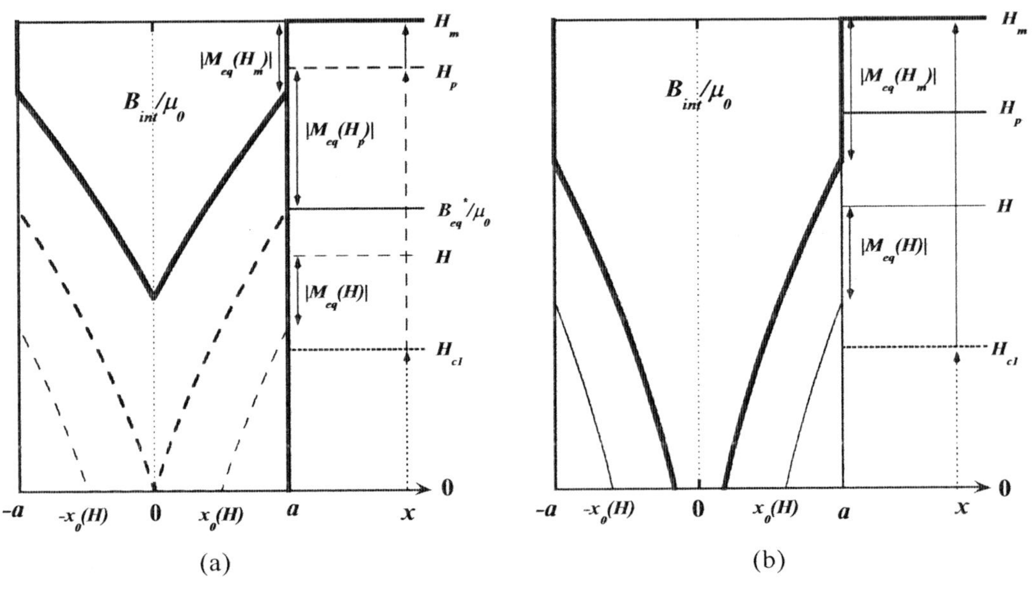


Figure 2. (a) Initial $B_{\rm int}(x,H)$-profile in *Region-I* and (b) in *Region-II*.

This equation was derived from Eqs. (1) and (2) for $J_{\rm c}^{\rm int} \leq 0$ and $B_{\rm int}(x,H) \geq 0$, and the boundary conditions of $B_{\rm vir}(a,H) = B_{\rm eq}(H)$ and $B_{\rm vir}(x_0,H) = 0$. Here $B_{\rm int}(x,H) = 0$ for $0 \leq x \leq x_0(H)$ and $B_{\rm int}(x,H) = B_{\rm vir}(x,H)$ for $x_0(H) \leq x \leq a$ as shown in Figs. 2 (a) and (b). We can divide the initial process into two cases by the range of $B_{\rm eq}^{\ *}$. The one is the *Region I* of $0 \leq B_{\rm eq}^{\ *} \leq B_{\rm eqm}$ corresponding to $x_0(H_{\rm m}) \leq 0$ which means $H_{\rm m} \geq H_{\rm p}$. Another one is the *Region II* of $B_{\rm eqm} \leq B_{\rm eq}^{\ *}$ corresponding to $x_0(H_{\rm m}) \geq 0$ which means $H_{\rm m} \leq H_{\rm p}$. The $B_{\rm int}(x,H)$-curves for *Region I* and *II* were shown in Fig. 2(a) and (b), respectively.

# Field Decreasing Process

On the field decreasing process after the above initial process, we must additionally divide the region *I* into *IA* and *IB*. In the *Region IA*, $B_{int}(0,H)$ is kept $B_{int}(0,H_m)$ at $H$ down to positive value of $H_d$ larger than $H_{c1}$. In the region *IB*, the $B_{int}(0,H)$ is kept $B_{int}(0,H_m)$ at $H$ down to negative value of $-H_f$ less than $-H_{c1}$. For the three regions of $B_{eq}^*$ (*IA*, *IB* and *II*), the expressions of $B_{int}(x,H)$ are obtained as shown in Figs. 3(a), (b) and (c).

*-Region IA-*

Fig. 3(a) shows the $B_{int}(x,H)$-profiles in *region IA* of $0 \le B_{eq}^* \le -B_0+\sqrt{\{B_0^2+K_m^2\}/2}$, where $B_{int}(x,H)=B_{vir}(x,H_m)$ for $0 \le x \le x_1(H)$, and $B_{int}(x,H)=B_{D1}(x,H)$ for $x_1(H) \le x \le a$ when $H_m \ge H \ge H_d$ as

$$B_{int}(x,H)=-B_0+\sqrt{K_m^2 - A - \frac{A}{a}\{x-2x_1(H)\}}=-B_0+\sqrt{K(H)^2 + A\left(1-\frac{x}{a}\right)} = B_{D1}(x,H) \qquad (4).$$

Here $x_1(H)=a\left\{1+\dfrac{K(H)^2-K_m^2}{2A}\right\}$, which was derived from Eq. (1) and (2) for $J_c^{int} \ge 0$ and $B_{int}(x,H) \ge 0$, and the boundary conditions of $B_{D1}(a,H)=B_{eq}(H)$ and $B_{D1}(x_1,H)=B_{vir}(x_1,H)$. The $H_d$ was defined by $B_{eq}(H_d)=-B_0+\sqrt{K_m^2-2A}$ from the relation of $B_{D1}(0,H_d)=B_{vir}(0,H_m)$.

When $H_d \ge H \ge H_{c1}$, $B_{int}(x,H)=B_{D1}(x,H)$ for $0 \le x \le a$. At the upper boundary of $-B_0+\sqrt{\{B_0^2+K_m^2\}/2}$ of the present $B_{eq}^*$-region of *IA*, $x_1(H_{c1})$ becomes 0.

When $H_{c1} \ge H \ge -H_{c1}$, $B_{int}(x,H)=B_{D1}(x,H_{c1})$ for $0 \le x \le a$. Thus, the $B_{int}(x,H)$ is independent of field in this $H$-region, where $B_{eq}(H_{c1})=\mu_0[H_{c1}+M_{eq}(H_{c1})]=0$ and $K(H_{c1})=B_0$ in the function of $B_{D1}(x,H_{c1})$. Same situation also holds in the following equations for $B_{D2}(x,H)$ and $B_{D3}(x,H)$.

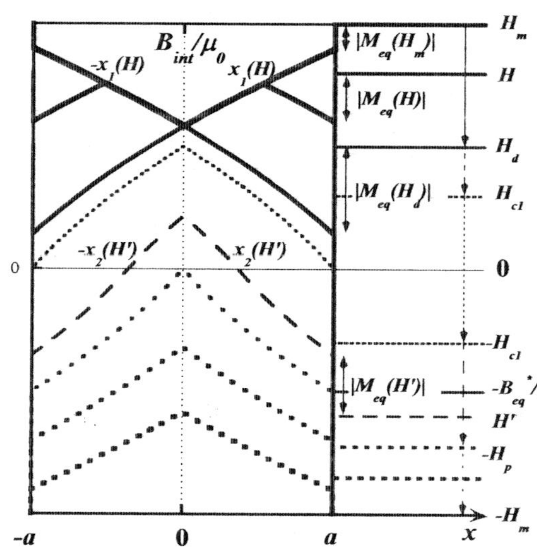


Figure 3(a). $B_{int}(x,H)$-profiles on field decreasing process in *Region IA*.

When $-H_{c1} \geq H \geq -H_p$,

$$B_{int}(x,H) = -B_0 + \sqrt{B_0^{\,2} - \frac{A}{a}\{x - x_2(H)\}} = -B_0 + \sqrt{2B_0^{\,2} - K(H)^2 + A\left(1 - \frac{x}{a}\right)} = B_{D2}(x,H) \quad (5),$$

for $0 \leq x \leq x_2(H)$ where $x_2(H) = a\left\{1 - \frac{K(H)^2 - B_0^{\,2}}{A}\right\}$, which was derived from Eq. (1) and (2) for $J_c^{\,int} \geq 0$ and $B_{int}(x, H) \geq 0$ with the boundary conditions of $B_{D2}(0,-H_{c1}) = B_{D1}(0, H_{c1})$ and $B_{D2}(x_2,H) = 0$.

For $x_2(H) \leq x \leq a$,

$$B_{int}(x,H) = B_0 - \sqrt{B_0^{\,2} + \frac{A}{a}\{x - x_2(H)\}} = B_0 - \sqrt{2B_0^{\,2} - K_m^{\,2} + A + \frac{A}{a}\{x - 2x_3(H)\}}$$

$$= B_0 - \sqrt{K(H)^2 - A\left(1 - \frac{x}{a}\right)} = B_{D3}(x,H) \quad (6),$$

where $x_3(H) = a\left\{1 + \frac{2B_0^{\,2} - K(H)^2 - K_m^{\,2}}{2A}\right\}$, which was derived from Eq. (2.1) and (2.2) for $J_c^{\,int} \geq 0$ and $B_{int}(x,H) \leq 0$, and the boundary conditions of $B_{D3}(a,H) = B_{eq}(H)$ and $B_{D3}(x_2,H) = 0$. When $-H_p \geq H \geq -H_m$, $B_{int}(x,H) = B_{D3}(x,H)$ for $0 \leq x \leq a$.

*-Region IB-*

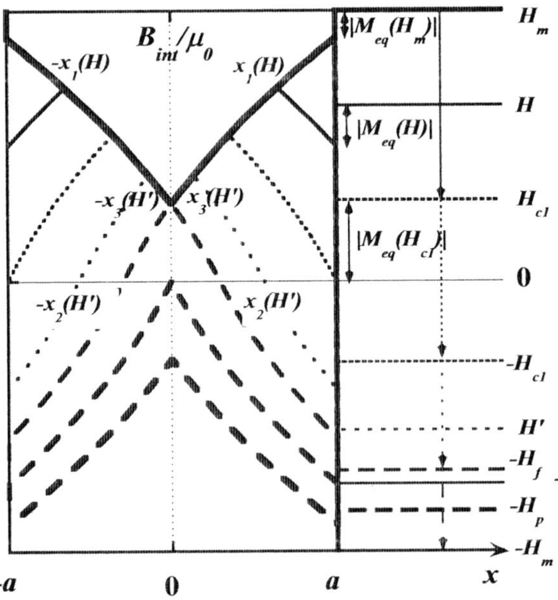


Figure 3(b). $B_{int}(x,H)$-profiles on decreasing process in the *Region IB*.

Figure 3(b). shows the $B_{int}(x,H)$-profiles in the *region IB* of $-B_0+\sqrt{\{B_0^2+K_m^2\}/2} \leq B_{eq}^* \leq B_{eqm}$, where $B_{int}(x,H)=B_{vir}(x,H_m)$ for $0 \leq x \leq x_1(H)$, and $B_{int}(x,H)=B_{D1}(x,H)$ for $x_1(H) \leq x \leq a$, when $H_m \geq H \geq H_{c1}$.

When $H_{c1} \geq H \geq -H_{c1}$, $B_{int}(x,H)=B_{vir}(x,H_m)$ for $0 \leq x \leq x_1(H_{c1})$, and $B_{int}(x,H)=B_{D1}(x,H_{c1})$ for $x_1(H_{c1}) \leq x \leq a$.

When $-H_{c1} \geq H \geq -H_f$, $B_{int}(x,H)=B_{vir}(x,H_m)$ for $0 \leq x \leq x_3(H)$, $B_{int}(x,H)=B_{D2}(x,H)$ for $x_3(H) \leq x \leq x_2(h)$, and $B_{int}(x,H)=B_{D3}(x,H)$ for $x_2(H) \leq x \leq a$. The $H_f$ is defind by $B_{eq}(-H_f) = B_0-\sqrt{2B_0-K_m^2+2A}$ from $B_{D2}(0,-H_f)=B_{vir}(0,H_m)$. The $x_3(-H_{c1})$ becomes 0 when $B_{eq}^*$ equals the lower boundary of $-B_0+\sqrt{\{B_0^2+K_m^2\}/2}$ in this *region IB*.

When $-H_f \geq H \geq -H_p$, $B_{int}(x,H)=B_{D2}(x,H)$ for $0 \leq x \leq x_2(H)$, and $B_{int}(x,H)=B_{D3}(x,H)$ for $x_2(H) \leq x \leq a$. When $-H_p \geq H \geq -H_m$, $B_{int}(x,H)=B_{D3}(x,H)$ for $0 \leq x \leq a$.

*-Region II-*

Fig. 3(c) shows the $B_{int}(x,H)$-profiles in the *region II* of $B_{eqm} \leq B_{eq}^*$, where $B_{int}(x,H)=0$ for $0 \leq x \leq x_0(H_m)$, $B_{int}(x,H)=B_{vir}(x,H_m)$ for $x_0(H_m) \leq x \leq x_1(H)$, and $B_{int}(x,H)=B_{D1}(x,h)$ for $x_1(H) \leq x \leq a$ when $H_m \geq H \geq H_{c1}$.

When $H_{c1} \geq H \geq -H_{c1}$, $B_{int}(x,H)=0$ for $0 \leq x \leq x_0(H_m)$, $B_{int}(x,H)=B_{vir}(x,H_m)$ for $x_0(H_m) \leq x \leq x_1(H_{c1})$, and $B_{int}(x,H)=B_{D1}(x, H_{c1})$ for $x_1(H_{c1}) \leq x \leq a$.

When $-H_{c1} \geq H \geq -H_m$, $B_{int}(x,H)=0$ for $0 \leq x \leq x_0(H_m)$, $B_{int}(x,H)=B_{vir}(x,H_m)$ for $x_0(H_m) \leq x \leq x_3(H)$, $B_{int}(x,H)=B_{D2}(x,H)$ for $x_3(H) \leq x \leq x_2(H)$, and $B_{int}(x,H)=B_{D3}(x,H)$ for $x_2(H) \leq x \leq a$.

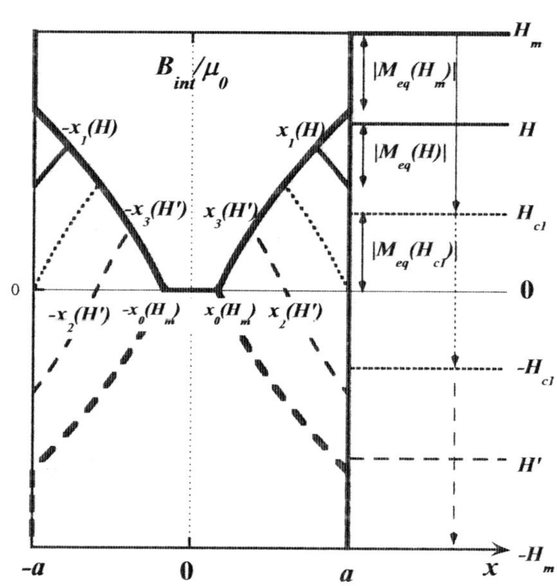


Figure 3(c). $B_{int}(x,H)$-profiles on decreasing process in the *region II* of $B_{eqm} \leq B_{eq}^*$.

## Averaged Magnetization

*Averaged magnetization* is obtained from

$$\mu_0 M(H) = \langle B_{\mathrm{int}}(x,H) \rangle - \mu_0 H \quad (7),$$

where $\langle\ \rangle$ means the spatial average. The spatially averaged flux density $\langle B_{\mathrm{int}}(x,H)\rangle$ for a $2a\times 2b$ orthorhombic long column can be obtained from the following integration [12],

$$\langle B_{\mathrm{int}}(x,H)\rangle = \left(\frac{1}{ab}\right)\int_0^a (2x+b-a)B_{int}(x,H)dx \quad (8).$$

This equation also gives the $\langle B_{\mathrm{int}}(x,H)\rangle$ of slab geometry with thickness $a$ when $a \ll b$, and polygonal long column geometry with the sides being tangent to a circle of radius $a$ when $a=b$. It should be noticed that $K(H)$ is simply represented by $K$ in the following equations.

## Initial Magnetization

Here we will give the expressions of *initial averaged magnetization* $M_{\mathrm{vir}}$ by the *Kimi. model* for two kinds of $B_{\mathrm{eq}}^*$-regions of *I* and *II*.

*-Region I-*

In the *Region I* of $0 \leq B_{\mathrm{eq}}^* \leq B_{\mathrm{eqm}}$, the initial magnetization after zero field cooling (ZFC) are given as

$$M^K{}_{V1} = -H \quad (9)$$

for $0 \leq H \leq H_{c1}$,

$$\mu_0 M^K{}_{V2} = -\mu_0 H + \frac{1}{3A}(B_0-K)^2(B_0+2K)$$
$$+\left(\frac{a}{b}\right)\frac{1}{15A^2}(B_0-K)^2[5A(B_0+2K)+(B_0-K)\{3B_0(B_0+3K)+8K^2\}] \quad (10)$$

for $H_{c1} \leq H \leq H_p$ and

$$\mu_0 M^K{}_{V3} = -\mu_0 H - B_0 + \frac{2}{3A}[K^3-(K^2-A)^{3/2}]$$
$$+\left(\frac{a}{b}\right)\frac{2}{15A^2}[(5A-4K^2)K^3+(A+4K^2)(K^2-A)^{3/2}] \quad (11)$$

for $H_p \leq H \leq H_m$.

*-Region II-*

In the *Region II* of $B_{eqm} \leq B_{eq}^*$ the initial magnetization is given by $M^K_{V1}$ for $0 \leq H \leq H_{c1}$, and $M^K_{V2}$ for $H_{c1} \leq H \leq H_m$.

Since the $M_{vir}(H_m)/H_m$ gives ZFC-DC-magnetic susceptibility under the external field of $H_m$, the $\chi_{DC}^{ZFC}$ are $M^K_{V1}(H_m)/H_m$=-1 for $0 \leq H_m \leq H_{c1}$, $M^K_{V2}(H_m)/H_m$ for $H_{c1} \leq H_m \leq H_p$, $M^K_{V3}(H_m)/H_m$ for $H_p \leq H_m << H_{irr}$ and $M_{eq}(H_m)/H_m = \chi_{eq}$ for $H_{irr} << H_m$.

## Magnetization Loops

On the field decreasing process after the above initial process, we must additionally divide the *Region I* into *IA* and *IB*. In region *IA*, the $B_{int}(0,H)$ is kept $B_{int}(0,H_m)$ at $H$ down to positive value of $H_d$ larger than $H_{c1}$. In *region IB*, the $B_{int}(0,H)$ is kept $B_{int}(0,H_m)$ at $H$ down to negative value of $-H_f$ smaller than $-H_{c1}$. For the three regions of $B_{eq}^*$ (*IA*, *IB* and *II*), the expressions of upper magnetization $M^K_U(H)$ on the field decreasing process are obtained from Figs. 3(a), (b) and (c).

The upper hysteresis curve by the *Kimi. model*, $M^K_U(H)$, are given as following for the field decreasing process from $H_m$ to $-H_m$. The lower hysteresis curve of $M^K_L(H)$ on the field increasing process from $-H_m$ to $H_m$ can be obtained from the transformation of $-M^K_U(-H)$ in the following equations.

*- Region IA -*

In this region of $0 \leq B_{eq}^* \leq -B_0 + \sqrt{(B_0^2 + K_m^2)/2}$, $M^K_U(H)$ are given by

$$\mu_0 M^K_{U1} = -\mu_0 H - B_0 - \frac{2}{3A}[K^3 + (K_m^2 - A)^{3/2} - \frac{1}{\sqrt{2}}(K^2 + K_m^2)^{3/2}]$$
$$-\left(\frac{a}{b}\right)\frac{2}{15A^2}[(5A+4K^2)K^3 - (A+4K_m^2)(K_m^2 - A)^{3/2}$$
$$-\frac{5}{\sqrt{2}}(A+K^2-K_m^2)(K^2+K_m^2)^{3/2}] \quad (12),$$

for $H_m \geq H \geq H_d$ , where $H_d$ was defined by $B_{eq}(H_d) = \sqrt{K_m^2 - 2A} - B_0$.

$$\mu_0 M^K_{U2} = -\mu_0 H - B_0 - \frac{2}{3A}[K^3 - (A+K^2)^{3/2}]$$
$$-\left(\frac{a}{b}\right)\frac{2}{15A^2}[(5A+4K^2)K^3 + (A-4K^2)(A+K^2)^{3/2}] \quad (13)$$

for $H_d \geq H \geq H_{c1}$,

$$\mu_0 M^K{}_{U3} = -\mu_0 H - B_0 - \frac{2}{3A}[B_0{}^3-(A+B_0{}^2)^{3/2}]$$
$$-\left(\frac{a}{b}\right)\frac{2}{15A^2}[(5A+4B_0{}^2)B_0{}^3+(A-4B_0{}^2)(A+B_0{}^2)^{3/2}] \qquad (14),$$

for $H_{c1} \geq H \geq -H_{c1}$,

$$\mu_0 M^K{}_{U4} = -\mu_0 H - B_0 - \frac{2}{3A}[K^3+3B_0(B_0{}^2-K^2)-(A+2B_0{}^2-K^2)^{3/2}]$$
$$-\left(\frac{a}{b}\right)\frac{2}{15A^2}[(15A+23B_0{}^2)B_0{}^3-\{15(A+2B_0{}^2)B_0-5(A+3B_0K)K+4K^3\}K^2$$
$$+(A-8B_0{}^2+4K^2)(A+2B_0{}^2-K^2)^{3/2}] \qquad (15)$$

for $-H_{c1} \geq H \geq -H_p$ and

$$\mu_0 M^K{}_{U5} = -\mu_0 H + B_0 - \frac{2}{3A}[K^3-(K^2-A)^{3/2}]$$
$$-\left(\frac{a}{b}\right)\frac{2}{15A^2}[(5A-4K^2)K^3+(A+4K^2)(K^2-A)^{3/2}] \qquad (16)$$

for $-H_p \geq H \geq -H_m$.

*- Region IB -*

In this region of $-B_0 + \sqrt{(B_0{}^2 + K_m{}^2)/2} \leq B_{eq}{}^* \leq B_{eqm}$, $M^K{}_U(H)$ are given by $\mu_0 M^K{}_{U1}$ for $H_m \geq H \geq H_{c1}$, and

$$\mu_0 M^K{}_{U6} = -\mu_0 H - B_0 - \frac{2}{3A}[B_0{}^3+(K_m{}^2-A)^{3/2}-\frac{1}{\sqrt{2}}(B_0{}^2+K_m{}^2)^{3/2}]$$
$$-\left(\frac{a}{b}\right)\frac{2}{15A^2}[(5A+4B_0{}^2)B_0{}^3-(A+4K_m{}^2)(K_m{}^2-A)^{3/2}-$$
$$\frac{5}{\sqrt{2}}(A+B_0{}^2-K_m{}^2)(B_0{}^2+K_m{}^2)^{3/2}] \qquad (17)$$

for $H_{c1} \geq H \geq -H_{c1}$,

$$\mu_0 M^K{}_{U7} = -\mu_0 H - B_0 - \frac{2}{3A}[K^3-3B_0(K^2-B_0{}^2)+(K_m{}^2-A)^{3/2}-\frac{1}{\sqrt{2}}(2B_0{}^2+K_m{}^2-K^2)^{3/2}]$$
$$-\left(\frac{a}{b}\right)\frac{2}{15A^2}[23B_0{}^5-4K^5+5A(3B_0{}^3-3B_0K^2+K^3)-15B_0K^2(2B_0{}^2-K^2)$$
$$-(A+4K_m{}^2)(K_m{}^2-A)^{3/2}-\frac{5}{\sqrt{2}}(A+2B_0{}^2-K^2-K_m{}^2)(2B_0{}^2-K^2+K_m{}^2)^{3/2}] \qquad (18)$$

for $-H_{c1} \geq H \geq -H_f$, where $H_f$ was defined by $B_{eq}(-H_f) = B_0 - \sqrt{2B_0 - K_m^2 + 2A}$.

The $M^K_U(H)$ are given by $\mu_0 M^K_{U4}$ for $-H_f \geq H \geq -H_p$, and $\mu_0 M^K_{U5}$ for $-H_p \geq H \geq -H_m$ as same in *Region IA*.

- *Region II* -

The expressions of $M^K_U(H)$ in this region of $B_{eqm} \leq B_{eq}^*$ are

$$\mu_0 M^K_{U8} = -\mu_0 H - \frac{1}{3A}[2K^3 - B_0(B_0^2 - 3K_m^2) - \sqrt{2}(K^2 + K_m^2)^{3/2}]$$
$$+\left(\frac{a}{b}\right)\frac{1}{15A^2}[3B_0^5 - 8K^5 + 5A(B_0^3 - 2K^3) - 5B_0K_m^2(3A + 2B_0^2 - 3K_m^2)$$
$$+5\sqrt{2}(A + K^2 - K_m^2)(K^2 + K_m^2)^{3/2}] \tag{19}$$

for $H_m \geq H \geq H_{c1}$,

$$\mu_0 M^K_{U9} = -\mu_0 H - \frac{1}{3A}[B_0(B_0^2 + 3K_m^2) - \sqrt{2}(B_0^2 + K_m^2)^{3/2}]$$
$$-\left(\frac{a}{b}\right)\frac{1}{3A^2}[(A + B_0^2 - K_m^2)\{B_0(B_0^2 + 3K_m^2) - \sqrt{2}(B_0^2 + K_m^2)^{3/2}\}] \tag{20}$$

for $H_{c1} \geq H \geq -H_{c1}$, and

$$\mu_0 M^K_{U10} = -\mu_0 H - \frac{1}{3A}[2K^3 + B_0\{5B_0^2 + 3(K_m^2 - 2K^2)\} - \sqrt{2}(2B_0^2 - K^2 + K_m^2)^{3/2}]$$
$$-\left(\frac{a}{b}\right)\frac{1}{15A^2}[43B_0^5 - 2K^2\{15B_0(2B_0^2 - K^2) + 4K^3\} + 5A\{5B_0^3 + 2K^2(K - 3B_0)\}$$
$$+5B_0K_m^2(3A + 2B_0^2 - 3K_m^2)$$
$$-5\sqrt{2}(A + 2B_0^2 - K^2 - K_m^2)(2B_0^2 - K^2 + K_m^2)^{3/2}] \tag{21}$$

for $-H_{c1} \geq H \geq -H_m$.

## Notes on the $M_{eq}$ for $J_c$ Estimation

The critical current density, $J_c$, and the pinning property of high $T_c$ superconductor, such as $MgB_2$, have been discussed extensively in the literature [13-18]. In these studies, the equation

$$|J_c(H)| = \frac{\Delta M(H)}{a\left(1-\dfrac{a}{3b}\right)}, \tag{22}$$

is assumed, where $\Delta M(H)$ is the difference between the upper and lower 'irreversible' magnetization curves of $[M_U(H)-M_L(H)]$ for an infinitely long orthorhombic column with a $2a\times 2b$ cross section. The external field $H$ is applied parallel to the long column axis. On the basis of Eq. (22), some authors have used the relation $|J_c(H)|=\Delta M(H)/a$ for a slab of $a<< b$, or $|J_c(H)|=3\Delta M(H)/(2a)$ for the tetragonal or cylinder case where $a=b$. However the validity of these equations is not so clear.

Eq. (22) has sometimes been confused with the result obtained from the Bean model [11],

$$|J_c| = \frac{\Delta M}{a\left(1-\dfrac{a}{3b}\right)} \tag{23}.$$

In the Bean model, the $|J_c|$ is spatially homogeneous and independent of $H$. Therefore, the field dependent $|J_c(H)|$ reported for $MgB_2$ [13-18] does not satisfy Bean's original assumption that $\Delta M$ is a constant at $B_{eq}^*/\mu_0=H_p+M_{eq}(H_p)$. It should also be noted that Eq. (23) is only applicable to the magnetic field within a range of $2H_p-H_m\leq H\leq H_m-2H_p$ when $H_p\leq H_m/2$.

The present extended *Kimishima*'s critical state model gives the $x$- and $H$-dependence of the *internal critical current density* $J_c^{int}$ in such a way that $|J_c^{int}(x,H)|=B_{eq}^*(B_{eq}^*+2B_0)/\{|B_{int}(x,H)|+B_0\}$. A similar equation to Eq. (22) can be derived from this model for the field dependent $|J_c^{eq}(H)|$, but with multiple conditions imposed. Here $J_c^{eq}$ is the critical current density at the inner sample surface at $x=a$. If $0\leq B_{eq}^*\leq -B_0+\sqrt{(B_0^2+K_m^2)/2}$ *(Region IA)*, $B_{int}(x,H)$ changes monotonously with $x$ in the half side of the sample in $0\leq x\leq a$ or $-a\leq x\leq 0$ on the field decreasing process in $H_{c1}\leq H\leq H_d$, and field increasing process in $H_p\leq H\leq H_m$. Then upper and lower irreversible magnetization $M_U$ and $M_L$ are given as following by $M^K_{U2}(H)$ and $-M^K_{U5}(-H)$ from Eqs. (2.13) and (2.16), respectively.

$$\mu_0 M_{U,L}(H) = -\mu_0 H - B_0 \mp \frac{2}{3A}\left(1+\frac{a}{b}\right)\left[K(H)^3 - \left\{K(H)^2 \pm A\right\}^{\frac{3}{2}}\right]$$
$$+\frac{4}{15A^2}\left(\frac{a}{b}\right)\left[\left\{K(H)^2 \pm A\right\}^{\frac{3}{2}}\left\{2K(H)^2 \mp 3A\right\} - 2K(H)^5\right] \tag{24}$$

where the upper and lower signs correspond to the equations for $M_U$ and $M_L$, respectively. This equation holds for $M_U$ on the demagnetized process in $H_{c1}\leq H\leq H_d$ after initial magnetization, where $H_{c1}$ is the lower critical field (smaller than $H_p$), and the $H_d$ (smaller than $H_m$) is defined by $B_{eq}(H_d)=\sqrt{K_m^2-2A}\,-B_0$. The $M_L$ on the magnetized process, recurred from the negative field region, is given by Eq. (24) for $H_p \leq H \leq H_m$. Here we define the equilibrium critical current density $|J_c^{eq}(H)|$ at an inner sample surface as

$$|J_c^{eq}(H)| = \frac{B_{eq}^*\left(B_{eq}^* + 2B_0\right)}{2\mu_0 a\left\{|B_{eq}(H)| + B_0\right\}} = \frac{A}{2\mu_0 a K(H)} \tag{25}$$

with $B_{int}(x,H)$ in Eq. (1) replaced by $B_{eq}(H)$. Replacing $\mu_0 a|J_c^{eq}(H)|/[2K(H)]=A/[2K(H)]^2$ by the parameter $X$, $\Delta M(=M_U-M_L)$ and $\Sigma M(=M_U+M_L)$ can be expanded by the series of $X$ as

$$\Delta M(H) = a|J_c^{eq}(H)|\left\{\left(1-\frac{a}{3b}\right)+\left(1-\frac{3a}{5b}\right)X^2+\frac{14}{3}\left(1-\frac{5a}{7b}\right)X^4+\bullet\bullet\bullet\bullet\right\}, \tag{26}$$

and

$$\Sigma M(H)/2 = M_{eq}(H) - a|J_c^{eq}(H)|\left\{\frac{1}{3}\left(1-\frac{a}{2b}\right)X-\left(1-\frac{2a}{3b}\right)X^3-6\left(1-\frac{3a}{4b}\right)X^5+\bullet\bullet\bullet\bullet\right\}, \tag{27}$$

for $H_p \leq H \leq H_d$. When $X<<1$, $|J_c^{eq}(H)|$ can be approximated as

$$|J_c^{eq}(H)| \sim \frac{\Delta M(H)}{a\left(1-\frac{a}{3b}\right)}. \tag{28}$$

This is similar to Eq. (22) except that the estimation of $J_c^{eq}(H)$ by this relation is only useful under the multiple conditions that $H_p \leq H \leq H_d$, $\mu_0 a|J_c^{eq}(H)|<<2K(H)$ and $0 \leq B_{eq}^* \leq -B_0 + \sqrt{(B_0^2+K_m^2)/2}$. Instead of using Eq. (22), it is recommended that you use the basic and universal expression of Eq. (25) for $J_c^{eq}(H)$, since it holds without any restriction. Though the $M_{eq}(H)$ is not correctly given by $\Sigma M(H)/2$ as shown in Eq. (27), it can be determined self-consistently by using Eqs. (26) and (27) expanded up to an adequate order of $X$. However, practically speaking, if experimental and fitted results of $\Sigma M(H)/2$ agree well with each other, $\Sigma M(H)/2$ can be considered as the good definition of $M_{eq}(H)$.

## Extended Critical State Model with Exponential Decay of *J* (*Exponential Kimishima Model*)

In the previous section, the author proposed an extended critical state model (*Kimishima model*) with the equilibrium magnetization $M_{eq}$ based on the well-known Kim's and Bean's critical state model, in which the absolute value of internal critical current density $J_c^{int}$ was expressed by

$$\mu_0 |J_c^{\rm int}(x,H)| = \frac{A}{2a[|B_{int}(x,H)|+B_0]} \tag{29},$$

where $A=B_{eq}^*(B_{eq}^*+2B_0)$.

Recently, the exponential field decay of $J_c$ was reported for Bi2223-system at relatively high temperatures [19,20], and some authors have presented the critical state model with exponential decay of $J_c$ [21-24], which gives more rapid decrease of magnetization to $M_{eq}$ at the finite field near $H_{irr}$. Though it looks like to give the $H_{irr}$ analytically, the physical meanings of characteristic parameters have not been clear in relation to those of usual critical state model.

Here a universal model with exponential decay of $J_c$ (*Exponential Kimishima model*) will be presented as following.

$$\mu_0 \quad |J_c^{\rm int}(x,H)| = \frac{G}{2aB_0} exp\left[-\frac{|B_{int}(x,H)|}{B_0}\right] \tag{30},$$

where $G=2B_0^2[exp(B_{eq}^*/B_0)-1]$. The definitions of $B_{eq}^*$ and $B_0$ are same as those in Eq. (1). Since the 2nd order expansion of $G$ for $(B_{eq}^*/B_0)$ equals $A$ of $B_{eq}^*(B_{eq}^*+2B_0)$ in equation (29) and the 1st order expansion of $exp[-|B_{int}|/B_0]$ for $(|B_{int}|/B_0)$ gives $B_0/[|B_{int}|+B_0]$, Eq. (30) can be approximated to Eq. (29) when $(B_{eq}^*/B_0)^3 \ll 1$ and $(|B_{int}|/B_0)^2 \ll 1$. Therefore, Eq. (30) is considered to be a new *universal* formula of $|J_c^{\rm int}|$ including Bean's, Kim's and exponential critical state. The difference between Eqs. (29) and (30) is the $J_c^{\rm int}$ behaviors at high field. When $|B_{int}| \gg B_0$, the $|J_c^{\rm int}|$ by Eq. (29) is proportional to $1/|B_{int}|$ as Anderson-Kim's model [10]. On the other hand, in Eq. (30), the $exp[-|B_{int}|/B_0]$-component becomes dominant for the decay of $|J_c^{\rm int}|$.

## Magnetic Flux Density Distribution

Expressions of internal magnetic flux density are obtained by the same procedures as in Section 2. On the initial magnetized process,

$$B_{vir}(x,H) = B_0 \ln[\frac{1}{2B_0^2}\{L(H)+G(\frac{x}{a}-1)\}] \tag{31},$$

where $x_0=a\{2B_0^2+G-L(H)\}/G$, $G=2B_0^2[exp(B_{eq}^*/B_0)-1]$ and $L(H)=2B_0^2 exp[|B_{eq}(H)|/B_0]$.

On the field decreasing process after the above initial process,

$$B_{D1}(x,H) = B_0 \ln[\frac{1}{2B_0^2}\{L(H)+G(1-\frac{x}{a})\}] \tag{32},$$

where $x_1(H)=a\{2G+L(H)-L_m\}/(2G)$,

$$B_{D2} = B_0 \ln[\frac{1}{2B_0^2}\{4B_0^2\text{-}L(H)+G(1-\frac{x}{a})\}] \quad (33),$$

where $x_2(H)=a\{2B_0^2+G\text{-}L(H)\}/G$, and

$$B_{D3} = B_0 \ln[\frac{1}{2B_0^2}\{L(H)+G(\frac{x}{a}-1)\}] \quad (34),$$

where $x_3(H)=a\{4B_0^2+2G\text{-}L(H)\text{-}L_m\}/(2G)$.

## Averaged Magnetization

Expressions of averaged magnetization are obtained by the same procedures as in the preceding *Kimi. model*. It should be noticed that $L(H)$ and $B_{eq}(H)$ are simply represented by $L$ and $B_{eq}$, respectively, in the following equations.

## Initial Magnetization

*- Region I -*

In this region of $0 \leq B_{eq}^* \leq B_{eqm}$, the initial magnetization after zero field cooling (ZFC) are given by *Exponential Kimishima* (*Exp. Kimi.*) *model* as

$$M^{EK}_{V1}=-H \quad (35)$$

for $0 \leq H \leq H_{c1}$,

$$\mu_0 M^{EK}_{V2} = -\mu_0 H - \frac{1}{G}[(B_0-B_{eq})L-2B_0^3]$$
$$+\left(\frac{a}{b}\right)\frac{1}{2G^2}[4B_0^3\{B_0^2+(G-2L)\}+L\{2B_{eq}(G-L)+B_0(3L-2G)\}] \quad (36)$$

for $H_{c1} \leq H \leq H_p$ and

$$\mu_0 M^{EK}_{V3} = -\mu_0 H - B_0 + B_{eq} + B_0(1-L/G)\ln\left(1-\frac{G}{L}\right)$$
$$-\left(\frac{a}{b}\right)\frac{B_0}{2}[1-2L/G+(2L/G)(1-L/G)\ln\left(1-\frac{G}{L}\right)] \quad (37)$$

for $H_p \leq H \leq H_m$.

*- Region II -*

In this region of $B_{eqm} \leq B_{eq}^*$, the initial magnetization is given by $M^{EK}{}_{V1}$ for $0 \leq H \leq H_{c1}$, and

$$M^{EK}{}_{V2} \text{ for } H_{c1} \leq H \leq H_m.$$

## Magnetization Loops

*- Region IA -* $(0 \leq B_{eq}^* \leq B_0 \ln[\{\exp(B_{eq}^*/B_0)+1\}/2])$

$$\mu_0 M^{EK}{}_{U1} = -\mu_0 H - B_0 + B_{eq}$$
$$+B_0[(1+L/G)\ln\left(\frac{L+L_m}{2L}\right)+(L_m/G-1)\ln\left\{\frac{L+L_m}{2(L_m-G)}\right\}]$$
$$-\left(\frac{a}{b}\right)\frac{B_0}{2G^2}[2(G-L_m)L_m\ln\left(\frac{L_m-G}{L}\right)$$
$$+(G+L-L_m)\{G-L-L_m-2(L+L_m)\ln\left(\frac{L+L_m}{2L}\right)\}] \quad (38),$$

for $H_m \geq H \geq H_d$ where $B_{eq}(H_d) = B_0 \ln[(L_m-2G)/(2B_0^2)]$,

$$\mu_0 M^{EK}{}_{U2} = -\mu_0 H - B_0 + B_{eq} + B_0(1+L/G)\ln(1+G/L)$$
$$-\left(\frac{a}{b}\right)\frac{B_0}{2G^2}[G(G+2L)-2L(G+L)\ln\left(1+\frac{G}{L}\right)] \quad (39)$$

for $H_d \geq H \geq H_{c1}$,

$$\mu_0 M^{EK}{}_{U3} = -\mu_0 H - B_0 + B_{eq}^* + 2B_0^2 B_{eq}^*/G$$
$$-\left(\frac{a}{b}\right)\frac{B_0}{2G^2}[G(G+4B_0^2)-4B_0 B_{eq}^*(2B_0^2+G)] \quad (40)$$

for $H_{c1} \geq H \geq -H_{c1}$,

$$\mu_0 M^{EK}{}_{U4} = -\mu_0 H - B_0 + \frac{B_0}{G}[4B_0^2-(2+B_{eq}/B_0)L-(4B_0^2+G-L)\ln\left(\frac{4B_0^2+G-L}{2B_0^2}\right)]$$
$$-\left(\frac{a}{b}\right)\frac{B_0}{2G^2}[24B_0^4+2L(L-G)B_{eq}/B_0+8B_0^2(2G-3L)+(G^2-6GL+6L^2)$$
$$-2(4B_0^2-L)(4B_0^2+G-L)\ln\left(\frac{4B_0^2+G-L}{2B_0^2}\right)] \quad (41)$$

for $-H_{c1} \geq H \geq -H_p$ and

$$\mu_0 M^{EK}{}_{U5} = -\mu_0 H + B_0 + B_{eq} - B_0(1-L/G)\ \ln\left(1-\frac{G}{L}\right)$$
$$+\left(\frac{a}{b}\right)\frac{B_0}{2G^2}\ [G(G-2L)+2L(G-L)\ \ln\left(1-\frac{G}{L}\right)] \tag{42}$$

for $-H_p \geq H \geq -H_m$.

- *Region IB* - $(B_0 \ln[\{\exp(B_{eq}{}^*/B_0)+1\}/2] \leq B_{eq}{}^* \leq B_{eqm})$

In this region, $M^{EK}{}_U(H)$ are given by $\mu_0 M^{EK}{}_{U1}$ for $H_m \geq H \geq H_{c1}$, and

$$\mu_0 M^{EK}{}_{U6} = -\mu_0 H - B_0 + \frac{B_0}{G}\ [(G-L_m)\ \ln\left(\frac{L_m - G}{2B_0{}^2}\right) + (2B_0{}^2 + L_m)\ln\left(\frac{2B_0{}^2 + L_m}{4B_0{}^2}\right)]$$
$$+\left(\frac{a}{b}\right)\frac{B_0}{2G^2}\ [2L_m(L_m-G)\ln\left(\frac{L_m - G}{2B_0{}^2}\right)$$
$$+(2B_0{}^2+G-L_m)(2B_0{}^2-G+L_m+2(2B_0{}^2+L_m)\ln\left(\frac{2B_0{}^2+L_m}{4B_0{}^2}\right)] \tag{43}$$

for $H_{c1} \geq H \geq -H_{c1}$,

$$\mu_0 M^{EK}{}_{U7} = -\mu_0 H - B_0 - \frac{B_0}{G}\ [4B_0{}^2 - (2+B_{eq}/B_0)L + (L_m - G)\ \ln\left(\frac{L_m - G}{2B_0{}^2}\right)$$
$$-\ (4B_0{}^2 - L + L_m)\ \ln\left(\frac{4B_0{}^2 - L + L_m}{4B_0{}^2}\right)]$$
$$-\left(\frac{a}{b}\right)\frac{B_0}{2G^2}\ [8B_0{}^2(B_0{}^2+G-2L)+2(B_{eq}/B_0)L(L-G)$$
$$+(G^2+5L^2+L_m{}^2-2G(2L+L_m)-2L_m(L_m-G)\ \ln\left(\frac{L_m - G}{2B_0{}^2}\right)$$
$$-2(4B_0{}^2+G-L-L_m)(4B_0{}^2-L+L_m)\ \ln\left(\frac{4B_0{}^2 - L + L_m}{4B_0{}^2}\right)] \tag{44}$$

for $-H_{c1} \geq H \geq -H_f$, where $H_f$ was defined by $B_{eq}(-H_f) = B_0 \ln[(4B_0{}^2+2G-L_m)/(2B_0{}^2)]$.

The $M^{EK}{}_U(H)$ are given by $\mu_0 M^{EK}{}_{U4}$ for $-H_f \geq H \geq -H_p$, and $\mu_0 M^{K}{}_{U5}$ for $-H_p \geq H \geq -H_m$ as same as those in *Region IA*.

- *Region II* - $(B_{eqm} \leq B_{eq}^{*})$

The expressions of $M^{K}{}_{U}(H)$ in this region are

$$\mu_0 M^{EK}{}_{U8} = -\mu_0 H + \frac{B_0}{G}[2B_0^2 - (B_{eq}/B_0)L - L_m + (L+L_m)\ln\left(\frac{L+L_m}{4B_0^2}\right)]$$
$$+\left(\frac{a}{b}\right)\frac{B_0}{2G^2}[4B_0^2(B_0^2+G-2L_m) - 2(B_{eq}/B_0)L(G+L) + L^2 + 2L_m(L_m-G)$$
$$+2(G+L-L_m)(L+L_m)\ln\left(\frac{L+L_m}{4B_0^2}\right)] \tag{45}$$

for $H_m \geq H \geq H_{c1}$,

$$\mu_0 M^{EK}{}_{U9} = -\mu_0 H + \frac{B_0}{G}[2B_0^2 - L_m + (2B_0^2+L_m)\ln\left(\frac{2B_0^2+L_m}{4B_0^2}\right)]$$
$$+\left(\frac{a}{b}\right)\frac{B_0}{2G^2}(2B_0^2+G-2L_m)[2B_0^2 - L_m + (2B_0^2+L_m)\ln\left(\frac{2B_0^2+L_m}{4B_0^2}\right)] \tag{46}$$

for $H_{c1} \geq H \geq -H_{c1}$, and

$$\mu_0 M^{EK}{}_{U10} = -\mu_0 H - \frac{B_0}{G}[2B_0^2 - (2+B_{eq}/B_0)L + L_m - (4B_0^2-L+L_m)\ln\left(\frac{4B_0^2-L+L_m}{4B_0^2}\right)]$$
$$-\left(\frac{a}{b}\right)\frac{B_0}{2G^2}[4B_0^2(B_0^2+G-4L+2L_m) + 2(B_{eq}/B_0)L(L-G) - L(4G-5L)$$
$$+2L_m(G-L_m) - 2(4B_0^2+G-L-L_m)(4B_0^2-L+L_m)\ln\left(\frac{4B_0^2-L+L_m}{4B_0^2}\right)] \tag{47}$$

for $-H_{c1} \geq H \geq -H_m$.

## Magnetization Analysis Of Sintered Bi2223 By Extended Critical State Model

The quantitative studies on the critical current density $J_c$ have been an important problem for the applications of high $T_c$ superconductors. The $J_c$-values of $Bi(Pb)_2Sr_2Ca_2Cu_3O_y$ (BSCCO-2223 or Bi2223) with higher $T_c$ of 110K have often been reported for poly- or single-crystal bulk sample [19,25-28], and Ag-sheathed tapes [29-32]. In these works on the Bi2223-system, the author paid attention to the exponential magnetic field dependence of $J_c$ at relatively high temperatures to discuss the pinning property of Bi2223. In the previous sections, the analytical expressions of DC magnetization $M$ were presented by the extended critical state model, including exponential decay of $J_c^{int}$. Here the fitting results to the experimental data of sintered Bi2223-system will be shown for intra- and inter-granular contribution [33].

The pressed powder mixtures of $Bi_2O_3$, $PbO_2$, $SrCO_3$, $CaCO_3$ and CuO with the mole ratio of 1.84:0.34:1.91:2.03:3.06, were sintered in the open furnace at 1068K for 15 hours. Then two times of sintering at 1116K for 80 hours were additionally performed. Successively, sintering was twice done at 1116K for 20 hours. In each interval of calcination, the samples were re-grounded and re-pressed. This method was empirically selected to obtain the single phase of Bi2223 in air, as shown in the Cu-$K\alpha$ powder x-ray diffraction patterns. The roughly parallel order of flake-like superconductive grains were confirmed by the SEM observations, after grinding and pressing several times in the calcination steps. The diameter of each grain distributed between about 3 and 5 $\mu$m with 1 $\mu$m thickness, and these grains accumulated in the blocks with the size of about $10^2\times 10^2\times 10\,\mu$m. The oxygen amount of 10.3 per molecule was obtained from the analysis by x-ray micro-analyzer. The magnetizations $M$ were measured by the Quantum Design's MPMS SQUID magnetometer between 10K and 90K under the external field between -5T and 5T. To discuss the intra- and inter-granular contribution, we measured the magnetizations of powdered sample and sintered bulk sample, respectively. Measured sintered bulk sample had the size of 4mm$\times$3mm$\times$0.3 mm, and powder sample was glued, pressed and cut to form a same size of bulk sample.

First, the magnetization expressions derived from Eq. (1) of the *Kimi. model*, were fitted to the experimental $\mu_0 M(H)$-curves after zero field cooled process, using the experimental $\mu_0 M_{eq}(H)$ of $\mu_0$ $(M_U+M_L)/2$. Some characteristic parameters were determined from the least square fitting by Gauss-Jordan method. The theoretical $\mu_0 M(H)$-curves were also calculated from equation (29) of the *Exp. Kimi.* model with these parameters. Thus, the theoretical curves of $\mu_0 M(H)$ and $\mu_0 a|J_c^{eq}(H)|$ were obtained from the above two models, respectively, where the $J_c^{eq}$ is the surface critical current density when the $B_{int}(x,H)$ in Eqs. (1) and (30) equals the surface flux density of $B_{int}(a,H)=B_{eq}(H)=\mu_0\{H+M_{eq}(H)\}$. Hereafter, $|J_c^{eq}(H)|$ is abbreviated as $J_c(H)$.

The analytical results for the $\mu_0 M(H)$-data of Bi2223 powder sample represent the intra-grain properties. In the sintered bulk samples, the $\mu_0 M(H)$-data under low field reflect the properties of inter-granular parts which cover or glue most of the grains. Therefore, we can discuss the analytical results for intra- and inter-granular part separately. Hereafter, we will use the suffix $g$ for the intra-granular quantities, and $j$ for inter-granular ones.

## Intra-Grain Magnetization Of Bi2223

For the temperatures of 10, 30, 50, 70 and 90K, the experimental $\mu_0 M_g(H)$-data and theoretical $\mu_0 M_g(H)$- and $\mu_0 aJ_{cg}(H)$-curves were depicted in Figs. 4(a)-(e), respectively. The numerical results are also given in Table 1. In these figures, the closed circles show the

$\mu_0 M_g(H)$-data, and the bold lines are theoretical $\mu_0 M_g^K(H)$ by the *Kimi. model*. The dotted lines are theoretical $\mu_0 M_g^{EK}(H)$ by the *Exp. Kimi. model*.

In Figs. 4(a)-(b), $\mu_0 M_g^K$ and $\mu_0 M_g^{EK}$ agreed well with each other in the field range of $|H| < H_{pg}$ , but the discrepancy between them increased as $|H|$ increased. In $|H| > H_{pg}$, the $\mu_0 M_g^K$ apparently fits to the experimental $\mu_0 M_g$ at 10K and 30K, and $\mu_0 M_g^{EK}$ better agrees with $\mu_0 M_g$ above 50K. Therefore. it was found that the $J_{cg}$ below 30K decayed being proportional to $1/(B_{eq}+B_0)$ of Kim type. On the other hand, the $J_{cg}$ had the exponential characteristics of $\exp(-B_{eq}/B_0)$ above 50K. The broken lines of $\mu_0 aJ_{cg}$ in Figs. 4(a)-(e) were obtained from Eq. (1) of the *Kimi. model* for 10K and 30K, and from Eq. (30) of the *Exp. Kimi. model* for 50, 70 and 90K.

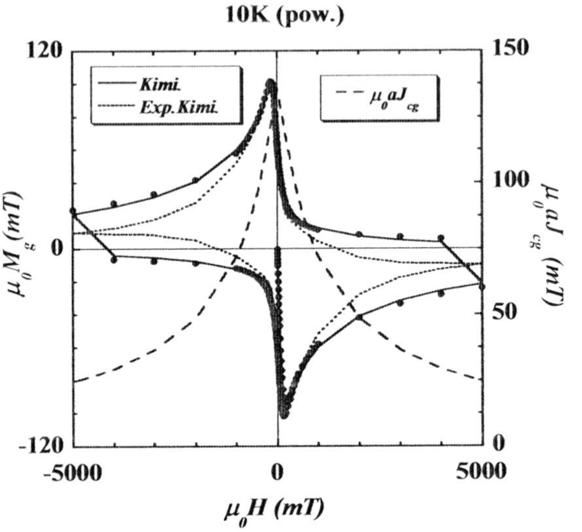


(a)

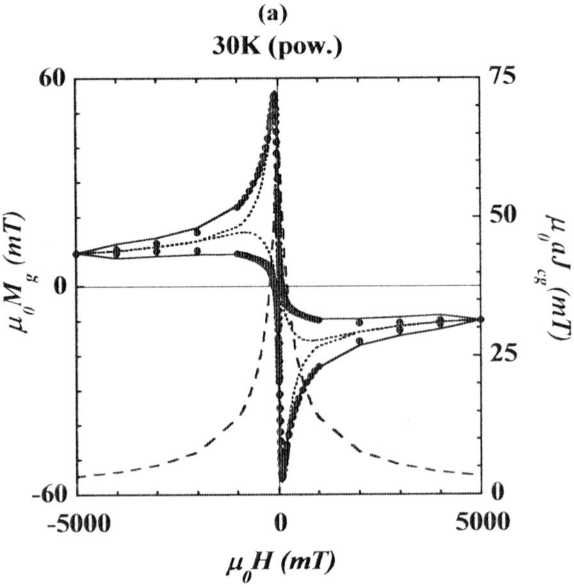


(b)

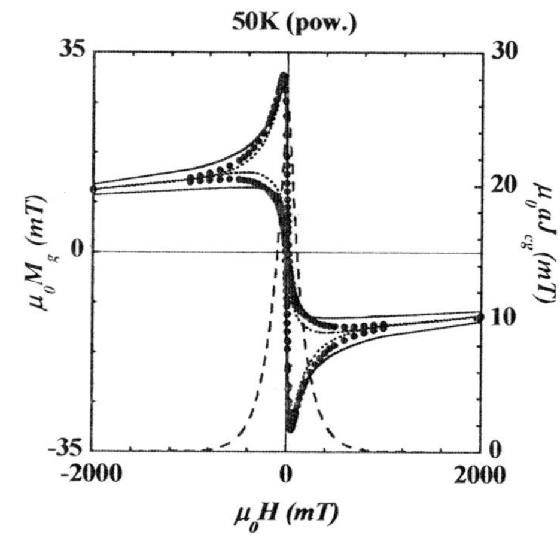


(c)

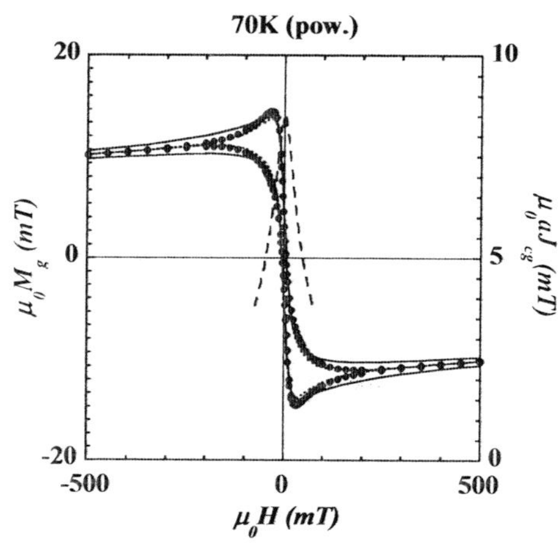


(d)

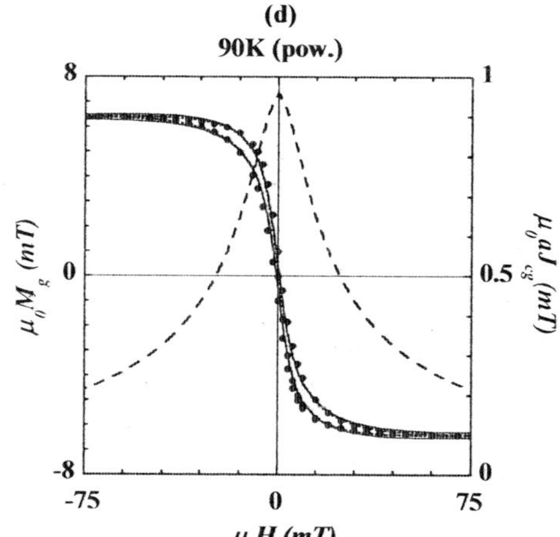


(e)

Figure 4. Experimental $\mu_0 M_g(H)$-data of powder sample and theoretical $\mu_0 M_g^K(H)$- and $\mu_0 M_g^{EK}(H)$ curves for the temperatures of (a) 10K, (b) 30K, (c) 50K, (d) 70K, and (e) 90K. The $\mu_0 aJ_{cg}(H)$-curves were intra-grain equilibrium critical current density at 0-field, calculated by Eq. (29) for (a) and (b), and Eq. (30) for (c)-(e).

Table 1. Intra-granular magnetic parameters of Bi2223-powder samples. The constant $c$ is the normalized factor in the least square fitting calculations.

| $T$ | [K] | 10 | 30 | 50 | 70 | 90 |
|---|---|---|---|---|---|---|
| $B_{eqg}^{*}$ | [mT] | 128. | 58 | 24 | 8.0 | 0.9 |
| $\mu_0 H_{pg}$ | [mT] | 163 | 85 | 50 | 20 | 4.8 |
| $B_{0\,g}$ | [mT] | 1104 | 272 | 154 | 50.4 | 19.8 |
| $\mu_0 H_{irr}^{g}$ | [T] | >5 | 5 | 2 | 0.6 | 0.12 |
| $\mu_0 aJ_{cg}(0)$ | [mT] | 136 | 64 | 26 | 8.6 | 0.96 |
| $c$ | | 0.19 | 0.17 | 0.16 | 0.16 | 0.15 |

## Inter-Grain Magnetization of Bi2223

The inter-grain magnetization of $\mu_0 M_j(H)$ were obtained from $\mu_0 M(H)$-data of sintered bulk sample under the field lower than $H_{irr}^{j}$, which were given in Table 2. Above $H_{irr}^{j}$, the experimental $\mu_0 M(H)$-data were essentially the same as the $\mu_0 M_g$ of powder specimen. All of theoretical $\mu_0 M_j(H)$- and $\mu_0 aJ_{cj}(H)$-curves in Figs. 5(a)-(e) were obtained from the fitting of $\mu_0 M^{K}(H)$-equations of the *Kimi. model*. The numerical results are also given in Table 2. As shown in Figs. 5(a)-(e), $\mu_0 M_j^{K}$ agreed well with experimental data at $|H| < H_{irr}^{j}$ for whole temperatures, meaning the $J_{cj}$ proportional to $1/[B_{eqj}(H)+B_{0j}]$.

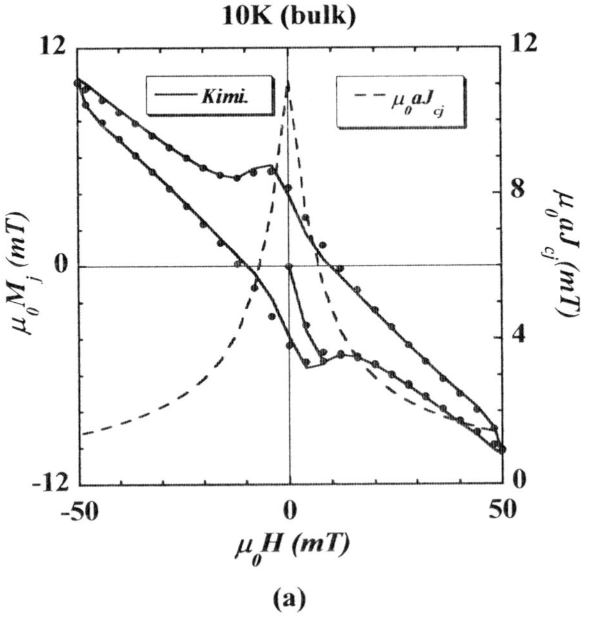


(a)

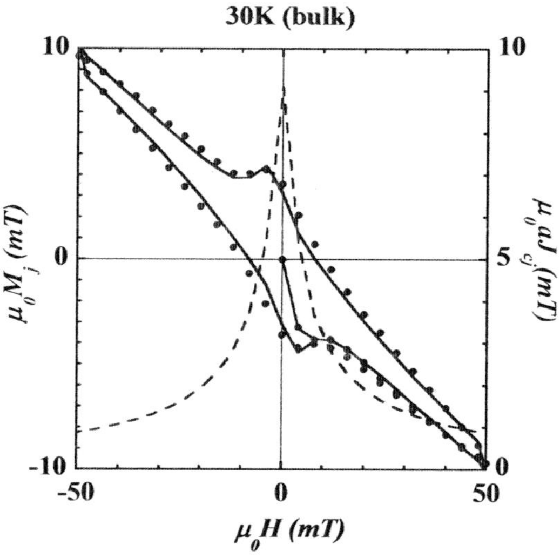
30K (bulk)
10
0
-10
-50
0
50
10
5
0
μ0Mj (mT)
μ0aJcj (mT)
μ0H (mT)

(b)

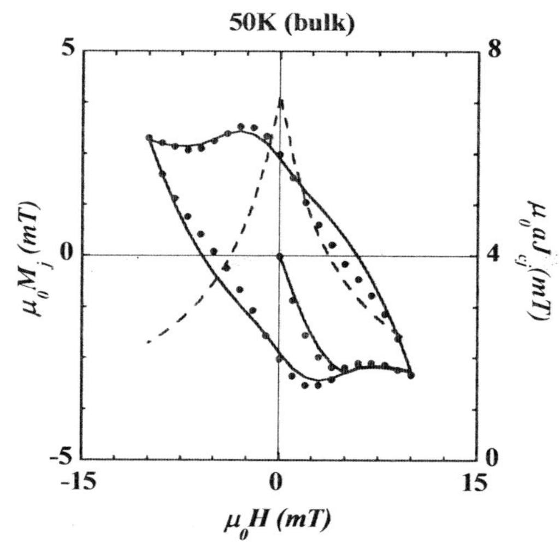
50K (bulk)
5
0
-5
-15
0
15
8
4
0
μ0Mj (mT)
μ0aJcj (mT)
μ0H (mT)

(c)

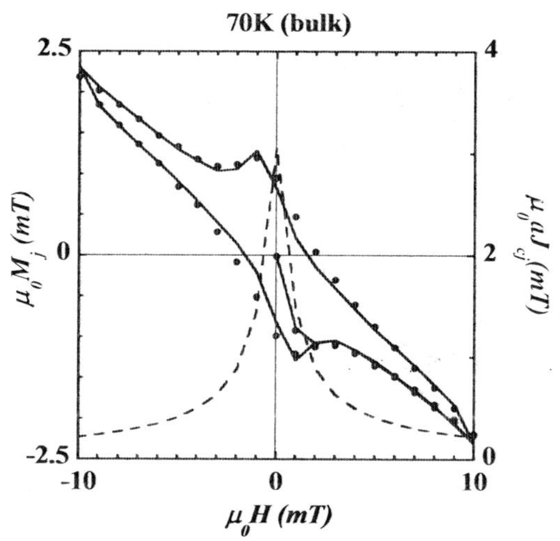
70K (bulk)
2.5
0
-2.5
-10
0
10
4
2
0
μ0Mj (mT)
μ0aJcj (mT)
μ0H (mT)

(d)

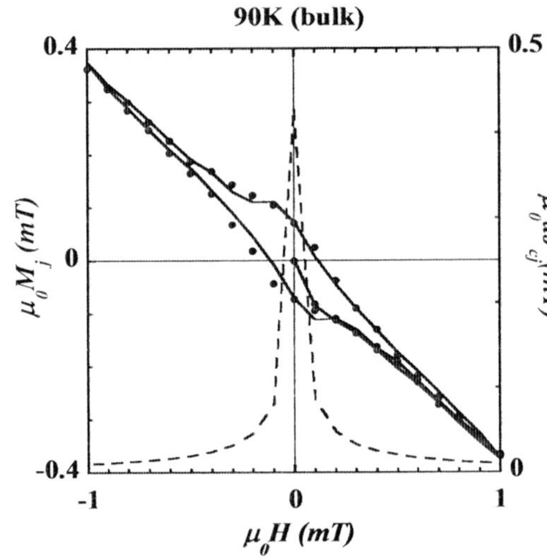


(e)

Figure 5. Experimental $\mu_0 M_g(H)$-data of sintered bulk sample and theoretical $\mu_0 M_g^{K}(H)$- curves for the temperatures of (a) 10K, (b) 30K, (c) 50K, (d) 70K, and (e) 90K. All $\mu_0 aJ_{cg}(H)$-curves were inter-grain equilibrium critical current density at 0-field, calculated by Eq. (1).

Table 2. Inter-granular magnetic parameters of Bi2223-bulk samples. The constant $c$ is the normalized factor in the least square fitting calculations.

| $T$ | [K] | 10 | 30 | 50 | 70 | 90 |
|---|---|---|---|---|---|---|
| $B_{eqj}^{*}$ | [mT] | 6.9 | 5.5 | 4.5 | 1.4 | 0.12 |
| $\mu_0 H_{pj}$ | [mT] | 12 | 8 | 5 | 1.8 | 0.2 |
| $B_{0j}$ | [mT] | 5.8 | 4.3 | 3.9 | 0.6 | 0.03 |
| $\mu_0 H_{irr}^{j}$ | [mT] | 100 | 50 | 30 | 10 | 1 |
| $\mu_0 aJ_{cj}(0)$ | [mT] | 11.1 | 9.1 | 7.1 | 3.1 | 1.4 |
| $c$ | | 1.02 | 1.02 | 0.92 | 0.87 | 0.47 |

## Temperature Dependence of Critical Current Density of Bi2223

For the high $T_c$ superconductors, the temperature dependence of $J_c$ has been discussed in relation to the GL-theory [27,34,35] or the flux creep [36-39]. The absolute value of $J_c$ had large ambiguity by the selection of thickness value $2a$. If we adopted the grain thickness of about 1 $\mu$m as $2a$, the $J_{cg}(0)$ and $J_{cj}(0)$ at 10K, they were estimated as the orders of $10^7$ and $10^6$A/cm$^2$, respectively, being much higher than the previously reported values [19,23]. When the thickness of grain-accumulated block of about 10 $\mu$m was used, the $J_{cg}(0)$ and $J_{cj}(0)$ decreased to about one tenth of the smaller values. While the measured sample thickness of $3\times 10^2$ $\mu$m gave the orders of $10^5$ and $10^4$A/cm$^2$ of $J_{cg}(0)$ and $J_{cj}(0)$ at 10K, respectively. So, to avoid this problem of absolute $J_c$ value, we will discuss the temperature dependence of $\mu_0 aJ_c$.

Tinkham [40] extended the Anderson theory for flux creep [41], and presented the equation of

$$J_c(T)=J_c(0)\{1-\alpha t-\beta t^2\} \tag{47}$$

where $t=T/T_c$, $\alpha$ and $\beta$ were constant for temperature. The $\alpha$ represents the order of magnitude of flux bundle creeping $kT/U(0)$, which is the ratio of thermal energy and pinning potential at 0K. Meanwhile the $\beta$ relates to the thermal decay of pinning potential $U(T)$ itself. The $\alpha$ and $\beta$ have the values between 0.1 and 1 for both of the high and low temperature superconductors. [37,38,40].

In Fig. 6, semi-log plots of $\mu_0 aJ_{c0}^{eq}$ vs. $T$ were given for intra- and inter-grains. The closed triangles between 80K and 100K are our previous $\mu_0 aJ_{c0j}^{eq}$ results from AC magnetic susceptibility, which gave the 81.6[mT]$\times(1-t)^{2.8}$-dependence near $T_{cj}$=104.6K [31]. As for the $\mu_0 aJ_{cj}(0)$, Eq. (47) were well fitted to the experimental result shown by closed squares between 10K and 80K, where the dotted theoretical curve was depicted when $t=T/T_{cj}$, $\mu_0 aJ_{cj}(0)$=11.6mT at 0K, $\alpha$=0.366 and $\beta$=1.045. These numerical results show that the creeping of flux bundle is dominant below $T\sim 40\text{K} \sim \alpha T_{cj}/\beta$ and thermal decay of pinning potential have greater weight in the intermediate $T$-region between about 40K and 80K for the suppression of $J_{cj}$ of Bi2223. At a very low temperature, below 10K, $\mu_0 aJ_{cj}(0)$ may have the $(1-t)$-characteristics as expected from the GL-theory with superconductive weak links [34].

Fitting of Eq. (47) for intra-grain $\mu_0 aJ_{cg}(0)$ gave the results of $\mu_0 aJ_{cg}(0)$=187mT at 0K, $\alpha$= 3.33 and $\beta$=3.93, but these large $\alpha$ - and $\beta$-values were unphysical [34]. It also had no power law dependence of $t^n$ or $(1-t)^m$, where $t=T/T_{cg}$ and $T_{cg}$=107.8K [36]. However, the $\mu_0 aJ_{c0g}^{eq}$ obeyed the exponential law of 234[mT]$\times$exp($-T$/21.8) between 10K and 70K, as shown in Fig. 6.

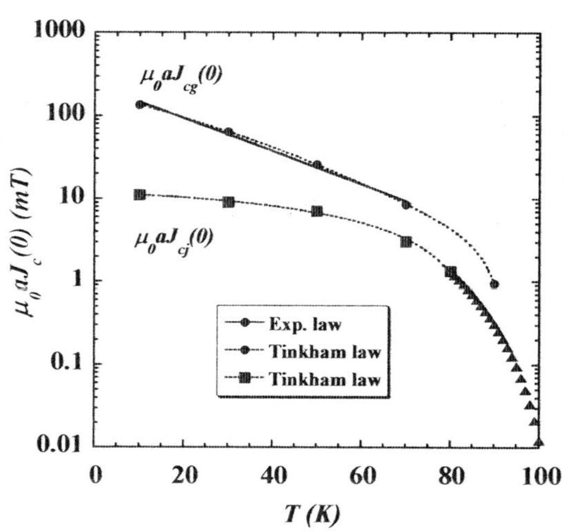


Figure 6. Semi log plots of temperature dependent $\mu_0 aJ_{cg}(0)$ and $\mu_0 aJ_{cj}(0)$. Dotted lines are the fitting curves of equation (47).

Feigel'man et al. [42] discussed the current relaxation due to the thermal activation process and gave the expression for the single vortex creep as

$$J_c = J_c(0)exp[-T/U_c \ln(t/\tau_0)] \quad (48),$$

where $U_c$ is the energy barrier for the single vortex pinning, and $\tau_0$ is the relaxation time of the single vortex creeping. The existence of a single vortex phase up to 70K near $T_{cg}$ is the interesting result for the Bi2223-system, because the power law was adequate for the $J_c$ of the other many high $T_c$-systems and the exponential decay of $J_c$ with temperature is a rare case at present.

## S-Doping Effects on Pinning Property of Bi2212 Single Crystals

Here the sulfur doping effect for $Bi_2Sr_2CaCu_2O_y$ (Bi2212) system is reported [43]. The enhancement of pinning property of high $T_c$ superconductor Bi2212 with highly electromagnetic anisotropy [44] has been a current problem for high field applications near liq. $N_2$ temperature of 77K. To improve the critical current $J_c$ and irreversible field $H_{irr}$, the creation of columnar pinning centers by heavy ions irradiation [45-48] or nano-rod composites [49] have been performed. These columnar defects supplied the larger $J_c$ or $H_{irr}$ at a low temperature below 77 K. The other promising method was carrier doping to the block double layers of Bi-O and reduction of anisotropy of Bi2212 by heavily Pb doping [50,51], which suppressed the flux flow and leaded to an increase in $J_c$ at relatively high temperatures. However, the improvement of pinning property has not been sufficient at 77 K. In the present study, sulfur-doping have been tried to the interstitial or $O^{2-}$ vacancy sites of Bi2212 single crystals. The enhancement of $J_c$ and $H_{irr}$ will be reported in this section on the basis of magnetization analysis by the extended critical state model.

Bi2212 single crystals were prepared by the self-flux (SF) method from the mixed powders of 99.9% $Bi_2O_3$, $SrCO_3$, $CaCO_3$ and CuO with the mole ratio of 2.4:2:1:2. The mixtures in the platinum crucible were pre-fired at 1093K for 8 hours and melted at 1323K for 8 hours in air. They were annealed from 1323K to 1143K for 87 hours and cooled to room temperature in the furnace. The sulfur-doping to a Bi2212 single crystal with the size of 5.1mm× 2.0mm× 0.6mm were performed in the vacuumed quartz cell at 1023K for 72 hours under the gaseous sulfur pressure of about 0.1MPa. The magnetizations were measured at 77K between −100mT and 100mT by the vibrating sample magnetometer and Quantum Design's MPMS SQUID magnetometer. The demagnetization correction was not done for all of the magnetization data because of the small magnitude near $T_c$.

In Fig. 7(a) and (b), temperature dependences of magnetic susceptibility under the field parallel to $c$-axis, $\chi_{\parallel c}$, and perpendicular to $c$-axis, $\chi_{\perp c}$, were illustrated, respectively, for pure and S-doped Bi2212. Here FC and ZFC means the warming up processes under the field of 10mT after 10mT field cooled and zero field cooled process, respectively. As shown in these figures, the magnitudes of negative susceptibility were considerably reduced, but the superconducting onset temperatures $T_c$ were unchanged as 90K by S-doping.

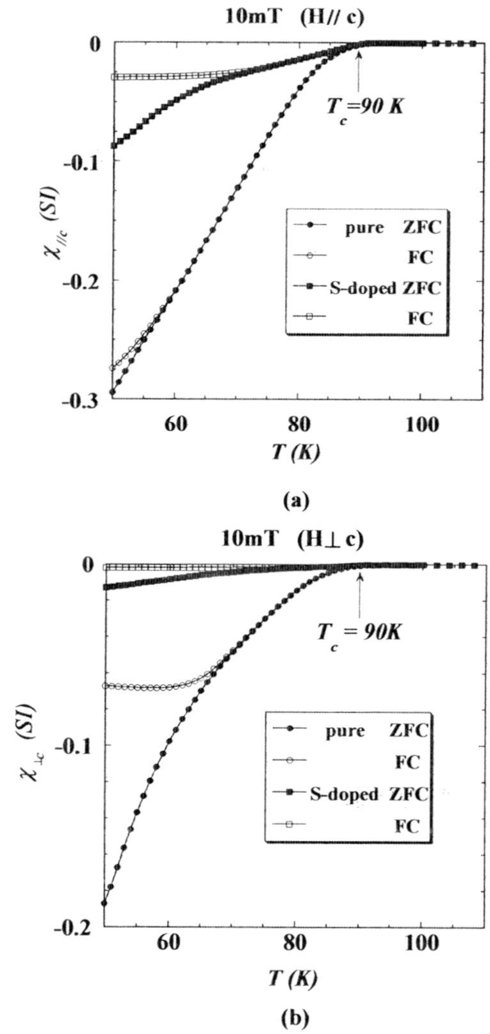


Figure 7. Temperature dependence of DC magnetic susceptibility under the field parallel to *c*-axis (a), and perpendicular to *c*-axis (b) for pure and S-doped Bi2212.

The irreversible temperature $T_{irr}$, at which the FC- and ZFC-$\chi$(T) separated from each other, increased from 60K to 75K in $\chi_{\parallel c}$ and from 70K to 85K in $\chi_{\perp c}$. These $\chi$(T)-results clearly show the enhancement of $J_c$ by S-doping instead of the reduction of superconductive volume fraction even near 77K.

The ZFC $M(H)$-curves at 77K are given in Fig. 7 and Fig. 8 under the field parallel and perpendicular to *c*-axis, respectively. In the parallel field case, the irreversible field $H_{irr}$, below which hysteretic $M(H)$-curves appear, increased from 2mT to 12mT by S-doping, as shown in Fig. 8. In the perpendicular field case, the $H_{irr}$ exceedingly increased from 10mT to 50mT, as shown in Fig. 9. Thus, the sulfur-doping can be considered to be worthy to produce strong pinning effects even at 77K for Bi2212. For the numerical estimation of $J_c$-enhancement, we will present the magnetization analysis by improved critical state model for the slab geometry as following.

Analysis of ZFC $M(H)$-curve perpendicular to $c$-axis at 77K were performed using Eqs. (9)~ (16). The parameters $B_{eq}^{*}$, $B_0$ and the normalization factor $c$ were evaluated from the least square fitting of the experimental $M(H)$-data to these equations. In this fitting, the equilibrium magnetization, $M_{eq}(H)$, was approximated to be $(M_U+M_L)/2$, and $H_p$ was determined as the field at which the initial magnetization $M_{vir}$ and $M_L$-curve meet each other. In Fig. 9, the theoretical curves are also depicted together with the $M_{\perp c}$-data, where the magnetic quantities were multiplied by $\mu_0$. The equilibrium critical current densities $|J_c^{eq}|$, estimated by Eq. (25) with $a$ of 0.6mm, are shown in Fig. 10 by $J_c$. The $J_c$ of S-doped crystal exceeded that of pure crystal near 1mT, being nearly constant at about $10^5$A/cm$^2$.

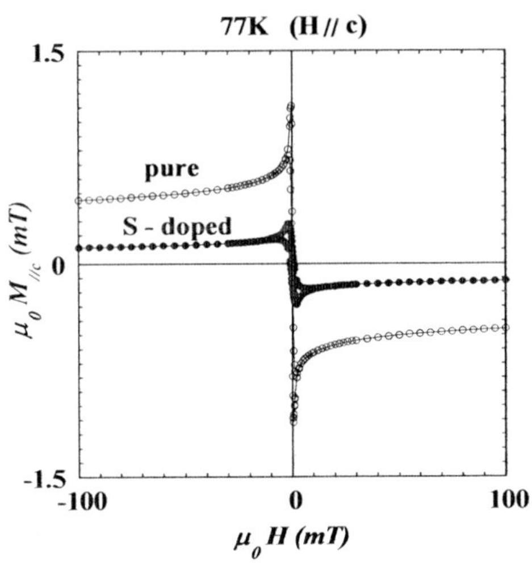


Figure 8. Magnetization curve at 77K under the field parallel to $c$-axis for pure and S-doped Bi2212.

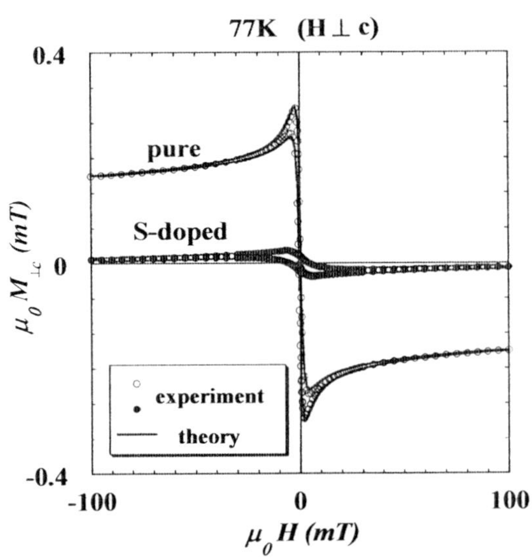


Figure 9. Magnetization curves at 77K under the field perpendicular to $c$-axis for pure and S-doped Bi2212. Fitted curves of magnetization are drawn by solid bold lines.

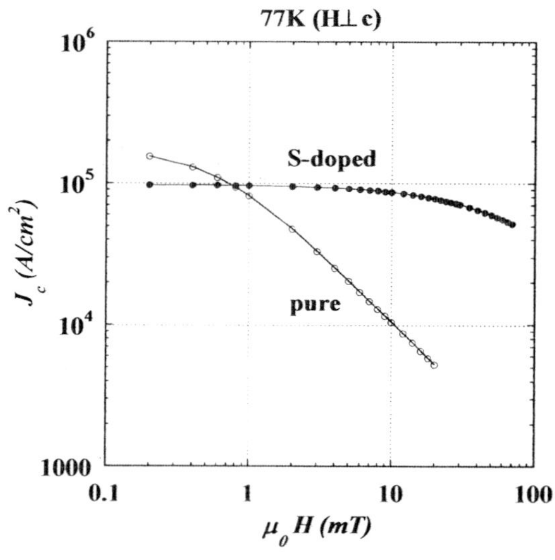


Figure 10. Field dependence of equilibrium critical current density under the field parallel to slab surface (perpendicular to *c*-axis).

Table 3. Magnetic parameters of pure and S-doped Bi2212 at 77K.

| | | ⊥ *c* (pure) | ⊥ *c* (S-doped) | ∥*c* (pure) | ∥*c*(S-doped) |
|---|---|---|---|---|---|
| $B_{eq}^*$ | [mT] | 0.35 | 0.37 | - | - |
| $\mu_0 H_p$ | [mT] | 0.8 | 0.6 | 2.0 | 2.0 |
| $B_0$ | [mT] | 0.59 | 80.1 | - | - |
| $\mu_0 H_{irr}$ | [mT] | 20 | 70 | 5 | 16 |
| $J_c$(10mT) | [$10^4$A/cm$^2$] | 1.1 | 8.7 | - | - |
| $c$ | | 0.74 | 0.04 | - | - |

The numerical results are given in Table 3. It can be seen that the $B_{eq}^*$ and $\mu_0 H_p$ are of the same order of magnitude in both pure and S-doped crystals under the field perpendicular to *c*-axis. On the other hand, the $B_0$ of S-doped crystal is much larger than that of pure crystal. This result also means that the S-doping effects suppress the decay of $J_c$ as described by the Bean model. The $J_c^{eq}$ and $\mu_0 H_{irr}$ in S-doped crystal at 77K are considerably larger than those by heavy ions irradiation [46], nano-rod composites [49] and heavily Pb-doping [50]. The very small value of $c$ in doped crystal is affected by the superconductive volume fraction and demagnetization effect. So one of the important problem is to realize the high $J_c$ doped crystal with high superconductive volume fraction $c$.

## Application of *Kimishima Model* to Mgb$_2$ Systems

In this section, an estimate of $J_c(H)$ for pure $MgB_2$ is given [52], where $J_c(H)$ is the abbreviated form of $|J_c^{eq}(H)|$. $MgB_2$ samples were prepared by sintering a mixed powder of Mg and B at a molar ratio of 1:2, pressing into a disk shape, wrapping in Ta-foil and enclosing in a vacuumed quartz cell. It was then fired in the furnace at 1173K for 2 hours. The results of powder x-ray diffraction showed a typical pattern of $MgB_2$, except for a few percentages of MgO. The $M(H)$-curves were measured at 5, 15, 25 and 35K under magnetic fields of between -5T and 5T by the Quantum Design's MPMS SQUID magnetometer.

Analysis of the zero field cooled $M(H)$-curve was performed using the $M(H)$-equations of (9)-(16) for slab geometry ($a/b \rightarrow 0$). In the first procedure, the parameters $B_{eq}^*$, $B_0$ and the normalization factor $c$ were evaluated from the least square fitting of the experimental $M(H)$-data to the theoretical $M(H)$-equations. In this fitting, the $M_{eq}(H)$ was approximated to be $\Sigma M(H)/2$, and $H_p$ was determined as the field at which the initial magnetization $M_{vir}$ and the $M_L$ curve meet each other. The numerical results are given in Table 4.

It can be seen that the $B_{eq}^*$, $\mu_0 H_p$ and $B_0$ are of the same order of magnitude. The values of $c$ obtained are affected by the superconductive volume fraction and demagnetization effect of the measured sample. Next, the $M_{eq}(H)$ was re-estimated self-consistently using Eqs. (26) and (27) with terms up to $X^2$ for $H_p \leq H \leq H_d$ and with the parameter values shown in Table 4. $M^K{}_{U3}(H)$ of Eq. (14) and $M^K{}_L(H)$ of $-M^K{}_{U4}(-H)$ where $M^K{}_{U4}(H)$ is given by Eq. (15), were applied for $H_{c1} \leq H \leq H_p$, in which case the $H_{c1}$ can be neglected. The values of $M_{eq}(H)$ thus obtained are small compared to $\Sigma M(H)/2$, even above $H_p$, as shown in Figs. 11(a)-(d), where experimental $M(H)$, $\Sigma M(H)/2$ and $\Delta M(H)$ values are shown by open circles, squares and triangles, respectively. Broken lines show the theoretical results of $M(H)$, $M_{eq}(H)$ and $aJ_c(H)$. Bold solid lines show the well fitted regions in the first trial by $M_{eq}(H) = \Sigma M(H)/2$, where the relationship $J_c(H) = \Delta M(H)/a$ also holds. All of the quantities presented are multiplied by $\mu_0$. Finally, the theoretical $M(H)$ and $aJ_c(H)$ values were calculated from the $M(H)$ equations of (9)-(16) and Eq. (25), respectively, using the re-estimated results of $M_{eq}(H)$.

In Figs. 11(a)-(d), the $aJ_c(H)$ begins to deviate from the $\Delta M(H)$ near and below $H_p$. We consider that the $aJ_c(H)$ curve from Eq. (25) is more reliable than that from $\Delta M(H)$, because the first estimated $B_{eq}^*$ and $B_0$, and re-estimated $M_{eq}(H)$ present good reproducibility for $M(H)$ data. Though the theoretical values of $M_U$ and $M_L$ could not be compared with the 5K data in Fig. 11(a) due to irregular flux jumps around the zero field, the agreement of the theoretical $M(H)$ predictions with the initial $M_{vir}$ data is satisfactory.

The theoretical results of $M(H)$ are rather consistent with the experimental ones for 5, 15, 25 and 35K. Therefore, it can be concluded that the improved expression given by Eq. (25) for $J_c(H)$ is useful even near and below $H_p$. The approximations of $J_c(H) = \Delta M(H)/a$ for a slab geometry and $J_c(H) = \Delta M(H)/[a(1-a/3b)]$ for orthorhombic columns should not be used to estimate $J_c(H)$, especially in the low field region. The zero field equilibrium critical current densities $\mu_0 aJ_c(0)$ of $A/(2B_0)$ from Eq. (25) are also given in Table 4. If we use the values of $a$ as about $10^{-4}$m (determined from the SEM measurements), $J_c(0)$ is estimated as about $3.3\times 10^5$ [A/cm$^2$] which is same order as that in previous reports [13-18].

The result that $B_{eq}^*$ and $B_0$ are of the same order of magnitude in Eq. (25), and the linear temperature dependence of $J_c(0)=3.69\times 10^5\,(1-T/36)$ [A/cm$^2$] shown from the $J_c(0)$-values in Table 4, indicate that the position- and time-decay of $J_c^{int}(x,H)$ is dominant and that the critical state is one of a quasi-equilibrium state of flux bundles [10] even below $H_p$. Being different from $MgB_2$, the high $T_c$ superconductor of Bi2223 has obviously larger $B_0$ than $B_{eq}^*$ together with the exponential temperature dependence of $J_c(0)$. This implies single vortex pinning as described for Bi2223-system.

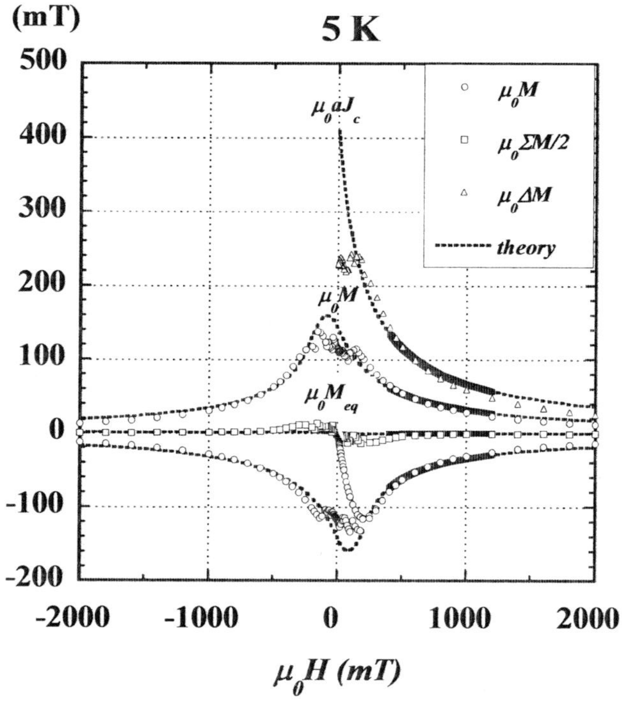
5 K
(mT)
500
400
300
200
100
0
-100
-200
-2000
-1000
0
1000
2000
μ0H (mT)
μ0aJc
μ0M
μ0Meq
μ0M
μ0ΣM/2
μ0ΔM
theory

(a)

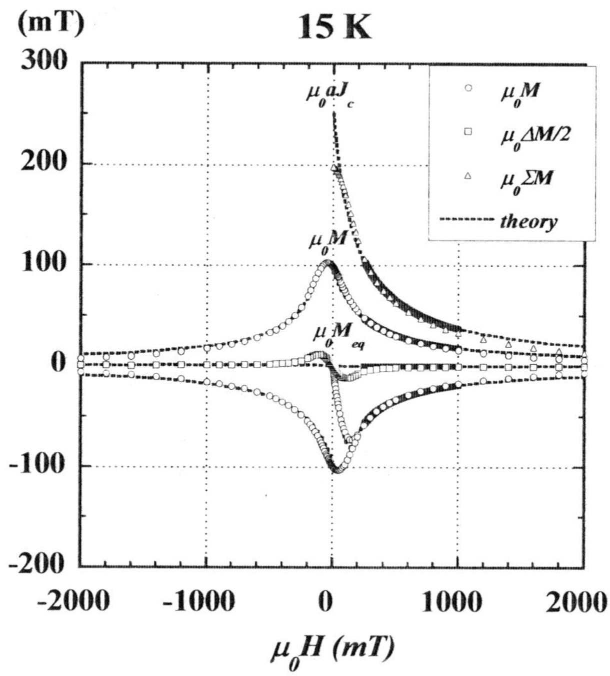
15 K
(mT)
300
200
100
0
-100
-200
-2000
-1000
0
1000
2000
μ0H (mT)
μ0aJc
μ0M
μ0Meq
μ0M
μ0ΔM/2
μ0ΣM
theory

(b)

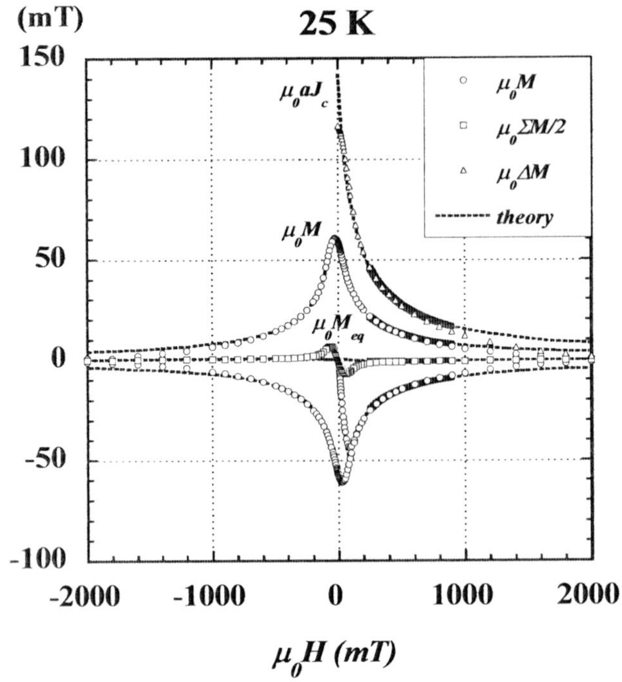


(c)

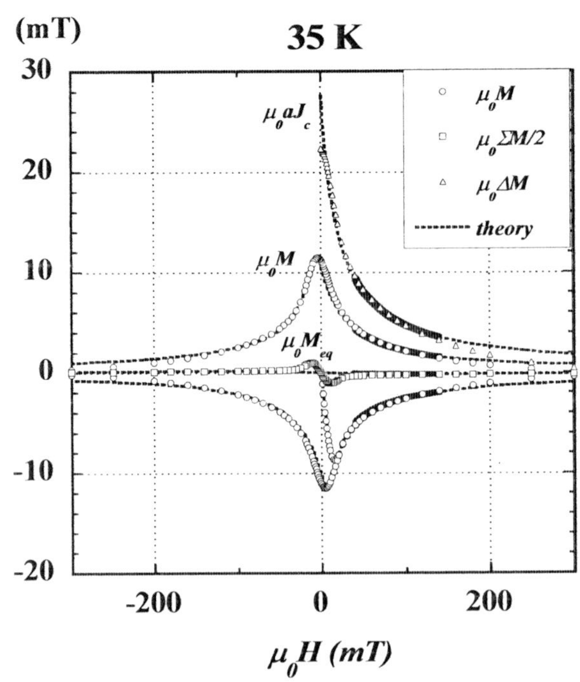


(d)

Figure 11. Magnetic data of $MgB_2$ at (a) 5K, (b) 15K, (c) 20K and (d) 35K.

Table 4. Magnetic parameters of $MgB_2$. $B_{eq}^*$ is the internal equilibrium flux density at $H$=$H_p$ above which $B_{int}$ is nonzero at the center of the sample. $B_0$ is the crossover parameter, $H_{irr}$ is the irreversible field and $J_c^{eq}(0)$ is the zero-field equilibrium critical current density. The constant $c$ is the normalized factor in the least square fitting calculations.

| **Temperature** | **[K]** | 5 | 15 | 25 | 35 |
|---|---|---|---|---|---|
| $B_{eq}^*$ | **[mT]** | 250 | 169 | 100 | 19 |
| $\mu_0 H_p$ | **[mT]** | 260 | 200 | 105 | 20 |
| $B_0$ | **[mT]** | 189 | 171 | 118 | 21 |
| $\mu_0 H_{irr}$ | **[T]** | >5 | 5 | 2.5 | 0.4 |
| $\mu_0 aJ_c(0)$ | **[mT]** | 415 | 251 | 138 | 25 |
| $c$ | | 1.04 | 1.23 | 1.18 | 1.01 |

## Lab$_6$-Doping Eefects on Pinning Property of Mgb$_2$

Here the La-doping effect was reported for the pinning property of $MgB_2$[53]. Since the superconductivity of $MgB_2$ with the $T_c$ of about 40K was discovered, the enhancement of the pinning property was studied by many authors. Especially, the critical current density $J_c$ and the irreversible field $H_{irr}$ at 20K have been important topics for the application of $MgB_2$ as the liquid hydrogen superconductor [52,54-63]. At 20K, pure $MgB_2$ generally has the $J_c$ of about $10^4$ A/cm$^2$ under 1 T and the $H_{irr}$ is about 4T [64]. As for the Ti-doping, a few or 10% Ti-doped $MgB_2$ showed the $J_c$-values larger than $10^5$A/cm$^2$ under 1 T and the $H_{irr}$ near 5T at 20K with small $T_c$-reductions [54-56]. The reason of pinning enhancement was considered was that the small (10nm) $MgB_2$ grains were surrounded by thin (1nm) $TiB_2$ layers which was good pinning centers by crystallographic matching with $MgB_2$.

On the other hand, the recent remarkable success was nano SiC-doping [57]. The 10% nano SiC-doped $MgB_2$ had large $J_c$ of $5\times 10^5$A/cm$^2$ under 1T and the $H_{irr}$ of 7T at 20K with scarce $T_c$-reduction. In this system, about 10nm inclusions of $Mg_2Si$, SiC and intra-grain defects were confirmed as the effective pinning centers. The nano-Si or C-doping also induced the effective pinning. In the nano Si-doped $MgB_2$ [58], the $Mg_2Si$ or Si inclusions with 10-30nm size enhanced the $J_c$ as large as that in the nano SiC-doped system at 1T and 20K, but it rapidly decayed in the higher field. In the nano carbon system [59,60], 5% nano C-doped $MgB_2$ had the $J_c$ of $10^5$A/cm$^2$ under 1T at 20K, but the $J_c$ disappeared at $H_{irr}$ of 5T being lower than the $H_{irr}$ of about 7T in the SiC-doped system. The effective pinning centers were formed by the uniformly distributed 5-10nm C, $Mg_2C_3$ and $MgB_2C_2$ precipitates in 100 nm $MgB_2$ grains [59]. Meanwhile, the formation of $MgAlB_4$ superstructure in $MgB_2$ by Al-doping [61] gave no improvement of pinning property, probably due to the reduction of density of state near the Fermi level which resulted in the decreasing of superconducting transition temperature. The Pb-doped $MgB_2$ with the impurity phases of Pb, MgO and $MgB_4$ also reduced $T_c$ of $MgB_2$, but the $|dH_{c2}/dT|$ increased with increasing the Pb content [62].

In the report on doping of $Y_2O_3$ nanoparticles [63], $J_c$ was enhanced to $2\times 10^5$A/cm$^2$ under 1 T and the $H_{irr}$ to 5.5T at 20K, producing regular distribution of nanometric (3-5nm) $YB_4$ particles in bulk $MgB_2$ by rapid formation technique. If the improvements of $J_c$ and $H_{irr}$

were due to the nano-$YB_4$ pinning centers, the same effects can also be expected for the formation of the other nano rare earth boride such as $LaB_6$.

The $LaB_6$-doped $MgB_2$ samples were sintered from the mixture of Mg, B and La powders with the mole ratio of Mg:B:La=1.2:2+6$x$:$x$ for the purpose of forming $(LaB_6)_x(MgB_2)_{1-x}$ mixed system, where $x$ was 0, 0.01, 0.02 and 0.03. The solid phase reactions of these powder mixtures were performed in the micro reactor cell at 1023 K under 40 MPa Ar-atmosphere for 30 hours.

In Fig. 12 the Cu$K\alpha$ x-ray powder diffraction (XRD) patterns are shown for the prepared samples. The intensities of (100) and (101) peaks of $MgB_2$ monotonously decreased with increasing $x$, and these peaks disappeared at $x$=0.03. On the other hand, the (110) peak intensity of $LaB_6$ increased with the increasing of La-content. For the $x$ = 0.01 and 0.02 samples, the half widths of diffraction peaks in Fig. 12 showed the production of 8~10 nm $LaB_6$ nano-particles in the $MgB_2$ phase with the small amount of MgO. So we investigated the pinning properties for the $x$=0.01 and 0.02 mainly. Another technique of rapid formation method [63] was also tried for La-doping in the vacuum quartz cell at 1123 K for 30 minutes. However, the $LaB_6$ phase became more dominant in the samples by this method.

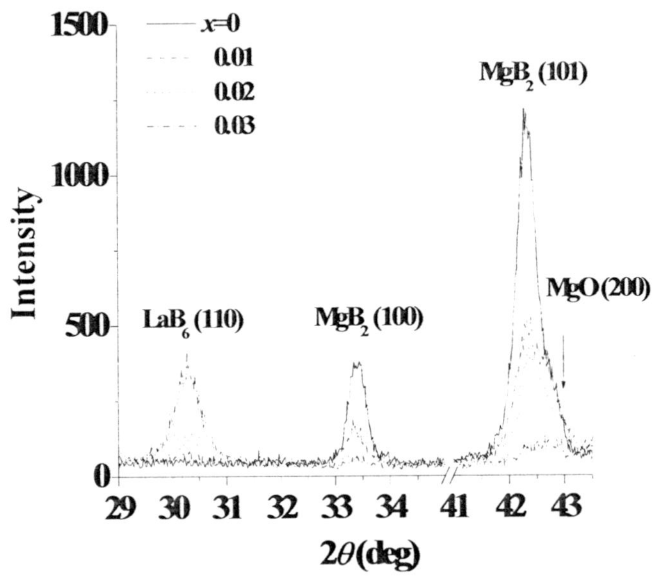


Figure 12. XRD patterns of $(LaB_6)_x(MgB_2)_{1-x}$ for $x$=0, 0.01, 0.02 and 0.03.

Electrical resistivities $\rho$ were measured by DC four terminals method for $x$=0, 0.01 and 0.02 samples produced in the micro reactor cell. The temperature dependence of normalized resistivities $\rho(T)/\rho(300K)$ was shown in Fig. 13. The superconducting transition temperatures $T_c$ were 38.4, 37.8 and 37.2K for $x$=0, 0.01 and 0.02 samples, respectively. The small amount of $|dT_c/dx|$ was preferable to improve the pinning property in this system.

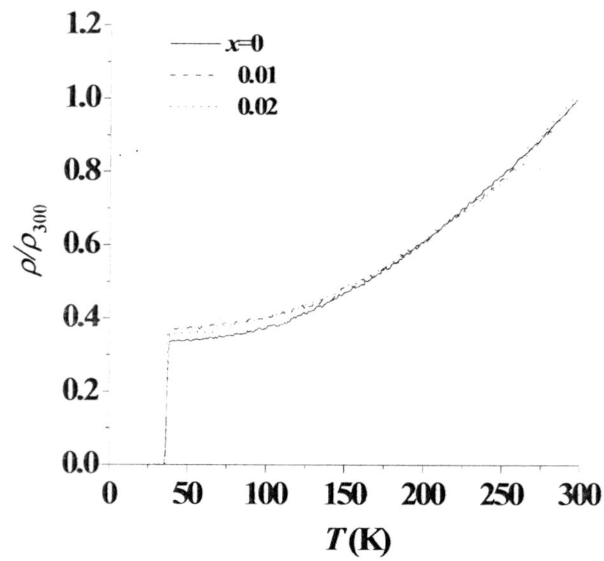


Figure 13. Temperature dependence of normalized resistivities of $x$=0, 0.01 and 0.02 samples.

The magnetization measurements were performed by the MPMS SQUID (Quantum Design) magnetometer at 5, 15, 25 and 35K up to the field of 5T. The experimental results were given in Figs. 14(a)-(c) for $x$=0, 0.01 and 0.02 samples, respectively. The $M(H)$-data and theoretical curves; obtained from the Eqs. (9)-(16), were depicted for $x$=0, 0.01 and 0.02 at 5, 15 and 25K, respectively, where the magnetic quantities were multiplied by vacuum permeability of $\mu_0=4\pi\times10^{-7}$ [Tesla m/A]. The numerical results were given in Table 5. The equilibrium critical current density $aJ_c$ (abbreviated form of $a|J_c^{eq}|$) at inner sample surface was estimated by Eq. (25). The all parameter-values of pure $MgB_2$ in Table 5 are larger than those in Table 4 mainly due to the difference in the value of $a$. However, it might suggest that the sintering in the high pressure oxygen atmosphere is useful to prepare the $MgB_2$ system with strong pinning centers.

If we use $10^{-4}$m as $a$, the $J_c$(1T)-value of pure $MgB_2$ was about $4\times10^4$A/cm$^2$ at 20K. In the present nano $LaB_6$-doped $(LaB_6)_x(MgB_2)_{1-x}$ system, $J_c$(1T)=$2\times10^4$A/cm$^2$ and $H_{irr}$=3.5T at 20K for $x$=0.01, and $J_c$(1T)=$6.5\times10^3$A/cm$^2$ and the $H_{irr}$=3.3T at 20K for $x$=0.02. Thus the enhancement of flux pinning could not be proved for the 10nm $LaB_6$-doped $MgB_2$. However, as shown in Table 5, the cross-over parameter $B_0$ of $x$=0.01 became larger than that of pure $MgB_2$. The large $B_0$ allows the mixed state of 2nd kind superconductor to have a Bean like critical state in which $J_c$ is constant for $H$ [10]. So this result means the decreasing of $J_c$ reduction rate against $H$ of $|d[J_c(H)/J_c(0)]/dH|$ in the $x$=0.01 sample for each temperature. Actually the $J_c(H)/J_c(0)$ of $x$=0.01 sample was larger than that of $x$=0 sample at 5 K as shown in Fig. 15.

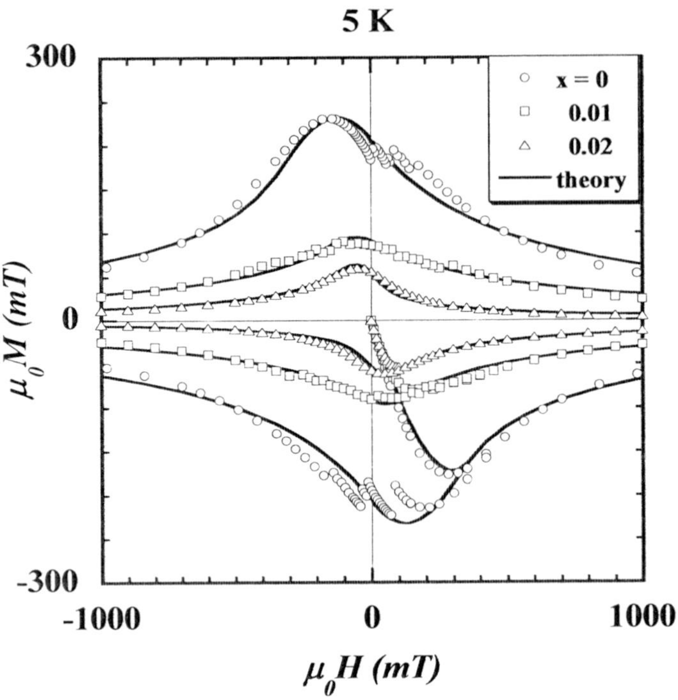
5 K
300
0
-300
-1000
0
1000
x = 0
0.01
0.02
theory
$\mu_0 M$ (mT)
$\mu_0 H$ (mT)

(a)

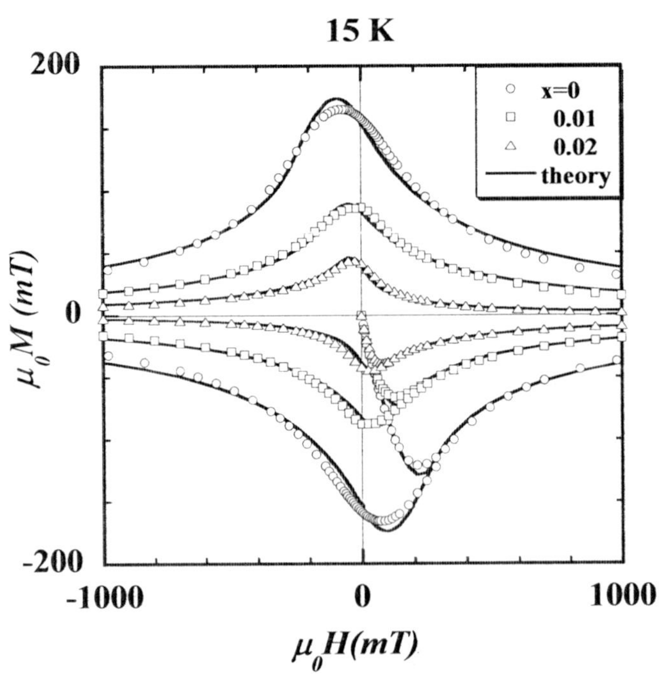
15 K
200
0
-200
-1000
0
1000
x=0
0.01
0.02
theory
$\mu_0 M$ (mT)
$\mu_0 H$(mT)

(b).

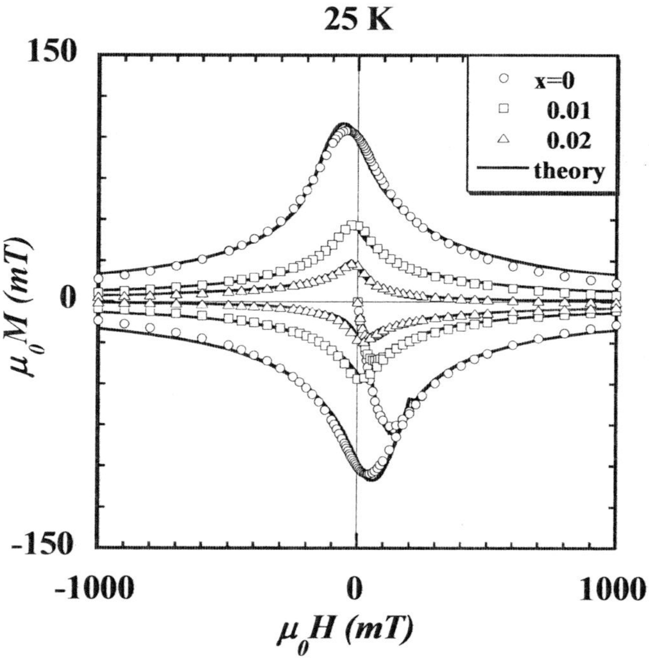


(c)

Figure 14. ZFC magnetization curves of $x$=0, 0.01 and 0.02 samples at 5K (a), 15K (b) and 25K (c). Bold solid lines are theoretical curves calculated using parameters in Table 5.

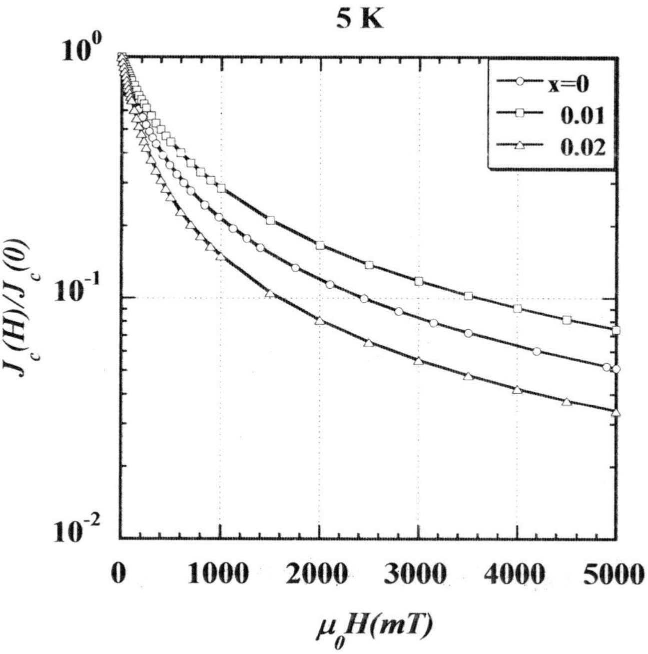


Figure 15. Field dependence of $J_c(H)/J_c(0)$ of $x$=0, 0.01 and 0.02 samples at 5K.

Table 5 Magnetic parameters and $J_c$ of $(LaB_6)_x(MgB_2)_{1-x}$.

| T [K] | | 5 | | | 15 | | | 25 | | |
|---|---|---|---|---|---|---|---|---|---|---|
| $x$ | | 0 | 0.01 | 0.02 | 0 | 0.01 | 0.02 | 0 | 0.01 | 0.02 |
| $B_{eq}$* | [mT] | 365 | 172 | 100 | 273 | 153 | 74 | 172 | 82 | 40 |
| $\mu_0H_p$ | [mT] | 420 | 180 | 100 | 300 | 164 | 80 | 210 | 88 | 50 |
| $B_0$ | [mT] | 271 | 401 | 176 | 201 | 218 | 146 | 133 | 138 | 101 |
| $\mu_0H_{irr}$ | [T] | >5 | >5 | 5 | 5 | 4.5 | 4 | 2.5 | 2.5 | 2.5 |
| $\mu_0aJ_c(0)$ | [mT] | 49 | 17 | 11 | 37 | 16 | 7.3 | 23 | 8.5 | 3.9 |
| $\mu_0aJ_c(1T)/J_c(0)$ | | 0.20 | 0.28 | 0.14 | 0.16 | 0.19 | 0.12 | 0.12 | 0.12 | 0.09 |
| $c$ | | 0.77 | 0.59 | 0.49 | 0.85 | 0.65 | 0.49 | 0.80 | 0.69 | 0.46 |

From the above results, it can be suggested that the nano-$LaB_6$ inclusions suppress the decay of $J_c$ by $H$ with no $T_c$- reduction. One of the origins of $J_c$ reduction by $LaB_6$-doping should be due to the insufficient growing of $MgB_2$ grains with the size of 20-30nm. Successive studies are needed to enhance the pinning property of $MgB_2$, using the $La_2O_3$ nano-particles as the starting materials.

## Pinning Enhancement in Ag-Doped Mgb$_2$

Here I report the pinning enhancement by Ag doping into $MgB_2$. Recently, the powder in tube (PIT) method to fabricate the superconducting wire or tape have been developed using Ag and Cu sheaths [65-68] in which the improved $J_c$'s of 2~3$\times 10^4$A/cm$^2$ under 1T at 20K were observed without $T_c$-reduction. By scanning electron microscopy (SEM) measurement, the formation of Mg-Ag and Mg-Cu alloys were confirmed between $MgB_2$ cores and the sheath materials of Ag and Cu, respectively [66]. Meanwhile the Fe- [69-72] or Ni- [73] clad $MgB_2$ tapes and wires showed higher $J_c$ and $H_{irr}$ than those of Ag- or Cu-clad ones, due to the lack of reaction between Fe or Ni and $MgB_2$. However, the magnetic elements such as Fe and Ni should not be adequate to the dopant for the enhancement of pinning property, but badly affect the superconducting state of host $MgB_2$. Therefore, the Ag or Cu could be one of the doping elements to form strong nano pinning centers in $MgB_2$.

The electronic structure calculations for noble metal diborides showed that the hole doped systems of $AgB_2$ and $AuB_2$ are potential candidates of BCS-type superconductors with 2~3 times higher $T_c$ than that in $MgB_2$ [74]. The solubility limit $x_L$ of Ag in $Mg_{1-x}Ag_xB_2$ single phase has been investigated by the chemical doping [75,76]. These studies revealed that $x_L$ was around 0.45%, and the lattice constants of $a$ and $c$ linearly decreased with increasing of $x$ in $0<x<0.45\%$. It means that the chemical pressure effect by the Ag-doping must be considered, except for the hole doping effect. As a result, only the reduction of $T_c$ was observed in this system. At $x>0.5$at%, the sintered sample had multi-phases of $Mg_{1-xL}Ag_{xL}B_2$ ($x_L$=0.0045) and residual Ag compounds or alloys.

The superconductive properties were also studied for $Ag_x(MgB_2)_{1-x}$ system [77,78], where x is the nominal mixing rate of Ag and $MgB_2$. Jin et al. [77] transferred the mixture of Ag powder with 1-10$\mu$m size and $MgB_2$ powder into Fe tubes under Ar atmosphere, and the Fe-clad ribbons of $MgB_2$ including 5at% Ag were fabricated by sintering at 900°C for 30min. Little effect on $T_c$ suggested the Ag is not incorporated into the $MgB_2$ superconductor phase.

However, the deterioration of $J_c$ was only observed in this case. Sun et al. [78] used the mixtures of commercial $MgB_2$ and Ag-powder in different mole ratios from 100:2.13 to 100:38.3. They were sintered in a vacuumed quartz ampoule at 950°C for 2h, and quenched to room temperature. In these samples, Ag reacted with $MgB_2$ to form AgMg and $MgB_4$. The $T_c$ remained at about 39 K, but the $J_c$ dropped from that of $MgB_2$ for all of the samples. Toulemonde et al. [79] synthesized $Mg_{0.9}Ag_{0.1}B_2$ by sintering in a cubic multi-anvil press at 3.5 GPa and 1000 °C for 1 h. The impurity phase of AgMg was formed in this sample, and no positive effect was observed for $T_c$, $J_c$, $H_{irr}$ and $H_{c2}$.

Here the author will present the first report of improvement of pinning property of $MgB_2$ by nano Ag-Mg alloy particles. The nominal composition of $Mg_{0.99}Ag_{0.01}B_2$ containing AgMg and $AgMg_4$ nano-particles showed remarkable enhancement of $J_c$ greater than that in pure $MgB_2$ at 15 and 20K.

The samples were sintered from the mixture of Mg-, B- and $Ag_2O$-powder with the mole ratio of Mg:B:$Ag_2O$=1.6(1-$x$):2:$x$/2, where $x$ was 0, 1 and 2%. The solid reaction of the powder mixture was performed in the electric furnace at 900°C for 30 minutes under Ar-atmosphere with 10% $H_2$ gas, followed by the rapid cooling in the furnace. In Fig. 16, the Cu$K\alpha$ -XRD patterns are shown for the prepared samples of $x$=0, 1 and 2%, where the (101) peak intensities of $MgB_2$-phase were normalized for each sample. For $x$=1 and 2% samples, Fig. 16 shows the production of impurity phase of AgMg and $AgMg_4$ of about 10-15nm size within the bulk $MgB_2$, except for small amounts of MgO. In Fig. 17, the transmission electron microscopy (TEM) image was given for $x$=1% sample, where the 10-20 nm particles of Ag-alloys (AgMg or $AgMg_4$) were observed in the $MgB_2$ matrices.

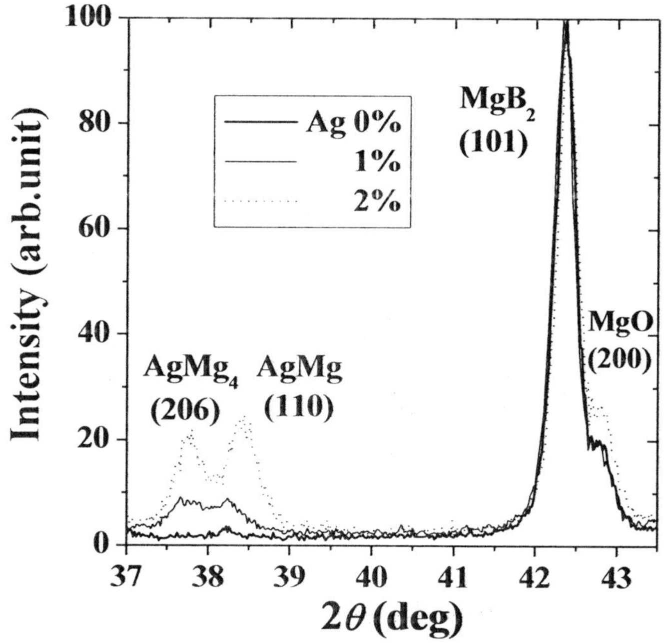


Figure 16. XRD patterns of nominal compounds of nominal $Mg_{1-x}Ag_xB_2$ for $x$=0, 1 and 2%.

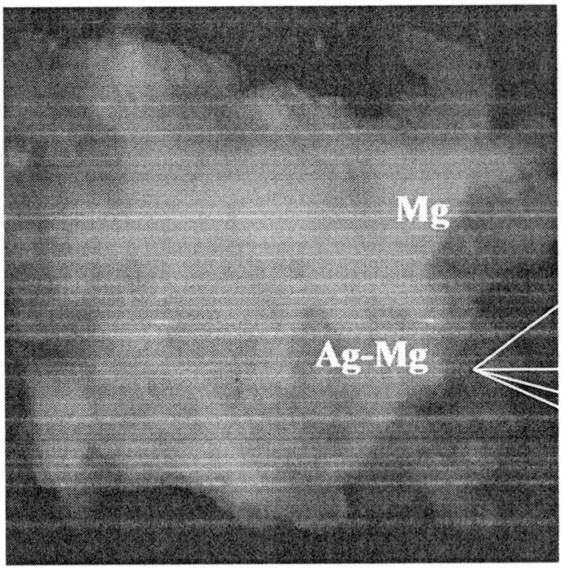


Figure 17. TEM image of Ag 1% sample. Nano-particles of Ag-Mg alloys were contained in the $MgB_2$ matrices.

Electrical resistivities $\rho$ were measured by DC four terminals method for $x$=0, 1 and 2% samples. The temperature dependence of resistivities $\rho$ ($T$) was shown in Fig. 18. The superconducting transition temperatures $T_c$ (mid points) were about 39K for all of the $x$ = 0, 1 and 2% of samples. Little change of $T_c$ was preferable to the pinning property in this system.

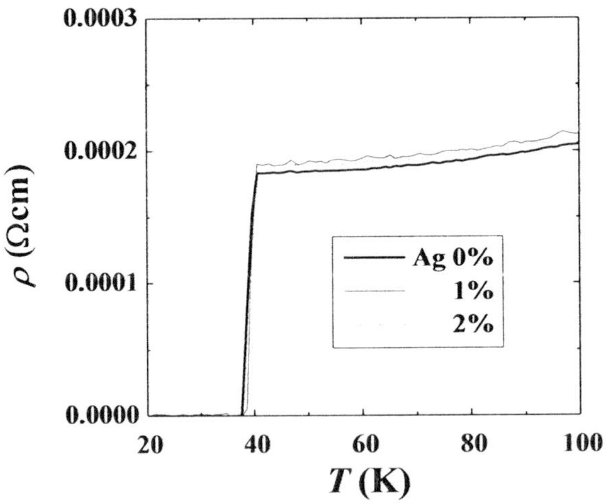


Figure 18. Temperature dependence of resistivities of Ag 0, 1 and 2% samples.

The magnetization measurements were performed by the MPMS SQUID (Quantum Design) magnetometer at 5, 15, 25 and 35K between the field of -5T and 5T. The Eqs. (9)-(16) for magnetization were fitted to the experimental data. The *equilibrium critical current density* $J_c(H)=|J_c^{eq}(H)|$ at inner sample surface was estimated by Eq. (25).

In Fig. 19(a)-(c), the *M*(*H*)-data at 5, 15, 20K and theoretical curves of *M* and $\mu_0 aJ_c$ were depicted for $x$=0, 1 and 2%, where all of the quantities were multiplied by $\mu_0$. As shown in this figure, the improvement of $J_c$ was visible to the $x$=1% sample, containing AgMg- and $AgMg_4$ nano-particles.

By the scanning electron microscopy (SEM), the accumulation of $MgB_2$ flakes with the thickness of about $10^{-4}$m was observed for all samples. If we use $10^{-4}$m as *a*, $J_c$'s at 20K and 1T was estimated as $2.3\times10^4$, $3.2\times10^4$ and $2.9\times10^4$A/cm$^2$ for $x$=0, 1 and 2% sample, respectively. Meanwhile, the irreversibly field $H_{irr}$'s at 20K were 4, 4 and 3.5 T for $x$=0, 1 and 2%, respectively. All of the numerical results were given in Table 6.

The Ag%-dependence of $T_c$, normalized by that of pure $MgB_2$, were shown in Fig. 20 with the results of other groups. The absence of $T_c$ reduction in our Ag1% and 2% samples means the coexistence of pure $MgB_2$-phase and nano-particles of AgMg and $AgMg_4$ alloy, because the large reduction of $T_c$ by the chemical pressure was observed for the solid solution of $Mg_{1-x}Ag_xB_2$ in the region of $0<x<x_L\sim0.45\%$ [75] as shown in Fig. 20. The substitution of Ag atoms into Mg sites in $MgB_2$ was realized by the sintering at a relatively low temperature of 700~950°C for a long time of 1~15 hours [75,76]. On the other hand, the Ag atoms should be unable to substitute with Mg atoms in $MgB_2$ by the rapid sintering at high temperature of 900~1000°C for 0.5~2 hours. This consideration is also supported by the similar $T_c$ results of refs.77 and 78 to ours.

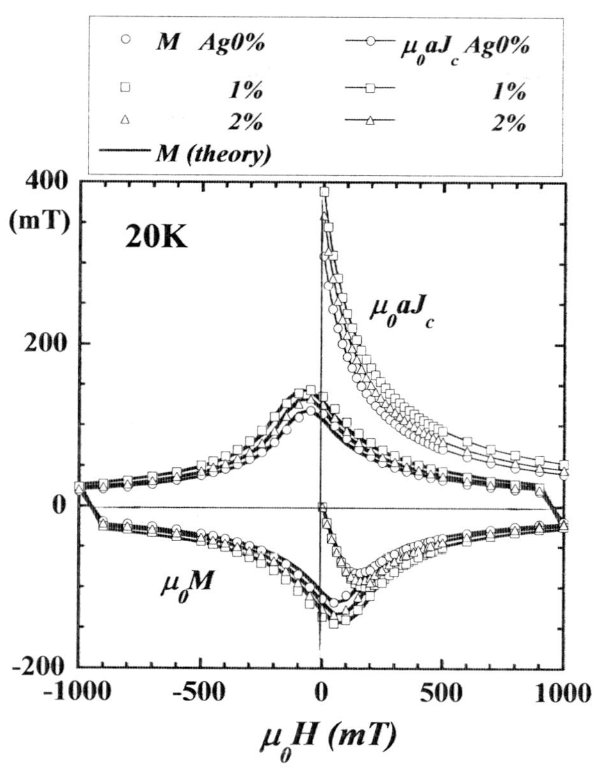


Figure 19. ZFC magnetization and $\mu_0 aJ_c$ curves of Ag 0, 1 and 2% samples at 20K. Bold lines are theoretical magnetization curves calculated using parameters in Table 6.

Table 6 Magnetic parameters and $aJ_c$ of nominal composition of $Ag_xMg_{1-x}B_2$.

| T [K] | 5 | | | 15 | | | 20 | | | 35 | | |
|---|---|---|---|---|---|---|---|---|---|---|---|---|
| Ag % | 0 | 1 | 2 | 0 | 1 | 2 | 0 | 1 | 2 | 0 | 1 | 2 |
| $B_{eq}$* [mT] | 276 | 331 | 320 | 233 | 284 | 261 | 189 | 227 | 209 | 38 | 41 | 38 |
| $B_0$ [mT] | 269 | 357 | 248 | 162 | 173 | 156 | 149 | 158 | 144 | 12 | 10 | 9 |
| $\mu_0H_{irr}$ [T] | >5 | >5 | >5 | 5 | 5 | 4.5 | 4 | 4 | 3.5 | 0.4 | 0.5 | 0.5 |
| $\mu_0 aJ_c(0)$ [mT] | 32 | 37 | 42 | 30 | 39 | 38 | 23 | 32 | 29 | 8 | 9.5 | 9.4 |
| $\mu_0 aJ_c(1T)$ [mT] | 7.0 | 10.2 | 8.3 | 4.4 | 6.1 | 5.1 | 3.2 | 4.2 | 3.6 | - | - | - |
| $c$ | 0.89 | 0.91 | 0.43 | 0.86 | 0.91 | 0.42 | 0.85 | 0.92 | 0.42 | 0.71 | 0.75 | 0.31 |

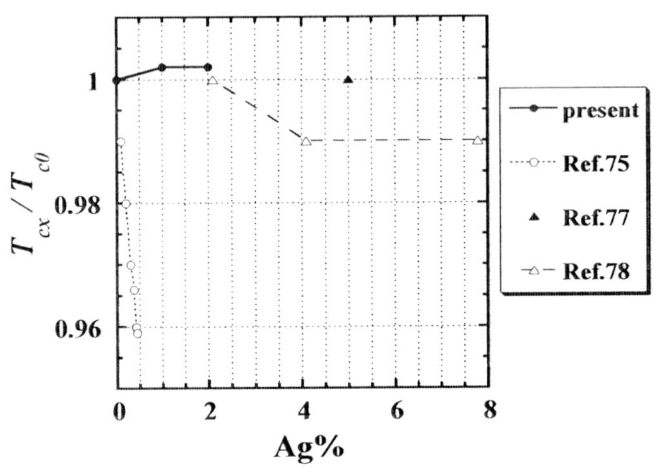


Figure 20. Ag%-dependence of $T_c$, being normalized by that of pure $MgB_2$.

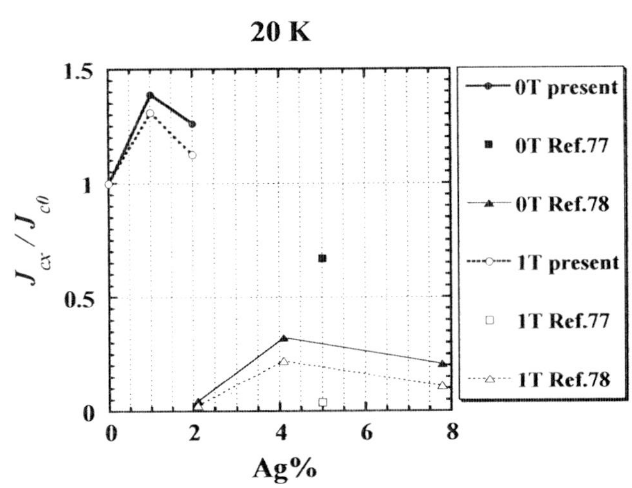


Figure 21. Ag%-dependence of $J_c$'s at 20 K under 0 and 1T, being normalized by that of pure $MgB_2$.

The Ag%-dependence of $J_c$'s at 20K under 0 and 1T, normalized by those of pure $MgB_2$, was shown in Fig. 21. The present report of enhanced $J_c$ at 15 and 20K is the first one for nominal $Mg_{1-x}Ag_xB_2$ system. The previous studies of the pinning property were performed for $Ag_x(MgB_2)_{1-x}$ or $Mg_{1-x}Ag_xB_2$ system with $x$ larger than 2% [77-79], and only negative results were obtained as shown in Fig. 21. Now we can expect that the nominal Ag content around 1% and 2% is a critical one for the pinning enhancement in nominal $Mg_{1-x}Ag_xB_2$ system by the existence of Ag-Mg nano-particles as the strong pinning centers. Poor $J_c$ results of $x$~2 % in Ref. 78 may be due to the other boride phase of $MgB_4$.

## Conclusion

In this chapter, the analytical expressions of magnetization by extended critical state model were presented by the author. First, $J_c$-formula of the *Kimishima model* was

$$\mu_0 \quad |J_c^{\text{int}}(x,H)| = \frac{B_{eq}^*(B_{eq}^* + 2B_0)}{2a[|B_{int}(x,H)| + B_0]} \tag{1},$$

which is one of the universal functions interconnecting the Bean and Anderson-Kim model. Second, the formula of $J_c$ by the *Exponential Kimishima model* was

$$\mu_0 |J_c^{\text{int}}(x,H)| = \frac{G}{2aB_0}\exp\left[-\frac{|B_{\text{int}}(x,H)|}{B_0}\right] \tag{30},$$

and $G=2B_0^2[\exp(B_{eq}^*/B_0)-1]$.

This equation was derived from the extended *Kimishima model* including exponential decay of $J_c$ near the irreversibly field $H_{irr}$.

As described in the previous sections, theoretical magnetization curves derived from Eq. (1) or Eq. (30) could be fitted to the experimental curves. Estimation of characteristic two parameters of $B_{eq}^*$ and $B_0$ for some high $T_c$ superconductors. The temperature dependences of $aJ_c$, and the parameters, $B_{eq}^*(T)$ and $B_0$ $(T)$, are also important for the comprehension of pinning properties of each high $T_c$ superconductor, especially for the types of pinned vortices. It will be possible to obtain phenomenological theory for the temperature dependences of magnetization, which will give the complete analytical formulation system for the magnetizations of type II high $T_c$ superconductors.

### Acknowledgements

The author expresses his thanks for the permission to reproduce his earlier articles of "Magnetization analysis of sintered Bi2223 using an extended critical state model" [33] Copyright (2004) by IOP, "Magnetization analysis of high $T_c$ superconductor by new critical state model with $M_{eq}$" [3] Copyright (2001), "Comments on magnetically estimated $J_c$ for MgB2 by critical state models" [52] Copyright (2002), "S-doping effects on pinning property of Bi2212 single crystals" [43] Copyright (2003) and "La-doping effects on pinning properties of $MgB_2$" [53] Copyright (2004) by Elsevier. He also thanks his collaborators,

Professor T. Kuramoto, Dr. M. Uehara, Dr. Y. Ichiyanagi and all of the students in his laboratory.

## References

[1] Kimishima, Y. *Cryogenics* 1998, 38, 763-775.
[2] Kimishima, Y.; Kamimura, K.; Ichiyanagi, Y. *Physica C* 2000, 329, 17-28.
[3] Kimishima, Y.; Ichiyanagi, Y. *Physica C* 2001, 353 , 111-122.
[4] Fietz, W. A.; Beasley, M. R.; Silcox, J.; Webb, W. W. *Phys. Rev.* 1964, 136, A335-A345.
[5] Stancil, D. D.; Schlesinger, T. E.; Stamper, A. K.; Wong, D. *J.Appl.Phys.* 1988, 64, 5899-5901.
[6] Krusin-Elbaum, L.; Malozemoff, A. P.; Cronemeyer, D. C.; Hpltzberg, F.;Clem, J. R.; Hao, Z. *J.Appl.Phys.* 1990, 67, 4670-4675.
[7] Indenbom, M. V.; Schuster, Th.; Kuhn, H.; Kronmüller, H.; Li, T. W.; Menkovsky, A. A. *Phys.Rev.B* 1995, 51, 15484-15491.
[8] Abrikosov, A. A. *Soviet Phys.JETP* 1950, 5, 1174-1182.
[9] Kim, Y. B.; Hempstead, C. F.; Strnad, A. R. *Phys.Rev.Lett.* 1962, 9, 306-309.
[10] Anderson, P. W.; Kim, Y. B. *Rev.Mod.Phys.* 1964, 36, 39-43.
[11] Bean, C. P. *Rev.Mod.Phys.* 1964, 36, 31-39.
[12] Chen, D.-X.; Goldfarb, R. B. *J.Appl.Phys.* 1989, 66, 2489-2500.
[13] Takano, Y.; Takeya, H.; Fujii, H.; Kumakura, H.; Hatano, T.; Togano, K. *Appl.Phys.Lett.* 2001, 78, 2914-2916.
[14] Bugoslavsky, Y.; Perkins, G. K.; Qi, X.; Cohen, L. F.; Caplin, A. D. *Nature* 2001, 410, 563-565.
[15] Joshi, Amish G.; Pillai, C. G. S.; Raj, P.; Malik, S. K. *Solid State Commun.* 2001, 118, 445-448.
[16] Eom, C. B.; Lee, M. K.; Choi, J. H.; Belenky, L.; Song, X.; Cooley, L. D.; Naus, M. T.; Patnaik, S.; Jiang, J.; Rikel, M.; Plyanskii, A.; Gurevich, A.; Cai, X. Y.; Bu, S. D.; Babcock, S. E.; Hellstrom, E. E.; Larbalestier, D. C.; Rogado, N.; Regan, K. A.; Hayward, M. A.; He, T.; Slusky, J. S.; Inumaru, K.; Haas, M. K.; Cava, R. J. *Nature* 2001, 411, 558-560.
[17] Bugoslavsky, Y.; Cohen, L. F.; Perkins, G .K.; Polichetti, M.; Tate, T. J.; Gwilliam, R.; Caplin, A. D. *Nature* 2001, 411, 561-563.
[18] Wen, H. H.; Li, S. L.; Zhao, Z. W.; Jin, H.; Ni, Y. M.; Kang, W. N.; Kim, H. J.; Choi, E. M.; Lee, S. I. *Phys.Rev.B* 2001, 64, 134505(1)-(5).
[19] Shimoyama, J.; Shimizu, K.; Ueda, K.; Horii, S.; Otzschi, K.; Kishio, K.; Chikumoto, N. *Physica C* 2002, 378-381, 457-461.
[20] Gioacchino, D. D.; Tripodi, P.; Gambardella, U. *Physica C* 2002, 372-376, 945-948.
[21] Ravi Kumar, G.; Chaddah, P. *Phys.Rev.B* 1989, 39, 4704-4707.
[22] Chen, D.-X.; Sanchez, A.; Munoz, J. S. *J.Appl.Phys.* 1990, 67, 3430-3437.
[23] Araújo-Moreira, F. M.; Ortiz, W. A.; de Lima, O. F. *J.Appl.Phys.* 1996, 80, 3390-3395.
[24] Hu, L.; Zhang, P.; Wang, J.; Teng, X.; Li, C.; Feng, Y.; Zhou, L.; Sulpice, A.; Mossonge, E.; Cao, L.; Ruan, K.; Li, X. *Physica C* 2003, 386, 131-137.

[25] van der Laan D. C., ; van Eck, H. J. N.; Schwarz, J.; ten Haken, B.; ten Kate H. H. J. *Physica C* 2002, 372-376, 1024-1027.
[26] Garnier, V.; Marinel, S.; Desgardin, G. *Physica C* 2002, 372-376, 1107-1110.
[27] Andrzejewski, B.;E. Guilmeau,; Simon, *Ch. Supercond.Sci.Technol.* 2001, 14, 904-909.
[28] Hari Babu, N.; Rajasekharan, T.; Seshu Bai, V. *Phys.Rev.B* 1998, 57, 1277-1283.
[29] Fujikami, J.; Kaneko, T.; Ayai, N.; Kobayashi, S.; Hayashi, K.; Takei, H.; Sato, K. *Physica C* 2002, 378-381, 1061-1064.
[30] Yoshizawa, S.; Hishinuma, Y.; Yamamoto, R.; Hirano, S.; Nishimura, A.; Matsumoto, A.; Kumakura, H. *Physica C* 2002, 372-376, 1232-1234.
[31] Maeda, H.; Inaba, T.; Sato, M.; Zhang, P. X.; Chen, W. P. *Physica C* 2001, 357-360, 1230-1233.
[32] Kawano, K.; Oota, A.; Fukuta, K.; Higuchi, T.; Fujimoto, H. *Cryogenics* 1999, 39, 853-858.
[33] Kimishima, Y.; Ichikawa, H.; Takano, S.; Kuramoto, T. *Supercond.Sci.Technol.* 2004, 17, S36- S41.
[34] Kimishima, Y.; Inagaki, K.; Tanabe, K.; Ichiyanagi, Y. *Supercond.Sci.Technol.* 2000, 13, 295-300.
[35] Tuszynski, J. A.; Sept, D.; Dixon, J. M. *Phys.Rev.B* 1999, 60, 10513-10521.
[36] Martin, S.; Fiory, A.T.; Fleming, R. M.; Espinosa, G. P.; Cooper, A.S. *Appl.Phys.Lett.* 1989, 54, 72-74.
[37] Tahara, S.; Anlage, S. M.; Halbritter, J.; Eom, C-B.; Fork, D. K.; Geballe, T. H.; Beasley, M.R. *Phys.Rev.B* 1990, 41, 11203-11208.
[38] Mannhart, J.; Chaudhari, P.; Dimos, D.; Tsuei, C. C.; McGuire, T. R. *Phys.Rev.Lett.* 1988, 21, 2476-2479.
[39] T. Fujiyoshi, K. Toko, T. Matsushita, K. Yamafuji, *Jpn.J.Appl.Phys.* 1989, 28, L1906-L1908.
[40] Tinkham, M. *Introduction to Superconductivity*.; MaGraw-Hill: New York, 1975.
[41] Anderson, P.W. *Phys.Rev.Lett.* 1962, 9, 309-311.
[42] Feigel'man, M. V.; Geshkenbein, V. B.; Vinokur, V. M. *Phys.Rev.B* 1991, 43, 6263-6265.
[43] Yorimasa, K.; Uehara, M.; Kuramoto, T.; Inoue, H.; Kimishima, Y. *Physica C* 2003, 392-396, 306-310.
[44] Kumakura, H.; Togano, J. K.; Maeda, H. *Appl.Phys.Lett.* 1991,58, 2830-2832.
[45] Gerhäuser, W.; Ries, G.; Neumüller, H. W.; Schmit, W.; Eibl, O.; Saemann-Ischenko, ; Klaumünzer S., *Phys.Rev.Lett.* 1992, 68, 879-882.
[46] Thompson, J. R.; Sun, Y. R.; Kerchner, H. R.; Christen, D. K.; Chakoumakos, B. C.; Marwick, A. D.; Civale, L.; Thompson, J. O. *Appl.Phys.Lett.* 1992, 60, 2306-2308.
[47] Kumakura, H.; Kitaguchi, J.; Togano, K.; Maeda, H.; Shimoyama, J.; Okayasu, S.; Kazumata, Y. *J.Appl.Phys.* 1993, 74, 451.
[48] Pradhan, A. K.; Roy, S. B.; Chaddah, P.; Kanjilal, D.; Chen, C.; Wanklyn, B. M. *Phys.Rev. B* 1996, 53, 2269-2272.
[49] Yang, P.; Lieber, C. M. *Science* 1996, 273, 1836-1840.
[50] Chong, I.; Hiroi, Z.; Izumi, M.; Shimoyama, J.; Nakayama, Y.; Kishio, K.; Terashima, T.; Bando, Y.; Takano, M. *Science* 1997, 276, 770-773.
[51] Motohashi, T.; Nakayama, Y.; Fujita, T.; Kitazawa, K.; Shimoyama, J.; Kishio, K. *Phys. Rev.B* 1999, 59, 14080-14086.

[52] Y. Kimishima,; M. Uehara,; T. Kuramoto,; Y. Ichiyanagi,; Y. Iriyama, K. Yorimasa, *Physica C* 2002, 377, 196-201.

[53] Kimishima, Y.; Uehara, M.; Kuramoto, T.; Takano, S.; Takami, S. *Physica C* 2004, 412-414, 402-406.

[54] Zhao, Y.; Feng, Y.; Huang, D. X.; Machi, T.; Cheng, C. H.; Nakao, K.; Chikumoto, N.; Fudamoto, Y.; Koshizuka, N.; Murakami, M. *Physica C* 2002, 378-381, 122-126.

[55] Zhao, Y.; Cheng, C. H.; Feng, Y.; Machi, T.; Huang, D. X.; Zhou, L.; Koshizuka, N.; Murakami, M. *Physica C* 2003, 386, 581-587.

[56] Anderson Jr., N. E.; Straszheim, W. E.; Bud'ko, S. L.; Canfield, P. C.; Finnemore, D. K.; Suplinskas, R. J. *Physica C* 2003, 390, 11-15.

[57] Dou, S. X.; Soltanian, S.; Horvat, J.; Wang, X. L.; Munroe, P.; Zhou, S. H.; Ionescu, M.; Liu, H. K.; Tomsic, M. *Appl.Phys.Lett.* 2002, 81, 3419-3421.

[58] Wang, X. L.; Zhou, S. H.; Qin, M. J.; Munroe, P. R.; Soltanian, S.; Liu, H. K.; Dou, S. X. *Physica C* 2003, 385, 461-465.

[59] Soltanian, S.; Horvat, J.; Wang, X. L.; Munroe, P.; Dou, S. X. *Physica C* 2003, 390, 185-190.

[60] Ribeiro, R. A.; Bud'ko, S. L.; Petrovic, C.; Canfield, P. C. *Physica C* 2003, 384, 227-236.

[61] Xiang, J. Y.; Zheng, D. N.; Li, J. Q.; Wen, H. H.; Zhao, Z. X. *Physica C* 2003, 386, 611-615.

[62] Gu, D. W.; Cai, Y. M.; Yau, J. K. F.; Cui, Y. G.; Wu, T.; Yuan, G. Q.; Shen, L. J.; Jin, X. *Physica C* 2003, 386, 643-647.

[63] Wang, J.; Bugoslavsky, Y.; Berenov, A.; Cowey, L.; Caplin, A. D.; Cohen, L. F.; J. L.; MacManus Driscoll,; Cooley, L. D.; Song, X.; Larbalestier, D. C. *Appl.Phys.Lett.* 2002, 81, 2026-2028.

[64] Canfield, P. C.; Bud'ko, S. L.; Finemore, D. K. *Physica C* 2003, 385, 1-7.

[65] Glowacki, B. A.; Majoros, M.; Vickers, M. E.; Zeimetz, B. *Physica C* 2002, 372-376, 1254-1257.

[66] Soltanian, S.; Wang, X. L.; Horvat, J.; Li, A. H.; Liu, H. K.; Dou, S. X. *Physica C* 2002, 382, 187-193.

[67] Zhou, S.; Pan, A. V.; Ionescu, M.; Liu, H.; Dou, S. *Supercond.Sci.Technol.* 2002, 15, 236-240.

[68] Martìnez, E.; Angurel, L. A.; Navarro, R. *Supercond.Sci.Technol.* 2002, 15, 1043-1047.

[69] Soltanian, S.; Wang, X. L.; Kusevic, I.; Babic, E.; Li, A. H.; Qin, M. J.; Horvat, J.; Liu, H. K.; Collings, E. W.; Lee, E.; Sumption, M. D.; Dou, S. X. *Physica C* 2001, 361, 84-90.

[70] Wang, X. L.; Soltanian, S.; Horvat, J.; Liu, A. H.; Qin, M. J.; Liu, H. K.; Dou, S. X. *Physica C* 2001, 361, 149-155.

[71] Glowacki, B. A.; Majoros,M.; Vickers, M.; Evetts, J. E. *Supercond.Sci.Technol.* 2001, 14, 193-199.

[72] Flükiger, R.; Suo, H. L.; Musolino, N.; Beneduce, C.; Toulemonde, P.; Lezza, P. *Physica C* 2003, 385, 286-305.

[73] Grasso, G.; Malagoli, A.; Ferdeghini, C.; Roncallo, S.; Braccini, V.; Siri, A.; Cimberle, A. S. *Appl.Phys.Lett.* 2001, 79, 230-232.

[74] Kwon, S. K.; Youn, S. J.; Kim, K. S.; Min, B. I. *cond-mat/0106483* 2001.

[75] Cheng, C. H.; Zhao, Y.; Wang, L.; Zhang, H. *Physica C* 2002, 378-381, 244-248.

[76] Cheng, C. H.; Zhao, Y.; Zhu, X. T.; Vowotny, J.; Sorrell, C. C.; Finlayson, T.; Zhang, H. *Physica C* 2003, 386, 588-592.
[77] Jin, S.; Mavoori, H.; Bower, C.; van Dover, R. B. *Nature* 2001, 411, 563-565.
[78] Sun, T.; Zhang, X. P.; Zhao, Y. G.; Shen, R.; Wang, K.; Zhang, L. W.; Cao, B. S.; Xiong, Y. H.; Li, P. J.; Wen, H. H. *Physica C* 2002, 382, 367-372.
[79] Toulemonde, P.; Musolino, N.; Flükiger, R. *Supercond.Sci.Technol.* 2003, 16, 231-236.

In: Superconductivity, Magnetism and Magnets
Editor: Lannie K. Tran pp. 83-102
ISBN 1-59454-845-5


*Chapter 3*

# Spontanous Vortex Phase, Pinning and Transport in Heterogenous Ferromagnetic-Superconducting Systems

***M. Amin Kayali***
Human Neuroimaging Laboratory, Baylor College of Medicine,
Houston, TX 77030, USA

**Abstract**

The past few years have witnessed an increasing interest in studying the static and dynamic phases of heterogeneous ferromagnetic-superconducting systems. Typically, a heterogeneous ferromagnetic-superconducting system consists of a superconductor placed in close proximity with a ferromagnet separated by a layer of insulator oxide. The layer of insulator oxide suppresses proximity effects and spin diffusion hence allowing for the coexistence of two competing order parameters, a feature that is special to these systems. The strong interaction between the ferromagnet and the superconductor gives raise to strong vortex pinning force which enhances the critical current of the superconductor. In this chapter, I will review the important theoretical developments in this area of research. The first part of this review will focus on the static properties of these systems; therefore, the general formulation of Londons theory for these systems will be presented. Afterward, I will apply the general theory to study few specific systems with emphasis on the appearance of spontaneous vortex phase and the important role played by the geometry and the magnetization distribution of the ferromagnet on the static and dynamic properties of the superconductor. In the second part, I will concentrate on the dynamic of vortices and antivortices in these systems. Therefore, I will review the anisotropic transport properties of a ferromagnetic-superconducting bilayer then I will study the dynamical properties of a system consisting of a superconducting thin film pierced by a square array of ferromagnetic columnar defects.

**PACS** 74.25.Dw,74.25.Ha,74.25.Op,74.25.Qt,74.25.Sv

## 1 Introduction

Dissipative behavior of type two superconductors (SC) is due to phase slippage, which may arise due to the motion of vortices or the existence of phase-slip centers or lines. To enhance the critical current of a superconductor the motion of vortices must be stopped; this occurs by what is known as vortex pinning. Therefore, it is very important both theoretically and experimentally to understand the pinning mechanisms and how to optimize pinning. In principle, most lattice defects create pinning potentials for flux lines in superconductors. The defects are either native to the superconductor's lattice or artificially introduced. Artificial pinning sites such as random columnar defects are typically created by heavy-ion irradiation [1]. The heavy-ion irradiation method creates random pinning centers. Vortex pinning and transport properties of superconductors with random pinning centers were well studied and thoroughly reviewed by Blatter *et.al.* [2]. It was shown later that flux pinning can be enhanced by employing a regular array of defects in the superconductor. Magnetization, critical current and resistivity measurements performed on systems consisting of a superconductor covered by a regular array of artificially created structural defects displayed commensurability effects [3], [4] that were not observed in systems with random pinning. The commensurability between the flux lattice and the defect array resulted in higher critical currents. Of particular interest are the results of the experiment by Martinoli *et.al* [5] who studied the transport properties of a SC film of periodically modulated thickness and observed oscillations of the critical current as a function of the applied magnetic field. It is worth mentioning that pinning in superconductors with artificial columnar defects such as those created by the heavy-ion irradiation is due to the multiply connected topology of the superconductor. The origin of pinning in superconductors with multiply connected topologies pertains to the variations in the effective value of the Ginzburg-Landau parameter $\kappa = \frac{\lambda}{\xi}$ in the neighborhood of the defect.

In recent years, the interest in heterogeneous ferromagnetic-superconducting systems (HFSS) has grown rapidly due to the preliminary experimental data which showed improved pinning strength in these systems. These systems typically consist of a superconductor placed in close proximity with a ferromagnet (FM). A thin layer of insulator oxide is sandwiched between the FM and SC to prevent proximity effects and spin diffusion, which both tend to destroy superconductivity. The interest in HFSS is motivated not only because of their technological importance but also for being a new class of physical systems in which two competing order parameters may coexist and possibly enhance one another. Several experimental groups [6], [7], [8], [9], [10] have investigated pinning in HFSS and found appreciable increases in the critical current. In most experiments, the superconductor is a continuous two dimensional sheet covered by a regular array of ferromagnetic dots. Each dot has a radius of the order of the SC penetration depth $\lambda$ and is magnetized in a direction either parallel or perpendicular to the SC. It is important to realize that pinning in HFSS does not appear due to the multiply connected topology, as occurs when the defect is established in the SC. It is rather due to the strongly inhomogeneous distribution of the magnetic field generated by the magnetic structure.

In 1957, Ginzburg [11] argued that superconductivity and ferromagnetism may coexist in systems of dimension $D \leq 2$ due to the large critical field of low-dimensional superconductors. However, proximity effects and spin diffusion pose serious problem for the coexistence of the two phenomena. To solve this problem, Lyuksyutov and Pokrovsky [12, 13] proposed to separate the ferromagnet and superconductor by inserting a thin layer of insulator oxide, which suppresses proximity effects. The FM and SC are now electronically separated; therefore, they interact solely via their magnetic fields without destroying one another's order parameter. Different realizations of HFSS were studied both experimentally and theoretically. Special consideration was given to the system of a superconducting thin film covered by a regular array of ferromagnetic nano-dots [6], [7], [8], [9], [10]. Such systems are usually prepared using electron-beam lithography and lift-off techniques [14]. The dots are magnetized either in the direction parallel or perpendicular to the SC film. These studies reported the observation of commensurability effects in these systems. The essence of these effects is that the dot array spontaneously creates vortices in the SC, which in turn will be pinned by the dots. Commensurability effects appear when the density of vortices in the SC is a fractional number of the density of dots in the FM.

Theoretically, many HFSS systems with different magnetization distributions were studied [15], [16], [17], [18], [19], [20], [21], [22], [23], [24], [25], [26], [27], [28]. Almost a decade ago, Marmorkos *et.al.* [17] studied the interaction between a ferromagnetic disk embedded in a bulk superconductor and showed that the disk generates several vortices in the SC. Their approach was to numerically solve the nonlinear Ginzburg-Landau equation with appropriate boundary conditions after taking into account the magnetic vector potential generated by the dot. Sasik and Hwa [23] studied a system consisting of a superconducting thin film covered by magnetic dots. They assumed the dots to be magnetized in a direction perpendicular to the SC film. They also simplified the problem by ignoring the real geometry of the dots by replacing them by point dipoles. This is the main drawback of their work since the geometry of the FM is of great importance in the study of the statics and dynamics of any HFSS. They were able to show that, in the absence of an external magnetic field, a spontaneous creation of superconducting vortices can take place if the stray magnetic field generated by the dot exceeds the field necessary to create vortices in the superconductor.

The ground state of a ferromagnetic-superconducting bilayer (FSB) was studied in [22]. The authors found that the homogeneous state, which is characterized by a magnetization directed perpendicular to the FM, can be unstable with respect to the creation of vortices in the SC. They showed that the weak interaction between the created vortices makes the structure with alternating domains an energy favorable ground state. They also considered the possibility of other structures, showing that the structure with alternating stripe domains costs less energy than other structures. Therefore, they speculated that a strong transport anisotropy can be viewed as an indirect observation of the stripe domain structure in the FSB.

Motivated by the current theoretical and experimental interest in the static and dynamical phases of the HFSS, I review some of the literature on static and transport properties of HFSS. This review is divided into two parts. The first part will begin by presenting the theory of HFSS within the London approximation then I present two specific example systems of HFSS and study their static properties. The first example is a SC film covered by a circular ferromagnetic dot whose magnetization is perpendicular to the SC plane while the second is for a dot with magnetization parallel to the SC film. In the second part I discuss the behavior of HFSS in the nonequilibrium regime. After a general introduction to vortex dynamics in HFSS, I summarized the transport properties of ferromagnetic-superconducting bilayers and then I present a study of the transport properties of a SC film pierced by and array of FM columnar defects.

# 2 Static Regime of HFSS

## 2.1 London Approximation and HFSS

Let the SC thin film be in the xy-plane and an infinitely thin ferromagnetic texture is placed on the top of it at a distance $a_0 \ll d \sim \xi$, where $a_0$ is an atomic dimension and $\xi$ is the SC coherence length. Thus the interaction between the ferromagnet and superconductor occurs via their magnetic fields, the theory to calculate the total magnetic field and screening currents is developed here. Let the magnetization distribution of the FM texture be $\mathbf{M}(\mathbf{r})$. The total magnetic field $\mathbf{B} = \mathbf{B}^s + \mathbf{B}^m$ is derived from the total vector potential $\mathbf{A} = \mathbf{A}^s + \mathbf{A}^m$, where the superscripts (s) and (m) refer to the superconducting and ferromagnetic parts respectively. The total vector potential is governed by Maxwell's equation

$$\nabla \times (\nabla \times \mathbf{A}) = \frac{4\pi}{c}(\mathbf{J}_s + \mathbf{J}_m) \tag{1}$$

and

$$\mathbf{J}^s = \frac{n_s \hbar e}{2m_s}(\nabla\varphi - \frac{4\pi e}{hc}\mathbf{A}) \tag{2}$$

$$\mathbf{J}^m = c\nabla \times \mathbf{M}. \tag{3}$$

Here $\hbar$ is Planck constant, $c$ is the speed of light, $n_s$ is the superconducting electron density, $m_s$ and $e$ are their mass and charge respectively, and $\varphi$ is the phase of the SC order parameter. For an SC thin film of thickness $d_s$, the effective penetration depth $\lambda$ is defined in terms of the London penetration depth $\lambda_L$ as $\lambda = \frac{\lambda_L^2}{d_s}$ with $\lambda_L = \sqrt{\frac{m_e c^2}{4\pi n_s e^2}}$

The presence of $N$-vortex in the superconductor, the phase gradient of the order parameter is

$$\nabla\varphi = \sum_{i=1}^{N} \nu_i \frac{\hat{z} \times (\mathbf{r} - \mathbf{r}_i)}{|\mathbf{r} - \mathbf{r}_i|^2} \tag{4}$$

where $\nu_i$ and $\rho_i$ are the vorticity and the location of the $i$-th vortex. Use of the superposition principle permits separation of the vortex part of the vector potential from that of the ferromagnet. The solution for the vortex part gives the Pearl solution for the vortex vector potential [30],[31],[32]

$$\mathbf{A}_v(\mathbf{r} - \mathbf{r}_0, z) = \frac{\phi_0}{2\pi} \sum_{i=1}^{N} \nu_i \frac{\hat{z} \times (\mathbf{r} - \mathbf{r}_i)}{|\mathbf{r} - \mathbf{r}_i|} \times \int_0^\infty \frac{J_1(q|\mathbf{r} - \mathbf{r}_i|)e^{-q|z|}}{1 + 2\lambda q} dq \tag{5}$$

where $q = |\mathbf{q}|$ is the amplitude of the 2D wave vector and $J_n(x)$ is the $n^{th}$ order Bessel function. The FM part of the vector potential is the solution of the following equation

$$\nabla \times (\nabla \times \mathbf{A}) = -\frac{1}{\lambda}\mathbf{A}\delta(z) + 4\pi\nabla \times [\mathbf{m}(\rho)\delta(z)] \tag{6}$$

where $\mathbf{m}(\rho)$ is the two-dimensional magnetization of the texture. The two-dimensional magnetization can be decomposed into three parts. The first part is perpendicular to the SC film $m^z$ and the second part $m^{||}$ is directed along the radial vector $\hat{\rho}$. The remaining part $m^{\perp}$ is directed along the unit vector $\hat{\rho} \times \hat{z}$ that is perpendicular to the plane spanned by $\hat{\rho}$ and $\hat{z}$. The solution of Eq.(6) is easily obtained using Fourier transform defined for any function $f(\mathbf{x})$ as

$$f(\mathbf{x}) = \frac{1}{(2\pi)^3} \int f_{\mathbf{k}} e^{-\imath \mathbf{k}\cdot\mathbf{x}} d^3\mathbf{k} \tag{7}$$

where $\mathbf{k} = (\mathbf{q}, k_z)$ is the $3D$ wave vector and $f_{\mathbf{k}}$ is the Fourier kernel. The solution of (6) can be represented by its Fourier kernel components [16] as:

$$A^{||}_{m\mathbf{k}} = -\frac{4\pi\imath m^{\perp}_{\mathbf{q}}}{k_z} e^{\imath k_z d} \tag{8}$$

$$A^{\perp}_{m\mathbf{k}} = -\frac{a^{\perp}_{\mathbf{q}}}{\lambda k^2} + \frac{4\pi\imath(k_z m^{||}_{\mathbf{q}} - q m_{\mathbf{q}z})}{k^2} e^{\imath k_z d} \tag{9}$$

where $\mathbf{a}_{m\mathbf{q}}$ is the Fourier transform of the vector potential at the SC film. Here $a^{||}_{m\mathbf{q}} = 0$ while

$$a^{\perp}_{\mathbf{q}} = -\frac{4\pi\lambda q(k_z m^{||}_{\mathbf{q}} - q m_{\mathbf{q}z})}{1 + 2\lambda q} e^{-qd} \tag{10}$$

Note that $A^{||}_{m\mathbf{k}}$ does not contain any information about the superconductor, so it is not affected by the presence of the superconductor in contrast to $A^{\perp}_{m\mathbf{k}}$. Now, the total magnetic field and energy of the HFSS can be calculated. The magnetic field's components can be calculated from the vector potential $\mathbf{A}$ from its definition $\mathbf{B} = \nabla \times \mathbf{A}$. This can be written in Fourier space as follows

$$B^z_{m\mathbf{k}} = \imath q A^{\perp}_{m\mathbf{k}}, \quad B^{||}_{m\mathbf{k}} = -\imath k_z A^{\perp}_{m\mathbf{k}}, \quad B^{\perp}_{m\mathbf{k}} = \imath k_z A^{||}_{m\mathbf{k}}. \tag{11}$$

The total energy of the HFSS is the sum of the vortex self energy $U_{sv}$, the vortex-vortex interaction $U_{vv}$, the energy of interaction between the ferromagnet and vortices $U_{mv}$ and the magnet self interaction $U_{mm}$. If there are $N$ vortices in the superconductor, then their total vector potential is the superposition of their individual potentials. This can be written in Fourier representation as follows

$$A_{v\mathbf{k}} = \frac{-2\imath\phi_0(\hat{z} \times \hat{q})F(\mathbf{q})}{k^2(1 + 2\lambda q)} \tag{12}$$

where $F(\mathbf{q}) = \sum_{j=1}^{N} \nu_j e^{\imath q r_j}$ with $r_j$ is the position of the $j$-th vortex and $\nu_j$ is its vorticity. The total energy of the HFSS can be written as

$$U = \int \left[ \frac{\mathbf{B}^2}{8\pi} + \frac{m_s n_s v_s^2}{2} - \mathbf{M} \cdot \mathbf{B} \right] d^3x \tag{13}$$

where $v_s$ is the superconducting carrier velocity defined as

$$v_s = \frac{\hbar}{2m_e}\nabla\varphi \tag{14}$$

The terms in the square bracket are as follows: the first term is the magnetic field energy, the second term is the kinetic energy due to the SC electrons and the last term is the Zeeman interaction term. Using Maxwell's equation $\nabla \times \mathbf{B} = (4\pi/c)\mathbf{J}$ with $\mathbf{B} = \nabla \times \mathbf{A}$, and integrating by parts then the magnetic energy can be rewritten as

$$\int \frac{\mathbf{B}^2}{8\pi} d^3x = \int \frac{\mathbf{J}\cdot\mathbf{A}}{2c} d^3x \tag{15}$$

This equation is gauge invariant because the total current is conserved $(\nabla \cdot \mathbf{J} = 0)$. I have omitted a surface term which arises from integration by parts. This is possible only because the vector potential and the currents decrease rapidly as $\mathbf{r}$ approaches infinity. The part due to the magnetic texture may be transformed to

$$\int \frac{\mathbf{J}_m \cdot \mathbf{A}}{2c} d^3x = \frac{1}{2}\int \mathbf{M}\cdot\mathbf{B}d^3x \tag{16}$$

Therefore, I rewrite (13) as

$$U = \int \left[\frac{n_s\hbar^2}{8m_s}(\nabla\varphi)^2 - \frac{n_s\hbar e}{4m_s c}\nabla\varphi\cdot\mathbf{A} - \frac{\mathbf{M}\cdot\mathbf{B}}{2}\right] d^3x \tag{17}$$

For the case when the SC and the FM thicknesses are infinitesimally small. Hence the total energy given by (17) must be modified by taking the 3D density of superconducting electrons in the SC film to be $n_s(\mathbf{r}) = n_s(\rho)\delta(z)$ and the 3D magnetization to be expressed in terms of the 2D magnetization $\mathbf{m}(\rho)$ of the FM as $\mathbf{M}(\mathbf{r}) = \mathbf{m}(\rho)\delta(z - D)$. In this case, the vortex self energy and the vortex-vortex interaction can be written as:

$$U_v = \frac{n_s\hbar^2}{8m_s}\int \nabla\varphi_{-\mathbf{q}}\cdot(\nabla\varphi_{\mathbf{q}} - \frac{2\pi}{\phi_0}\mathbf{a}_{v\mathbf{q}})\frac{d^2q}{4\pi^2} \tag{18}$$

The interaction energy between superconducting vortices and ferromagnet is

$$U_{mv} = -\frac{n_s\hbar e}{4m_s c}\int \nabla\varphi_{-\mathbf{q}}\cdot\mathbf{a}_{m\mathbf{q}}\frac{d^2q}{4\pi^2} - \frac{1}{2}\int \mathbf{m}_{-\mathbf{q}}\cdot b_{v\mathbf{q}}\frac{d^2q}{4\pi^2} \tag{19}$$

The ferromagnet self interaction energy $U_{mm}$ is

$$U_{mm} = -\frac{1}{2}\int \mathbf{m}_{-\mathbf{q}}\cdot b_{m\mathbf{q}}\frac{d^2q}{4\pi^2} \tag{20}$$

where $b_{m\mathbf{q}}$ and $b_{v\mathbf{q}}$ used in Eq.(18)-(20) are the Fourier representation of the magnetic fields generated by the FM and the vortex evaluated at the surface of the superconductor. The total energy $U$ is the sum $U = U_v + U_{mv} + U_{mm}$. This is a many variable function with its variables being the locations of the N vortices. Therefore, the distribution of vortices which minimizes the total energy of the system must be found. In most cases the total energy of the system is a very complex function of the locations of vortices. Hence the minimal energy configuration must be found numerically. In what follows I will review two examples of HFSS focussing only on their static properties. In these examples, the system consists of a superconducting thin filmm interacting with a ferromagnetic structure in its close proximity.

## 2.2 Circular Magnetic Dots

An infinitely thin circular ferromagnetic disk whose radius is $R$ is grown on the top of the SC film and its magnetization is prepared to be either parallel or perpendicular to the plane of the dot as in Fig. 5. The separation between the SC and the FM (The thickness of the insulator oxide layer) is $d \ll \lambda$.

### 2.2.1 Perpendicular Magnetization

The magnetization of the dot is $\mathbf{M}(\mathbf{r}) = m\theta(R-\rho)\delta(z-d)\hat{z}$ where $m$ is the two-dimensional magnetization, $\theta(x)$ is the step function and $\delta(x)$ is Dirac delta function. the magnetization current $\mathbf{J}_M = c\nabla \times \mathbf{M}$ correspond to a current flowing along a ring of radius $R$. Using (8), (9) and (10) gives

$$
\begin{aligned}
A^{\perp}_{m\mathbf{k}} &= -\frac{8\pi^2 \imath m R J_1(qR)}{k^2} \\
&\times \left[\frac{2\lambda q e^{-qd}}{1+2\lambda q} + (e^{\imath k_z d} - e^{-qd})\right]
\end{aligned}
\tag{21}
$$

Therefore, the components of the magnetic field produced by the ferromagnet in the presence of the superconductor are:

$$
B^z_m(\rho, z) = 4\pi m\lambda R \int_0^\infty \frac{q^2 J_1(qR) J_0(q\rho) e^{-q|z|}}{1+2\lambda q} dq \tag{22}
$$

$$
\begin{aligned}
B^\rho_m(\rho, z) &= 2\pi m R \int_0^\infty q J_1(qR) J_1(q\rho) e^{-q|z|} \\
&\times \left[\frac{2\lambda q}{1+2\lambda q}\theta(z) + \theta(z-d) - \theta(z)\right] dq
\end{aligned}
\tag{23}
$$

Erdin et.al argued that a vortex, if appears, has to be centered at the dot's center. They also showed that in the presence of a SC vortex the z-component of the total magnetic field generated by the system changes sign at some finite $\rho = R_c$ and $z = 0$. Therefore, they proposed that such a sign change could be used to detect SC vortices confined by the magnetic dot.

The presence of a vortex changes the energy of the hybrid system by an amount of

$$
\Delta = \epsilon_0 \ln(\frac{\lambda}{\xi}) - \epsilon_m R \int_0^\infty \frac{J_1(qR)}{1+2\lambda q} dq \tag{24}
$$

where $[\epsilon_0 = \phi_0^2/16\pi^2\lambda]$ is the energy scale for the vortex self interaction and with $\epsilon_m = m\phi_0$ is energy scale for the FM-SC interaction. The appearance of the vortex is energy favorable when $\Delta \leq 0$. Hence a curve in the plane of two dimensional variables $R/\lambda$ and $\epsilon_m/\epsilon_0$ can be found. This curve separates between regimes with $(\Delta < 0)$ and without $(\Delta > 0)$ vortices.

The value of the ratio $\delta_m = \frac{\epsilon_m}{\epsilon_0}$ determines how favorable is the spontaneous creation of vortices in the SC. To study the control parameters of $\delta_m$, we put $m = n_m d_m g \mu_B S$, $\epsilon_0 = \phi_0^2/16\pi^2\lambda$ then

$$\delta_m = gS\frac{n_m d_m}{n_s(T)d_s} \tag{25}$$

where $g$ is Lande factor and $mu_B = \hbar e/2m_e c$ is Bohr's magneton. Here $S$ is the elementary spin of the FM, while $n_m$ and $n_s$ are the electronic density for the FM and SC respectively with $d_m$ the thickness of the FM film and similarly $d_s$ is the thickness of the SC. Note that, close to the superconducting transition temperature $T_s$ the SC electrons density becomes very small making $\delta_m$ very large.

Although this argument is presented for the case when a single SC vortex is present. Erdin et. al. studied the ground state of this system for when two or three vortices exist in the SC and showed that similar phase diagrams exist for these cases. However, studying the distribution of vortices in the ground state of such system becomes a hard optimization problem when the number of vortices is large.

### 2.2.2 Parallel Magnetization

Here the system is exactly as before except for the magnetization of the dot which is taken to be directed along $\hat{x}$ instead of $\hat{z}$. So the magnetization of the dot is $\mathbf{M}(\mathbf{r}) = m\theta(R - \rho)\delta(z - d)\hat{x}$ The Fourier transform of the dot's magnetization is

$$\mathbf{M}_{\mathbf{k}} = \frac{2\pi m R}{q} J_1(qR) e^{\imath k_z d}\hat{x} \tag{26}$$

Using (8) and (9), the vector potential due to the dot in the presence of the superconductor is

$$A^{\perp}_{m\mathbf{k}} = \frac{8\pi^2 m_0 R}{k_z^2 + q^2} J_1(qR)\cos(\phi_q) \times \left(\frac{\imath k_z e^{\imath k_z d}}{q} + \frac{e^{-qd}}{1 + 2\lambda q}\right) \tag{27}$$

$$A^{||}_{m\mathbf{k}} = \frac{8\pi^2 \imath m_0 R}{k_z q} J_1(qR)\sin(\phi_q) e^{\imath k_z d} \tag{28}$$

The Fourier transform of the dot's vector potential at the surface of the superconductor is

$$a^{\perp}_{m\mathbf{q}} = -\frac{8\pi^2 m_0 \lambda R}{1 + 2\lambda q} J_1(qR)\cos(\phi_q) e^{-qd} \tag{29}$$

The magnetic field has three components $B^{||}$, $B^{\perp}$ and $B^z$; of particular interest to us here are $B^{||}$ and $B^z$. The inverse Fourier transforms of $B^{||}_{m\mathbf{k}} = -\imath k_z A^{\perp}_{m\mathbf{k}}$ and $B^z_{m\mathbf{k}} = \imath q A^{\perp}_{m\mathbf{k}}$ are

$$B^z_m(\rho,\phi,z) = 2\pi m_0 R\cos(\phi)\int_0^\infty qJ_1(qR)J_1(q\rho)\times \left(\frac{e^{-q(|z|+d)}}{1+2\lambda q} - sign(d-z)e^{-q|d-z|}\right)dq \tag{30}$$

$$B^\rho_m(\rho,\phi,z) = 2\pi m_0 R\cos(\phi)\int_0^\infty qJ_1(qR)J_0(q\rho)\times \left(2\delta(z-d) - qe^{-q|d-z|} - \frac{qe^{-q(|z|+d)}}{1+2\lambda q}sign(z)\right) \tag{31}$$

where the $sign(x)$ is $+1$ for $x > 0$ and $-1$ for $x < 0$. Note that $B^\rho_m$ and $B^z_m$ both have discontinuities at $z = 0$ and $z = d$. Therefore, extra care is needed to calculate the field components at the SC surface.

The magnetic field generated by the magnetic dot in the absence of the superconductor is equivalent to the field of a ring of radius $R$ with a magnetic charge distribution proportional to $\cos(\phi)$. The $z$-component of the dot's magnetic field vanishes at the center of the dot then it increases to reach a maximum at $\rho = R$. At large distances $\rho > R$, it decays as $1/\rho^3$. The component $B^\rho_m$ is maximum at $\rho = 0$. The magnetic field of the dot has all three components but only $B^z_m$ and $B^\rho_m$ are of interest for two reasons. First, $B^\phi_m$ does not contain any information about the superconductor, it is exactly equal to its value in the absence of the SC. Second, the vortex's magnetic field has only $B^\rho_v$ and $B^z_v$ components, so it is not affected by $B^\phi_m$.

Both $B^z_m$ and $B^\rho_m$ have maxima at $\phi = 0$ and minimum at $\phi = \pi$. If flux lines are to appear in the SC due to the field of the dot, then a vortex-antivortex pair would appear with the vortex at some $\rho = \rho_0$ and another at $\rho = -\rho_0$, with $\rho_0$ to be determined.

The total energy for a system of magnetic dot with in-plane magnetization interacting with $N$ single quanta flux lines is

$$U_N = N\epsilon_0\ln(\frac{\lambda}{\xi}) + \frac{1}{2}\sum_{i\neq j}^{N}\nu_i\nu_j\epsilon_{vv}(|\rho_i-\rho_j|) + \sum_{i=1}^{N}\nu_i\epsilon_{mv}(\rho_i) \tag{32}$$

where $\nu_i = +1$ for a vortex and $\nu_i = -1$ for antivortex. Here $\epsilon_{vv}$ is the vortex-vortex interaction which can be expressed in terms of Neumann's function $H_0$ and the Struve function $Y_0$ [33] as

$$\epsilon_{vv}(|\rho_i-\rho_j|) = \frac{\epsilon_0}{\pi}\left(H_0(\frac{|\rho_i-\rho_j|}{2\lambda}) - Y_0(\frac{|\rho_i-\rho_j|}{2\lambda})\right) \tag{33}$$

The interaction energy between a flux line and the magnetic dot can be calculated using (29) together with the phase gradient and magnetic field due to the flux line at $\rho_i$. It is

$$\epsilon_{mv}(\rho_i) = -\epsilon_m R\cos(\phi_i)\int_0^\infty \frac{J_1(qR)J_1(q\rho_0)}{1+2\lambda q}dq \tag{34}$$

If the SC has only one vortex-antivortex pair then it is logical to assume that the vortex and antivortex will be located on the x-axis since the magnetic field is strongest on the x-axis with respect to its angular distribution. Therefore, the case when the vortex is at $x = -\rho_0$ and the antivortex at $x = +\rho_0$ is studied. The total energy for a vortex-antivortex pair coupled to a magnetic dot is

$$E_{N=2} = 2\epsilon_0 \ln(\frac{\lambda}{\xi}) - 4\epsilon_0\lambda \int_0^\infty \frac{J_0(2q\rho_0)}{1+2\lambda q} dq - 2\epsilon_m R \int_0^\infty \frac{J_1(qR)J_1(q\rho_0)}{1+2\lambda q} dq + E_0 \quad (35)$$

The first and second terms in (35) represent the energy of the flux lines without the magnetic dot. The third term is the interaction energy between the flux lines and the magnetic dot, and $E_0$ is the magnetic dot self interaction energy.

The energy (35) is numerically minimized with respect to $\rho_0$ to find that $E_{N=2}$ is a minimum at $\rho_0 = R$. The presence of the vortex-antivortex pair changes the energy of the system by an amount $\Delta_1$ given by

$$\Delta_1 = 2\epsilon_0 \ln(\frac{\lambda}{\xi}) - 4\epsilon_0\lambda \int_0^\infty \frac{J_0(2qR)}{1+2\lambda q} dq - 2m_0\phi_0 R \int_0^\infty \frac{J_1(qR)J_1(qR)}{1+2\lambda q} dq \quad (36)$$

The creation of a vortex-antivortex pair is energy favorable if the energy of the overall system is lowered by such process. The curve $\Delta_1 = 0$ is depicted in Fig.1 separates between a regime without magnetic flux in the SC from another with one vortex-antivortex pair.

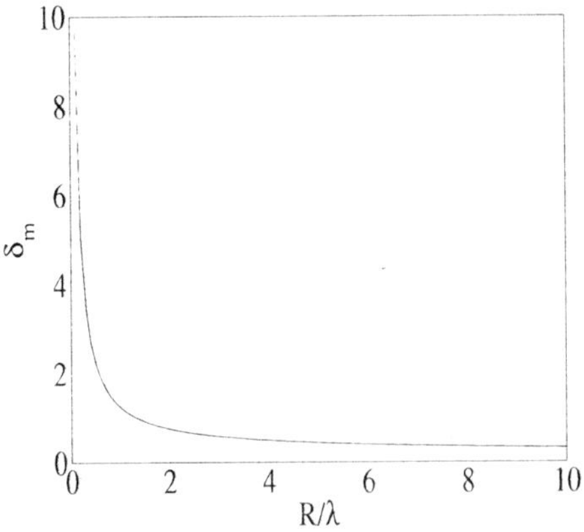


Figure 1: Phase diagram for one vortex-antivortex pair creation by a magnetic dot with in-plane magnetization.

The region above the curve represents a phase with one vortex-antivortex pair while that under the curve refers to a system without any flux line.

Next, consider the creation of two vortex-antivortex pairs coupled to the FM dot. Extra caution in handling the energy's angular dependence must be used. The results above show that the first pair appears with the vortex and antivortex situated along the $x$-axis with their centers under the edge of the dot. For two vortex-antivortex pairs, two vortices will appear on one side of the dot and two antivortices will be on the other side as shown in Fig.2.

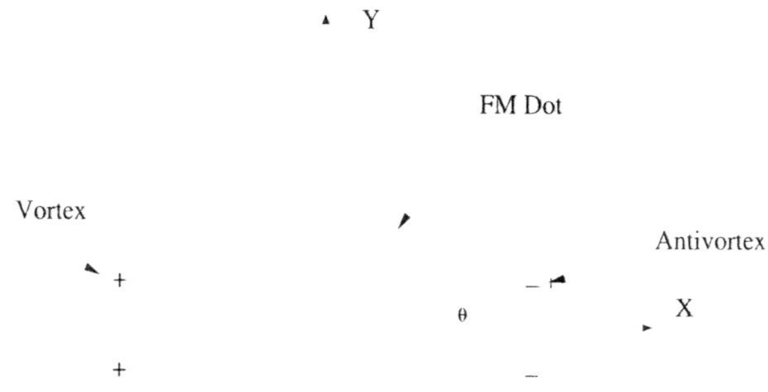


Figure 2: Illustration of two vortex-antivortex pairs in a superconducting thin film coupled to a ferromagnetic dot with in-plane magnetization.

Minimizing the energy for the system shown in Fig.2 showed that the vortices and antivortices centers are exactly under the boundary of the dot $\rho_0 = R$ with angular inclination $13° < \theta < 17°$, as illustrated in Fig.3.

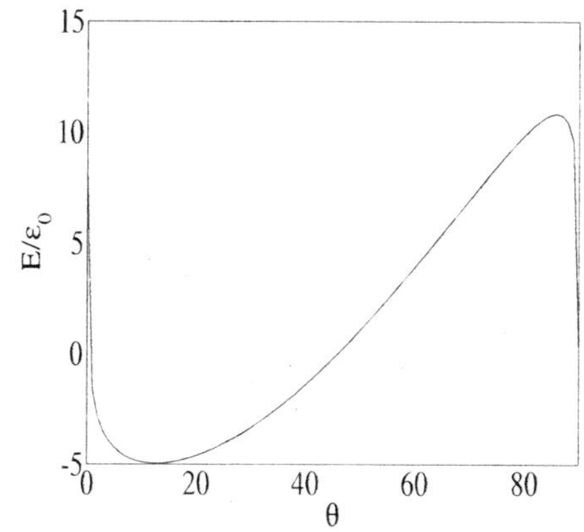


Figure 3: The angular dependence of the energy of two vortex-antivortex pairs created by a magnetic dot with in-plane magnetization.

The creation of more vortex-antivortex pairs is possible for higher values of $m$; however, finding the equilibrium distribution of flux lines around the dots has not been investigated. Circular dots with finite thickness were considered by Milosevic *et.al.* [19], [20], [21]; however, no qualitative differences with the studies reviewed above were found.

Many other studies used geometries other than the circular one. For example, effects of the magnetic dot's shape anisotropy on the ground state of the SC film were studied in [26]. Spontaneous vortex phase in FM-SC bilayer system was studied in [22]. The interaction between vortices in SC film and an array of magnetic dipole arrays placed outside the SC film was also investigated in [23], [24], [25]. The interaction between superconducting vortices in SC film pierced by a ferromagnetic columnar defect was recently studied in [27].

## 3 Dynamic Regime of HFSS

In the previous section, I briefly reviewed within the London approximation a general formulation for hybrid ferromagnetic-superconducting system. In this section, I will briefly discuss a phenomenological approach to study these hybrid systems in the presence of externally applied transport current flowing in the superconductor.

Let us consider a generic hybrid ferromagnetic-superconducting system that consists of a SC thin film and a ferromagnetic structure in its close proximity. If the magnetization of the ferromagnet is above a threshold value then vortices will spontaneously appear in the ground state of the system. If an external electric current flow in the SC, when the current reaches a critical value, vortices (antivortices) begin to move due to none zero net force acting on them. The forces acting on a moving vortex (antivortex) are Magnus force, viscous force, pinning force and the usual inertial force. Let the coordinates of the vortex center be $(\rho, z)$. Then the inertial force on the vortex is $\mathbf{f}_i = m_v\ddot{\rho}$ where $m_v$ is the mass of the vortex. Inertial force is typically ignored due to its smallness compared to other forces exerted on the vortex. Magnus force is defined as

$$\mathbf{f}_m = \pi n_s \hbar d_s(\mathbf{v}_s - \dot{\mathbf{r}}) \times \hat{z} \tag{37}$$

where $n_s$ is the SC electron density, $\mathbf{v}_s$ is the velocity of the SC electrons and $\dot{\rho}$ is the vortex velocity. The viscous (friction) force is $\mathbf{f}_f = -\eta\dot{\rho}$ where the $\eta$ is the Bardeen-Stephen drag coefficient [34] and is defined as

$$\eta = \frac{\phi_0 H_{c2} d_s}{\rho_n c^2} \tag{38}$$

where $H_{c2}$ is the upper critical magnetic field of the SC and $\rho_n$ is the resistivity of the SC sample when it is in the normal state. Putting it all together, the equation of motion for the vortex in our HFSS is

$$\eta\dot{\rho} = \pi n_s \hbar d_s(\mathbf{v}_s - \dot{\mathbf{r}}) \times \hat{z} + \mathbf{f}_p(\rho) \tag{39}$$

where $\mathbf{f}_p(\rho)$ is the pinning force felt by the vortex. If we assume the clean limit for the SC film, then the form and strength of pinning force will depend mainly on the properties of the ferromagnet and the distribution of other vortices and antivortices in the SC film. The pinning force exerted on a vortex in the SC is derived from a pining potential that is the sum of the energy of its interaction with the ferromagnetic structure and all other vortices and antivortices in the SC. After this brief introduction about the motion of vortices in superconductors and the explanation of various forces acting on vortices, I will present detailed studies of transport in HFSS. The first example will summarize the study reported in [35] by Kayali and Pokrovsky who considered the transport properties of hybrid ferromagnetic-superconducting bilayer. The second example will report new results on vortex dynamics in SC film pierced by ferromagnetic columnar defect.

### 3.1 Vortex Dynamics in Ferromagnetic Superconducting Bilayer

In [22], the authors studied the equilibrium structure of a FM-SC bilayer (FSB). They have proved that it represents a two-dimensional periodic stripe domain structure consisting of two equivalent sub-lattices, in which both the magnetization $m_z(\mathbf{r})$ and the vortex density $n_v(\mathbf{r})$ alternate. Thus, they predicted spontaneous violation of the translational and rotational symmetry in the bilayer. In [35], Kayali and Pokrovsky showed that the FSB exhibits strong anisotropy of the transport properties: the bilayer may be superconducting for the current parallel to the domain walls, and resistive when the current is perpendicular to them.

Periodic pinning forces in the direction parallel to the stripes do not appear in continuously distributed vortices, their reappearance is associated with the discreteness of the vortex lattice. Therefore, the theory [22] must be modified to incorporate the discreteness effects. Assume that the saturation magnetization per unit area of the FM film is $m$ and its width is $L$. The energy necessary to create a single Pearl vortex [30] in the superconductor is $\epsilon_{v0} = \epsilon_0 \ln(\frac{\lambda}{\xi})$,

It was shown in [22] that the interaction between the superconducting vortices and the magnetization in the stripe structure renormalizes the single-vortex energy to the value $\tilde{\epsilon_v} = \epsilon_{v0} - m\phi_0$ which must be negative to allow development of the stripes. The density of the superconducting vortices increases when approaching the domain walls and in the continuous approximation [22] it can be expressed as

$$n_v(x) = \frac{\pi\tilde{m}}{L\phi_0}\frac{1}{\sin(\frac{\pi x}{L})} \tag{40}$$

where $\tilde{m} = m - \frac{\epsilon_{v0}}{\phi_0}$ is the renormalized magnetization of the FM stripe. and $L$ is the stripe domain width. The vortices spontaneously appear in the superconductor. Kayali and Pokrovsky showed that vortices and antivortices line themselves up in straight chains as in Fig. 4.

They calculated the force between two chains of vortices falls off exponentially with the distance separating the two chains. They also showed that in the vicinity of the superconducting transition temperature $T_s$ and the vortex disappearance temperature $T_v$, pinning force in the direction parallel to the chains goes to zero very fast.

By assuming that vortices in one domain moves with the same velocity, they were able to solve the equations of motion for vortices and antivortices for the driving current direction parallel and perpendicular to the domains. They calculated the parallel and perpendicular critical currents and showed that the parallel critical current is much higher than the perpendicular one, at least in a vicinity of the two transition temperatures. This strong transport anisotropy is due to the fact that, for perpendicular current, the induced motion

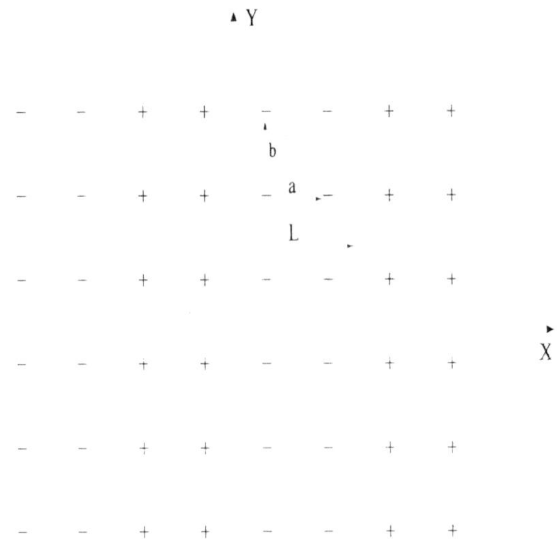


Figure 4: Schematic vortex distribution in the FM-SC bilayer. The sign $\pm$ refers to the vorticity of the trapped flux.

is a Goldstone mode and it is specific to a system of mobile domains and vortices. The ratio of the parallel to perpendicular critical current to be in the range $10^2 \div 10^4$ close to the superconducting transition temperature $T_s$ and to the vortex disappearance temperature $T_v$. The anisotropy decreases rapidly when the temperature goes from the ends of this interval, reaching a minimum somewhere inside it. The anisotropy can be destroyed by a rather weak magnetic field perpendicular to the bilayer. This anisotropic transport behavior might serve as a diagnostic tool to discover spontaneous topological structures in magnetic-superconducting systems.

### 3.2 Vortex Dynamics in SC film pierced by Ferromagnetic Columnar Defect

In [27], I studied the static properties of a superconducting thin film pierced by an array of ferromagnetic columnar defects (FCD). The FCD is assumed to have a length of $2L$ and a finite radius $R$ and its magnetization is $\mathbf{M} = M\hat{z}$. It was found that when the magnetization of the FCD is above a threshold value then vortices are spontaneously created in the ground state of the system. These vortices are strongly bound to the FCD and the interaction energy between the vortex and the FCD is given by

$$\begin{aligned} U(\rho) &= -2M\phi_0 R \int_0^\infty \frac{J_0(q\rho)J_1(qR)(1-e^{-\frac{qL}{2}})}{q(1+2\lambda q)} dq \\ &\quad + U_p(\rho) \end{aligned} \tag{41}$$

where $U_p(\rho)$ is the Mkrtchyan and Schmidt pinning energy and is given by

$$U_p(\rho) = \begin{cases} -\epsilon_0 \ln(\frac{R}{\sqrt{2}\xi}), & \rho < R \\ \epsilon_0 \ln\left[1 - \left(\frac{\sqrt{2}R}{\sqrt{2}\rho+\xi}\right)^2\right], & R < \rho < \lambda \end{cases}$$

This energy appear due to the presence of the cavity in SC film caused by the penetration of the FCD. In equilibrium, the lowest energy is reached when $\rho = 0$. The pinning force exerted by the FCD on a vortex at $\rho_0$ in the SC is given by the gradient of the interaction energy between the FCD and the vortex (41) is

$$\begin{aligned} F_\rho &= -\frac{\partial U_p(\rho)}{\partial \rho}|_{\rho=\rho_0} \\ & -2\epsilon_m \frac{R}{\lambda} \int_0^\infty \frac{J_1(q\rho_0) J_1(qR)(1 - e^{-qL/2})}{1 + 2\lambda q} dq \end{aligned} \tag{42}$$

Note that if $\rho_0 < R$, then the pinning force is purely of a magnetic origin, while when $\rho_0 > R$, the pinning force is both magnetic and non-magnetic. The non-magnetic part is due to the Mkrtchyan and Schmidt pinning potential, while the magnetic part is specific to our problem. At small distances $\rho < \lambda$, the magnetic pinning force behaves as $F_\rho \propto \rho$ if $\rho < R$, while it behaves as $1/\rho$ if $R < \rho < \lambda$. At large distances $\rho \gg \lambda$, the pinning force decays as $1/\rho^2$. The solution of (39) can not be attained analytically with the full form of $\mathbf{F}_p$. However, the analysis of $\mathbf{F}_p$ shows that $\mathbf{F}_p$ is maximum at $r = R$. Therefore, one can solve (39) for $0 \le \rho \le R$ by approximating the pinning force by a linear restoring force $\mathbf{F}_p = -k\rho$ with $k = \epsilon_m/2\lambda^2$. Therefore, the vortex motion is now equivalent to a forced two-dimensional harmonic motion that can be solved exactly for $0 \le \rho \le R$. Let a DC external current $J_0$ flow in the SC along the $x$-axis, then the solution of (39) is

$$x(t) = \frac{F_0}{k} \sin(\omega_0 t) e^{-\frac{\omega_0}{\alpha} t} \tag{43}$$

$$y(t) = \frac{F_0}{k} \left[1 - \cos(\omega_0 t) e^{-\frac{\omega_0}{\alpha} t}\right] \tag{44}$$

where $F_0 = \phi_0 J_0 d_s / c$, $\alpha = \eta'/\eta$ with $\eta' = \pi n_s \hbar d_s$ and $\omega_0 = \frac{\eta' k}{\eta^2 + \eta'^2}$. Note that real values of $\alpha$ are of the order of $10^{-2}$; therefore, the appearance of $\alpha$ in the exponential leads to strong damping of the oscillation part of the vortex motion. It is clear that the vortex motion along the $x$-axis will be suppressed fast, but the motion along the $y$-axis proceed until it stops at $y_\infty = F_0/k$. The pinning force acting on the vortex at distances $r > R$ has power less than linear. The critical value of the current $J_c$ is reached when $y_\infty = R$. This is because when $y_\infty \ge R$, the vortex escapes the pinning potential of the columnar defect. Therefore, the critical current for this case is

$$J_c = \frac{M_0 R c}{2\lambda d_s} \tag{45}$$

Taking $M_0 = 40$ G, $R = 500$ nm, $\lambda = 1000$ nm and $d_s = 30$ nm, then (45) yields $J_c \sim 10^8 - 10^{10}$ A/cm$^2$.

In the remainder of this article, I study the dynamical properties of a superconducting thin film pierced by a square array of ferromagnetic columnar defects. The analysis in this case must account for the presence of all other vortices in the SC and the periodic pinning force produced by the ferromagnet. The FCD is infinitely long and its radius is $R \ll a$, where $a$ is the spacing between two neighboring columnar defects. The FCD is uniformly magnetized along its symmetry axis with a constant magnetization $M_0$.

I showed in [27] that if $M$ is larger than a critical value, then vortices and antivortices are spontaneously created in the superconductor to minimize the energy of the whole system. Here I will assume that $M \ll M_c$ and that an external magnetic field $B_0 = \phi_0/a^2$ is applied to the superconductor along its $c$-axis. The magnetic field creates vortices in the superconductor such that each lattice cell has only one vortex. The minimum energy configuration for vortices is that when vortices form a square lattice that coincide with the FCD array [27]. Although larger values of the external field lead to configurations with more than one vortex per FCD, I will confine the treatment here to the ideal case of one vortex per FCD.

I will assume that the array unit vectors **a** and **b** are directed along the axis $\hat{x}$ and $\hat{y}$ respectively. If an external electric current with density **J** flows in the superconductor, then vortices begin to move. Assume that all vortices in the film move together such that one can neglect the force among them. The equations of motion for a vortex are

$$\eta\dot{x} = -F_0\sin(\theta) + \eta'\dot{y} - f_x^p(x,y) \quad (46)$$

$$\eta\dot{y} = F_0\cos(\theta) - \eta'\dot{x} - f_y^p(x,y) \quad (47)$$

where $\theta$ is the angle between the applied current and one side of the FCD array, while $f_x^p(x,y)$ and $f_y^p(x,y)$ are the components of the pinning force along $x$ and $y$ respectively. The result of [27] showed that if $R \ll a$ and $a < 4\pi\lambda$, then $f_x^p(x,y)$ and $f_y^p(x,y)$ can be approximated as

$$f_x^p(x,y) = -\mathcal{F}_0 \sum_{n,m}{}' \frac{n\sin(\frac{2\pi}{a}(nx+my))}{n^2+m^2} \quad (48)$$

$$f_y^p(x,y) = -\mathcal{F}_0 \sum_{n,m}{}' \frac{m\sin(\frac{2\pi}{a}(nx+my))}{n^2+m^2} \quad (49)$$

where the $\sum'$ is the sum without the $n = m = 0$ term. $\mathcal{F}_0 = \frac{M\phi_0R^2}{2\lambda a}$ is the scale of the magnetic pinning force. Equations (46) and (47) together with (48) and (49) form a set of first order nonlinear differential equations whose analytical solution is hard to attain. Therefore, I resort numerical methods to solve equations (46) and (47).

The numerical solution of equations (46) and (47) uses the Runge-Kutta algorithm with a time step $\Delta t = 10^{-3}$ in units of $\eta R/\mathcal{F}_0$ with the initial conditions $x(0) = 0$ and $y(0) = 0$. The sum in (48) and (49) is cut at $n_{max} = a/\xi$ and $m_{max} = a/\xi$. In the numerical calculations, I take $\lambda = 1000$ nm, $a = 4000$ nm, $\xi = 100$ nm, $R = 500$ nm, while $M = 40$ G. Therefore, the numerical value used for $n_{max}$ and $m_{max}$ is $a/\xi = 40$. The numerical solution of the (46) and (47) in the case of a current applied to the superconductor at $\theta = 45°$ with respect to one side of the FM array is shown in Fig. 5.

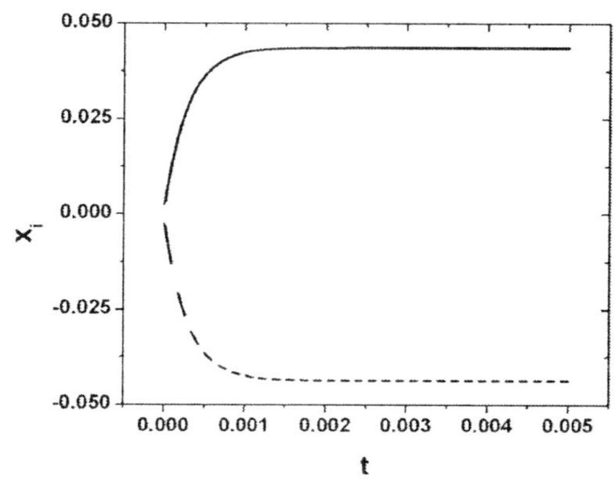


Figure 5: The coordinates $x$ (dashed) and $y$ (solid) for the vortex driven by a DC current applied at $\theta = 45°$ with respect to the $x$-axis.

Fig. 5 shows that the vortex moves both in $x$ and $y$ direction until $(x_\infty, y_\infty)$ when the driving force becomes equal to the pinning force. The solution shown in Fig. 5 tells us that the linear approximation of the pinning force works well even in this case provided the applied current is less than a critical value. However, once the applied current exceeds a critical value, the linear approximation becomes invalid. The validity of the linear approximation depends solely on the ratio $R/a$. The case considered here assume a ratio of $R/a = 1/8$ then the valley of the pinning potential has an effective radius close to $R$. If the ratio $R/a$ is made larger either by increasing $R$ or decreasing $a$, then the effective radius of the pinning potential increases well above $R$. Because the pinning force acting on the vortex is not rotationally invariant, then one expects the critical current to depend on the angle at which the current is applied to the superconductor.

The dependence of the critical current on the angle $\theta$ at which the current is applied with respect to one side of the FM array is studied by solving (46) and (47) for each value of $\theta$. I will consider here that $\theta$ is the angle measured between the applied current and the $x$-axis. The angle $\theta$ is varied between $\theta = 0°$ and $\theta = 90°$ with an angular step size $\Delta\theta = 15°$. The angular dependance of the critical current is illustrated in Fig. 6.

Fig. 6 shows that the maximum critical current is obtained when the driving current is applied along either side of the FM array and it reaches a minimum value when it is at $\theta = 45°$.

To conclude, I have shown that the motion of the vortex in a region of radius $R$ around the defect can be modeled by a forced two-dimensional harmonic motion whose fundamental frequency and damping coefficient were found. The value of the critical current of the superconductor in the presence of the pinning potential of the array is in the range of $J_c \sim 10^8 - 10^{10}$ A/cm$^2$. The angular dependance of the SC critical current for a system of SC film pierced by an array of columnar defects a minimum value at some critical angle $\theta_c$ whose value is $\theta_c = 45°$ for square array of FCD.

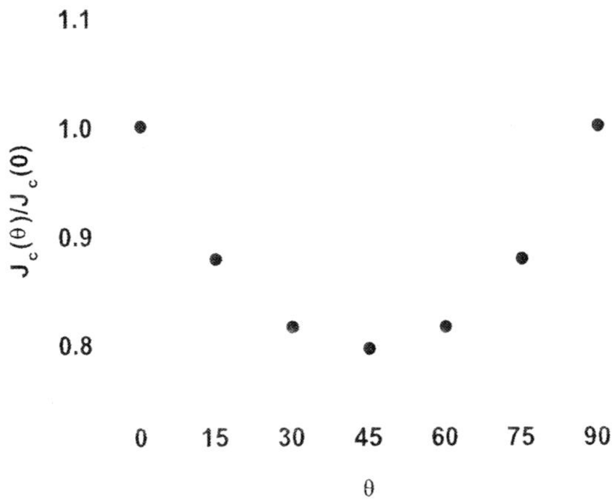


Figure 6: The angular dependance of the critical current for a superconducting thin film pierced by a square array of FCD. The angles $\theta$ is the angle between the direction of the current and the $x$-axis. The critical current is measured in unites of $J_c(0)$.

# 4 Summary

The usage of ferromagnetic structures to pin vortices in superconductors has been proven to be an efficient method to increase the critical current of the superconductors. Despite all the experimental and theoretical progress achieved in studying these systems, the subject is still in its infancy and will remain attractive for both experimentalists and theorists in superconductivity. This attraction stems not only due to their possible technological applications but also from being new class of systems where two competing order parameters coexist without being destroying each other. This coexistence is the essence of all interesting and peculiar static and dynamical phases that appear in these systems. Due to space and time limitations, there is a great deal of literature that I kept out of this short review. Among those topics that I left out are the proximity effects in HFSS for which I refer the readers to the recent review article by Lyuksyutov and Pokrovsky [36].

## Acknowledgement

I thank P. R. Montague for his support and encouragement and for giving me the time to write this review.

# References

[1] L. Civale, A. D. Marwick, T. K. Worthington, M. A. Kirk, J. R. Thompson, L. Krusin-Elbaum, Y. Sun, J. R. Clem, and F. Holtzberg, *Phys. Rev. Lett.* **67**, 648 (1991).

[2] G. Blatter, M. V. Feigel'man, V. B. Geshkenbein, A. I. Larkin and V. M. Vinokur, *Rev. Mod. Phys.* **66**, 1125 (1994).

[3] M. Baert, V. V. Metlushko, R. Jonckheere, V. V. Moshchalkov, and Y. Bruynseraede, *Phys. Rev. Lett.* **74**, 3269 (1995).

[4] Y. Nozaki, Y. Otani, K. Runge, H. Miyajima, B. Pannetier, J. P. Nozieres, and G. Fillion, *J. Phys.* **79**, 8571 (1996).

[5] P. Martinoli, M. Nsabimana, G.A. Racine, H. Beck, and R. Clem, *Helv. Phys. Acta* **56**, 765 (1983).

[6] J. I. Martin, M. Velez, J. Nogues, and I. K. Schuller, *Phys. Rev. Lett.* **79**, 1929, (1997).

[7] D. J. Morgan and J. B. Ketterson, *Phys. Rev. Lett.* **80**, 3614(1998).

[8] M. J. Van Bael, J. Bekaert, K. Temst, L. Van Look, V. V. Moshchalkov, Y. Bruynseraede, G. D. Howells, A. N. Grigorenko, S. J. Bending, G. Borghs, *Phys. Rev. Lett.* **86**, 155 (2001).

[9] M. Lange, M. J. Van Bael, L. Van Look, K. Temst, J. Swerts, G. Guntherodt, V. V. Moshchalkov, Y. Bruynseraede, *Europhys. Lett.* **51**, 110 (2001).

[10] S. J. Bending, G. D. Howells, A. N. Grigorenko, M. J. Van Bael, J. Bekaert, K. Temst, L. Van Look, V. V. Moshchalkov, Y. Bruynseraede, G. Borghs, R. G. Humphreys, *Physica C* **332**, 20 (2000);

[11] V. L. Ginzburg, *Zh. E'ksp. Teor. Fiz.* **31**, 202 (1957) [*Sov. Phys. JETP* **4**, 153 (1957)].

[12] I. F. Lyuksyutov and V. L. Pokrovsky, *Phys. Rev. Lett.* **81**, 2344 (1998).

[13] I. F. Lyuksyutov and V. L. Pokrovsky, *cond-mat/9903312*.

[14] M. J. Van Bael, K. Temst, V. V. Moshchalkov, Y. Bruynseraede, *Phys. Rev. B* **59**, 14674 (1999).

[15] L. N. Bulaevskii, A. I. Buzdin, M. L. Kulic and S. V. Panyukov, *Adv. Phys.* **34**, 175 (1985).

[16] S. Erdin, M. A. Kayali, I. F. Lyuksyutov and V. L. Pokrovsky, *Phys. Rev. B* **66**, 014414 (2002).

[17] I. K. Marmorkos, A. Matulis and F. M. Peeters, *Phys. Rev. B* **53**, 2677 (1996).

[18] M. A. Kayali, *Phys. Lett. A* **298**, 432 (2002).

[19] M. V. Milosevic, S. V. Yampolskii and F. M. Peeters, *Phys. Rev. B* **66**, 174519 (2002).

[20] M. V. Milosevic and F. M. Peeters, *Phys. Rev. B* **68**, 094510 (2003).

[21] M. V. Milosevic and F. M. Peeters, *Phys. Rev. B* **69**, 104522 (2004).

[22] S. Erdin, I. Lyuksyutov, V. Pokrovsky and V. Vinokur, *Phys. Rev. Lett.* **88**, 017001 (2002).

[23] R. Sasik, T. Hwa, *cond-mat/0003462*.

[24] J. E. Santos, E. Frey and F. Schwabl, *Phys. Rev. B* **63**,4439 (2001).

[25] G. Carneiro, *Physica C* **404**, 78 (2004).

[26] M. A. Kayali, *Phys. Rev. B* **69**, 012505 (2004).

[27] M. A. Kayali, *Phys. Rev. B* **71**, 024515 (2005).

[28] L. E. Helseth, R. W. Hansen, E. I. Il,yashenko, M. Baziljevich and T. H. Johansen, *Phys. Rev. B* **64**, 174406 (2002).

[29] G. S. Mkrtchyan and V. V. Schmidt, *Sov. Phys.* **JETP** 34 (1972) 195.

[30] J. Pearl, *J. Appl. Phys. Lett.* **5**, 65 (1964).

[31] P. G. de Gennes, *Superconductivity of Metals and Alloys* (Addison-Wesley, New York, 1989).

[32] A. A. Abrikosov, *Introduction to the Theory of Metals* (North-Holland, Amsterdam, 1986).

[33] I. S. Gradshteyn and I. M. Ryzhik, *Table of Integrals, Series, and Products* (Academic Press, Boston,1994), 5th ed.

[34] J. Bardeen and M. J. Stephen, *Phys. Rev.* **A 140**, 1197 (1965).

[35] M. A. Kayali and V. L. Pokrovsky, *Phys. Rev. B* **69**, 018413 (2004).

[36] I. F. Lyuksyutov and V. L. Pokrovsky, *cond-mat/0409137*.

In: New Research on Superconductivity and Magnetism
Editor: Lannie K. Tran, pp. 103-142
ISBN: 1-60021-540-8


*Chapter 4*

# SCANNING SQUID MAGNETIC IMAGING FOR NONDESTRUCTIVE EVALUATION: FROM MICRON-SCALE RESEARCH TO PRACTICAL LARGE-SCALE APPLICATION

***K. Isawa***
R & D Center, Tohoku Electric Power Co., Inc.
2-1 Nakayama 7-chome, Aoba-ku, Sendai, 981-0952 JAPAN

## Abstract

Superconducting magnetic sensor SQUID is the most sensitive magnetic sensing element known that has high spatial resolution. Accordingly, scanning SQUID magnetic imaging is currently being developed for applications to NDE (nondestructive evaluation). Laboratory-scale scanning SQUID microscopes (SSM) are useful for academic research as well as for industrial applications that require high-resolution measurements. This paper firstly reviews current knowledge of (a) temperature dependence of local magnetic distribution in high-Tc superconductors by SSM, and (b) high-resolution measurements using SSM on stress-induced magnetism in austenitic. For practical applications, a cost-effective NDE method that is highly accurate is desired. SQUID-NDE techniques are also reviewed here because they were recently proved capable of detecting damage such as fatigue, deformation, and cracks in materials, especially metals. Considering the demand in industry, conventional stationary SQUID-NDE methods are practically useless for inspecting large structures or material components. To answer this demand, an advanced SQUID-NDE system, called a *traveling SQUID-NDE*, was recently developed. This traveling SQUID-NDE can continuously "travel" over the surface of an object during testing without the need for magnetic shielding. And finally, this system is also reviewed here.

# 1 High-Resolution Experiments using a Scanning Squid Microscope

In this section, recent data from high-resolution experiments using scanning SQUID microscope (SSM) are reviewed to demonstrate the capability of SQUID used mainly for academic research. SSM is a 'microscope' that can produce a magnetic image at micron-scale resolution. The spatial resolution of SSM depends on the diameter of the pickup coil integrated into the SQUID body, and thus, the highest spatial resolution is currently ~10μm. Here, first, the principle of SQUID is briefly described, and then, current SSM research on high-Tc superconductors as well as on stainless steel is highlighted.

## 1.1 SQUID Sensor

This section briefly describes how SQUID works as a magnetic sensor with unparalleled sensitivity. SQUID is the abbreviation for Superconducting Quantum Interference Device that operates based on the Josephson effect phenomena and is used to measure variations in magnetic flux with the highest sensitivity to date. Figure 1-1 schematically illustrates this principle. Typically, a DC-SQUID is a ring of superconductors interrupted by two Josephson junctions. A Josephson junction is where electrons can tunnel from one superconducting region to another region separated by an insulating region (i.e., superconducting-insulator-superconducting, SIS). In a low-Tc SQUID, the Josephson junction is usually a tunneling junction (i.e., superconducting-normal-superconducting, SNS). When flux is induced into the SQUID loop, additional screening currents are generated. The SQUID voltage will change sinusoidally with a period of flux quantum, $\phi_0$ ($\phi_0$=2.07 × $10^{-15}$ Wb) when bias current is applied marginally (i.e., at the boundary level between superconducting and nonsuperconducting) up to non-superconducting level. Because such a nonlinear relation is not preferable in a magnetic sensor, a feedback circuit called Flux-Locked-Loop (FLL) [1,2] is usually used to obtain a linear relationship between the output SQUID voltage and the sensing magnetic field. Details of the fundamentals and fabrication of SQUID sensors are reported elsewhere [1].In brief, a SQUID is the most sensitive magnetic field sensing element known that utilizes superconductivity and that can sense and visualize flux quantum, $\phi_0$. Accordingly, SQUID can be applied to academic research as well as to industrial applications. Among the magnetic imaging methods that utilize SQUID, scanning SQUID microscopy (SSM) can measure local magnetic intensities and can visualize magnetic distributions at the micron scale [3]. In the next section, recent SSM data is reviewed to show how SQUID is sensitive while maintaining high spatial resolution.

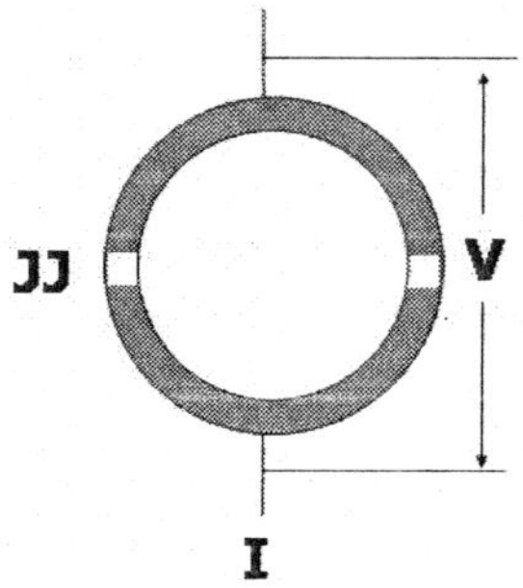


Fig.1-1. Schematic of superconducting ring. A DC-SQUID contains two Josephson junctions (JJ). For a loop of a superconducting ring interrupted by a JJ, Josephson predicted the possibility of electrons tunneling from one superconducting region to another where separated by a resistive weak-link, i.e., a JJ [1]. SQUID uses the Josephson effect to measure variations in magnetic flux with the highest sensitivity to date.

## 1.2 Scanning SQUID Microscopy (SSM)

SSM is the highest resolution SQUID magnetic imaging technique currently known. This section first summarizes the specifications of SSM, and then reviews observations of vortexes in high-$T_c$ superconductors using SSM, and finally reviews results from high-resolution measurements using SSM for stainless steel, mainly type-304L data.

Fig.1-2. Photo of commercially available scanning SQUID microscope (SSM) (Seiko Instruments, SQM-2000S) for high-resolution experiments. Inset: photo of the pick-up coil, 10μm in diameter.

Figure 1-2 shows a commercially available scanning SQUID microscope (Seiko Instruments, SQM-2000S) typically used in SSM measurements [4]. The spatial resolution of this microscope depends on the size of the pick-up coil connected to the DC-SQUID magnetometer. The pick-up coil, which was mounted at the tip of a flexible cantilever, was 10μm in diameter. This pick-up coil senses the magnetic flux density perpendicular to the surface of the object during scanning. Because the magnetic flux sensitivity of this SQUID microscope was better than $5\mu\phi_0 Hz^{-1/2}$, this SSM has acme-level spatial resolution and magnetic sensitivity. Note that, the background magnetic field was kept below approximately 1μT by using a permalloy shielding chamber.

## 1.3 Observation of Vortexes below and above Critical Temperature $T_c$ in High-$T_c$ Superconductors

As pointed out in section 1.2, SSM is so sensitive that it is useful for academic research. For example, SSM can be used to determine the distribution of local magnetic moment and the temperature dependence of vortices patterns. Because spatial resolution and magnetic sensitivity of SSM is superior to any other magnetic sensor, SSM is used as a convenient diagnostic tool to determine the magnetic phase diagram, especially for combinatorial thin films. Hasegawa et al. reported a phase diagram of combinatorial thin films, such as $(Nd,Sr)MnO_3$ (colossal magnetic resistive oxide) and $TiO_2$:Co (first reported transparent magnetic oxide) [5]. Using SSM measurements, Iguchi et al. [6] successfully detected a fractional diamagnetic state in $La_{2-x}Sr_xCuO_4$ (LSCO) thin films, even at temperatures much higher than the critical temperature $T_c$. Encouraged by this finding, independent groups reexamined observations of the fractional diamagnetic state above $T_c$ for LSCO, because the local diamagnetism above $T_c$ could be a precursor to the superconducting Meissner state and also be related to a pseudogap in LSCO. However, this local diamagnetism above $T_c$ has not been confirmed in successive measurements on bulky LSCO single crystals [7, 8]. To resolve this controversy, local magnetic images of hole-doped LSCO single crystals utilizing the SSM technique and SSM images of electron-doped systems were investigated.

### 1.3.1 SSM Measurements for Hole-Doped LSCO Systems

High-quality, large LSCO single crystals have been grown by using a traveling solvent floating zone (TSFZ) method by Prof. Yamada's group at the Institute for Materials Research, Tohoku University [9]. Flat surfaces (ab-plane) were obtained by mechanically polishing the crystals. At a fixed distance, usually less than 2 μm, the sample surface was scanned by a pickup coil (10 μm in diameter) on the cantilever.

Figure 1-3 shows a typical example of the scanning SQUID microscope image for an optimally doped region, x=0.14 in LSCO, measured under *Bex*=0, where *Bex* is the applied external magnetic field. At 3 K, the vortex patterns were clearly observed, and agreed well with the data for LSCO crystals [8]. Roughly, 70 vortices were identified in the image. In general, positions of vortices are determined by the pinning potential and/or the sample cooling conditions.

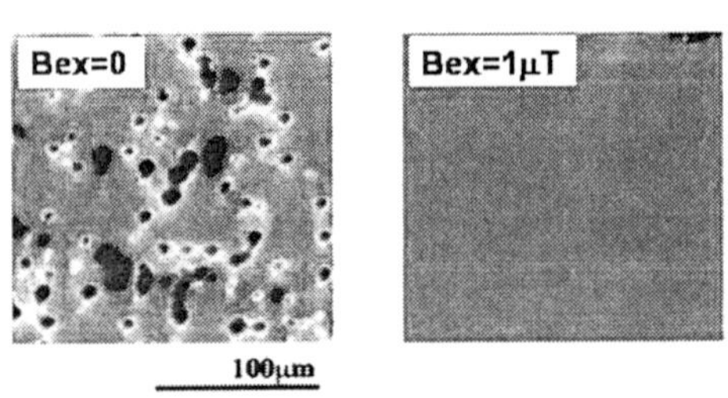


Fig. 1-3. Typical example of an SSM image of an optimally doped region of LSCO crystals (x=0.14) measured under *Bex*=0, where *Bex* is the applied external magnetic field. Empirically, the vortices were not trapped in the crystals when cooled while subjected to *Bex*=□1μT applied perpendicular to the surface so as to cancel the remnant environmental magnetic field. When *Bex*=0, the background magnetic field was thus shielded down to ~1μT by using a permalloy chamber.

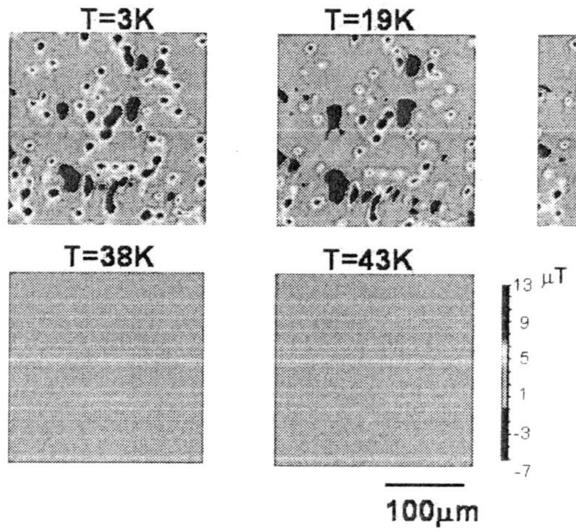


Fig. 1-4. Temperature dependence of SSM images for LSCO (x=0.14). Vortex patterns persisted up to 31 K, and completely disappeared at 38 K. No diamagnetic domains emerged in these magnetic images above $T_c$ for the LSCO crystals.

As clearly seen in Fig. 1-4, at this doping level (x=0.14), vortex patterns persisted up to 31 K, though they totally disappeared at 38 K. Similar tendencies were commonly observed for samples in the doping range between x=0.14 and 0.16. Namely, the vortex patterns for the optimally doped LSCO samples with x=0.14-0.16 were similar in their magnetic behavior. No diamagnetic domains emerged in their magnetic images above $T_c$ for the crystals. Based on empirical results, the vortices were not trapped in the crystals when cooled in *Bex*=+1μT applied perpendicular to the surface so as to cancel the remnant environmental magnetic field (see right image in Fig.1-3). In overdoped LSCO crystals [10], however, the vortices aligned in a quasi-1D manner with an arbitrary angle with respect to the crystallographic axes. Furthermore, Kasai et al. [10] found that Sr/La composition was modulated with a periodicity of ~200μm, similar to the periodicity of the striped magnetic pattern detected in their SSM results on LSCO. Kasai et al. concluded that the striped magnetic patterns originated from the spatial fluctuation of the chemical composition, resulting in sinusoidal modulation of $T_c$ ($\Delta T_c$<2K).

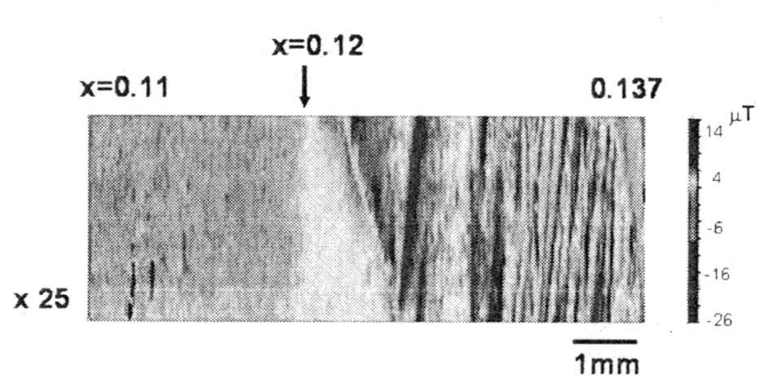


Fig. 1-5. SSM image of an underdoped LSCO single crystal, which had a distribution in carrier concentration along the direction of crystal growth, measured at 3 K for *Bex*=0. Spatial gradient of Sr in the crystal ranged from x=0.11 to 0.137 from left to right.

Figure 1-5 shows an SSM image of an underdoped LSCO single crystal, which had a distribution in carrier concentration along the direction of crystal growth, measured at 3K under *Bex*=0 [7]. The concentration gradient increased consecutively from x=0.11 to 0.137. The scanning area was 8000×1000μm. Interestingly, the SSM image changed at around x=0.12, namely, superconductivity was slightly suppressed. Consistently, the dip in $T_c$ at x~0.12 was clearly evident in the phase diagram of the concentration gradient crystals [9]. Possible origins of this change in the SSM image include inhomogeneity of the crystal and/or change in local magnetism due to spatial concentration gradient. Quite recently, similar measurements were carried out for lightly-doped (x<0.08) LSCO crystals near the boundary between superconducting (SC)/nonsuperconducting (spin glass, SG). For LSCO, nonsuperconductive SG phase appears at low Sr concentration, namely, the boundary of SC/SG is around at x=0.02~0.05 [11]. A high-quality large single LSCO crystals (ab-plan) with a spatial Sr concentration gradient of 0.04<x<0.08 was also grown by the TSFZ method.

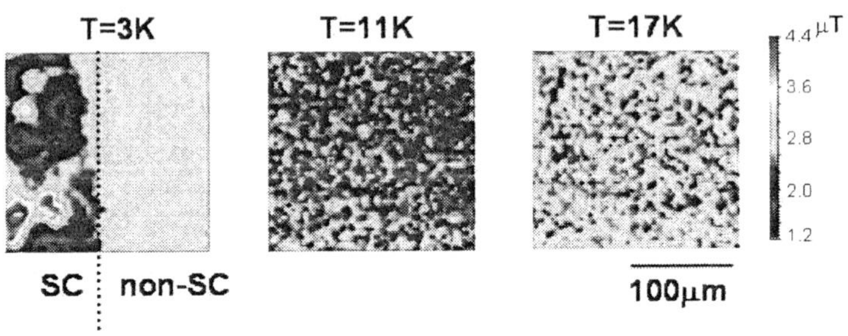


Figure 1-6. Temperature dependence of SSM images near the boundary of SC/SG for an LSCO crystal with spatial Sr concentration gradient from 0.04<x<0.08.

Figure 1-6 shows the temperature dependence of an SSM image near the boundary of SC/SG for this LSCO crystal. The scanning area was 200 × 200μm. At 3K, diamagnetic domains emerged in the magnetic images. Half of the left-hand image indicates the superconducting Meissner state, while the right-hand image clearly indicates the nonsuperconducting state. Using only the SSM data shown in Fig.1-6, it is unclear if the detected nonsuperconducting region is SG, however. The superconducting region was not confirmed at much higher temperature, such as 11K or 17K, although the scanning area was exactly the same as that at 3K, indicating that the SC/SG boundary depends on temperature. This temperature-dependency is more clearly confirmed when the scanning area is extended as follows. At first, the superconducting Meissner state is clearly confirmed by the data shown in Fig.1-7 for the LSCO crystal (ab-plane) with a spatial Sr concentration gradient of 0.04<x<0.08. At 6K, diamagnetic signals emerged only on the left side of the SSM image, clearly indicating the superconducting/nonsuperconducting (SC/non-SC) boundary.

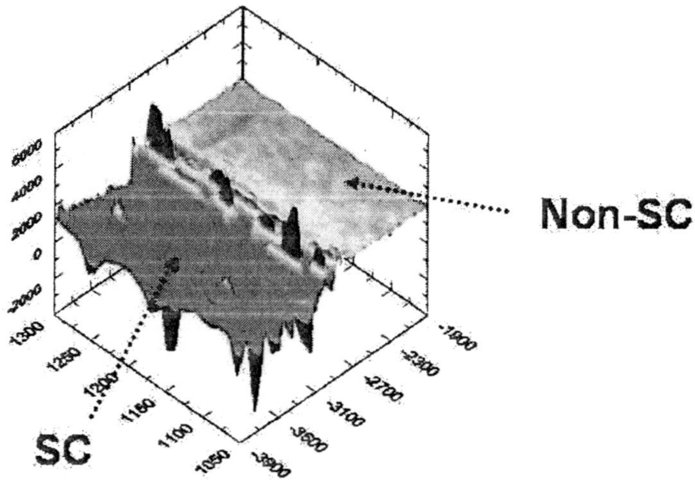


Fig.1-7. 3D SSM image of LSCO crystal near SC/SG with Sr gradient for a wide scanning area (2000 × 250μm) at 6K. For clarity, the y-axis is elongated by a factor of 4. The superconducting Meissner state is clearly visualized only on the left half of the plot.

For the same scanning area, the temperature dependence was then examined. Figure 1-8 shows the temperature dependence of the SSM images for a scanning area of 2000 × 250μm of an LSCO crystal (ab-plane) with spatial Sr concentration gradient of $0.04<x<0.08$ (gradually decreasing from left to right in the image). Clearly evident in this image is that the superconducting region decreased with increasing temperature. In conclusion, SSM measurements can be used to visualize SC/non-SC boundaries, including the SC/SG boundary.

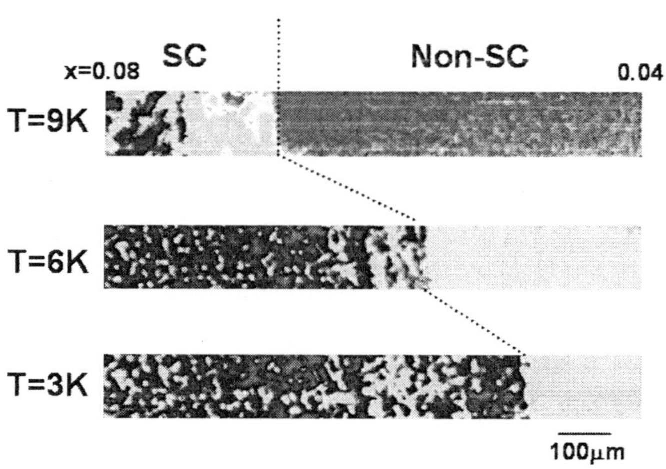


Fig. 1-8. Temperature dependence of SSM images of LSCO crystal near SC/SG with Sr gradient for a wide scanning area (2000 × 250μm). The SC/SG boundary shift left side with increasing the temperature measured.

### 1.3.2 SSM Measurements for Electron-Doped Systems

As complementary studies on hole-doped systems, studies on electron-doped systems provide important data about the mechanism of high-$T_c$ superconductivity. Similarities and differences between electron-doped and hole-doped systems have been reported [12,13]. To obtain SSM images of electron-doped systems, high-quality large electron-doped $Nd_{2-x}Ce_xCuO_4$ (NCCO) single crystals with x=0.15 and 0.17 were grown by using the TSFZ technique [14]. Superconducting NCCO crystals were obtained by post-annealing in a reducing atmosphere. The $T_c$ of these NCCO(x=0.15) and NCCO(x=0.17) crystals was 25K and 13K, respectively. Flat surfaces (ab-planes) of the crystals were obtained by mechanically polishing to a thickness of ~1mm. Two types of sample stages were used to control an external magnetic field *Bex* applied to the crystal: a nonmagnetic stage, where *Bex* from -100 to100μT can be applied by an external coil, and a magnetic stage (made by using a Au-based Ni plating technique), where crystals can be field-cooled under *Bex*>20μT.

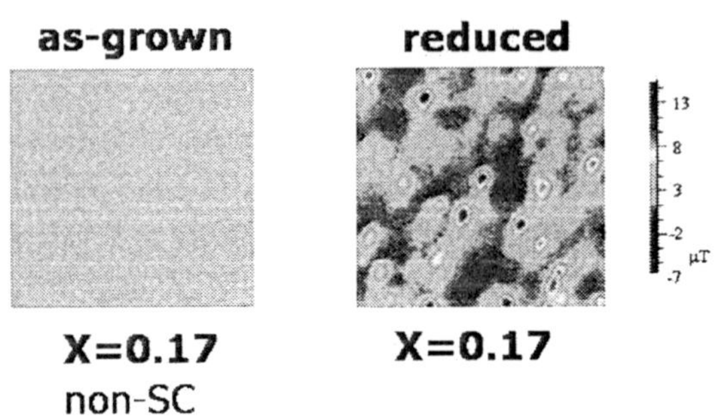


Fig.1-9. SSM image of a 200 × 200 μm area of as-grown and reduced NCCO(x=0.17) single crystals imaged at 3K under no applied external field, *Bex*=0.

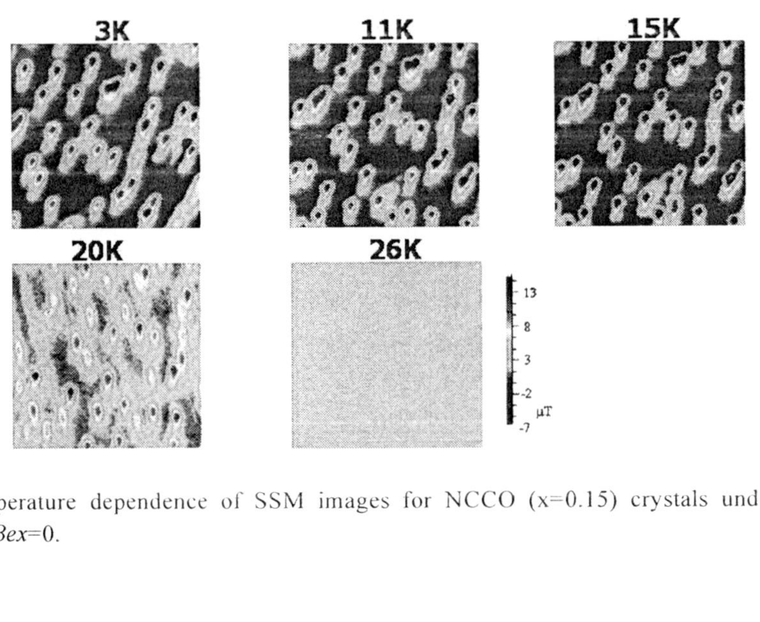


Fig.1-10. Temperature dependence of SSM images for NCCO (x=0.15) crystals under no applied external field, *Bex*=0.

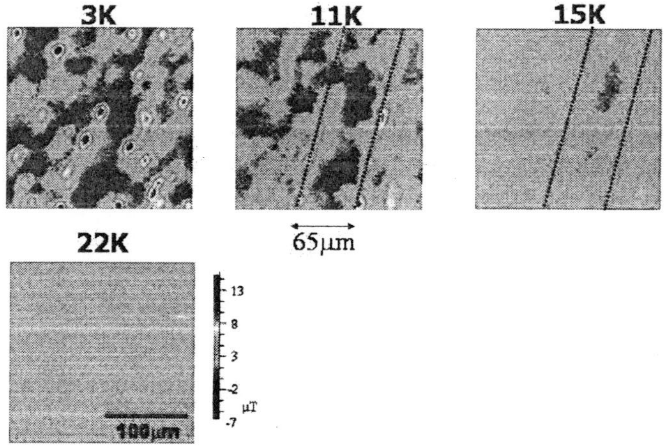


Fig.1-11. Temperature dependence of SSM images for NCCO (x=0.17) crystals under no applied external field, *Bex*=0. At 11K, vortices aligned in a quasi-1D manner (dashed lines) at intervals of about 65μm. Generally, magnetic patterns depend on the geometrical configuration between the pick-up coil and crystal. The direction of the striped pattern was not identified yet.

Figure 1-9 shows a typical SSM image for NCCO at T=3K measured under *Bex*=0. For the NCCO(x=0.17) crystals, vortex patterns were clearly evident when as-grown crystals were post-annealed in a reducing atmosphere. Figures 1-10 and 1-11 show SSM images for NCCO (x=0.15) and NCCO (x=0.17) crystals, respectively, at various temperatures and under *Bex*=0. The vortex patterns persisted up to just below $T_c$, and then totally disappeared at temperatures higher than $T_c$. For the NCCO bulky crystals, the diamagnetic precursor state above $T_c$ as previously reported for LSCO thin films [6] was not detected. These vortex patterns are similar to the magnetic behavior at temperatures much lower than $T_c$, as seen in Figs.1-10 and 1-11. A notable feature of vortex patterns is evident near $T_c$ only for the overdoped NCCO (x=0.17) crystals, namely, the vortices aligned in a quasi-1D manner at T=11K. Such alignment yields 'stripe-shaped' magnetic (hereafter simply called *striped magnetic*) images as depicted by a darker contrast (blue area). This blue area corresponds to a lower magnetic flux density perpendicular to the sample surface. Typical spatial periodicity of the striped magnetic region was ~ 65μm (dashed lines in Fig.1-11). Unfortunately, prior to the SSM measurements, the direction of the crystals was not determined, because the main interest of the measurements was the possible observation of local diamagnetism above $T_c$ in electron-doped NCCO single crystals. Unexpectedly, such unusual striped magnetic patterns were detected for NCCO crystals [15]. In $Bi_2Sr_2Ca(Cu,Co)_2O_{8+\delta}$ [16], similar quasi-1D vortex patterns were previously reported by Zhao et al. [16]. They found that the vortex assemblies in Co-doped Bi2212 form a quasi-1D structure along a certain crystallographic direction, either (010) or (110). The vortex patterns for the NCCO (x=0.15) and NCCO (x=0.17) crystals were totally different when the crystals were field-cooled under *Bex*>0, compared with those at *Bex*=0 [15]. Using a magnetic sample stage, the temperature dependence of the vortex patterns for the NCCO(x=0.15) crystal was also investigated (see Fig.1-12). Based on the SSM data at 42K shown in Fig.1-12, the estimated *Bex* was >20μT. The quasi-1D vortex alignment commonly observed at temperatures below $T_c$, however, totally vanished at temperatures higher than $T_c$ [15]. Furthermore, the periodicity of the striped magnetic patterns for the NCCO (x=0.15) crystal were similar to that observed for the NCCO(x=0.17) crystal (dashed lines in Fig.1-11). For instance, the interval marked by α and α' in Fig.1-12 was estimated to be ~65μm at 22K. In addition to this interval, a much shorter one (approximately six times shorter) was also confirmed, namely, the interval marked by β and β'. Note that the length of β and β' is comparable to that of the limit of the spatial resolution of the SSM apparatus of ~10μm. If *Bex* is increased, the periodicity of the striped magnetic patterns might decrease. This possibility is reasonable because the density of the magnetic field in the crystal is much higher when *Bex* is stronger than when *Bex* is weaker. The configuration between the pick-up coil and the crystal surface affects the geometric orientation of the striped patterns.

In NCCO crystals, a post-annealing procedure in a reducing atmosphere is essential to induce superconductivity. However, a secondary $(Nd,Ce)_2O_3$ phase with a volume fraction of 1-3% is produced due to decomposition of the NCCO matrix during the reduction. Such a secondary phase region has a spatial extent well above 1 μm parallel to the $CuO_2$ planes of NCCO, although 'stripe-shaped' distribution of the $(Nd,Ce)_2O_3$ has not been reported yet. Because $(Nd,Ce)_2O_3$ is paramagnetic even at 1.9K [17], further study is needed to determine if the $(Nd,Ce)_2O_3$ phase affected the present SSM measurements on NCCO. Such a secondary phase, however, is not distributed on the top of the surface of the measured NCCO crystal. Furthermore, the spatial distribution of $(Nd,Ce)_2O_3$ phase [17] is much shorter than 10μm, which is the limit of the spatial resolution of the SSM used in that study. Local magnetic images of the ab-plane of NCCO single crystals above and below $T_c$ were obtained by SSM measurements. For an NCCO(x=0.15) crystal, unusual striped magnetic patterns in the images were observed below $T_c$, only when the crystals were field-cooled with $Bex > 20\mu T$ [15]. No diamagnetic state was confirmed for the bulky NCCO crystals at temperatures higher than $T_c$.

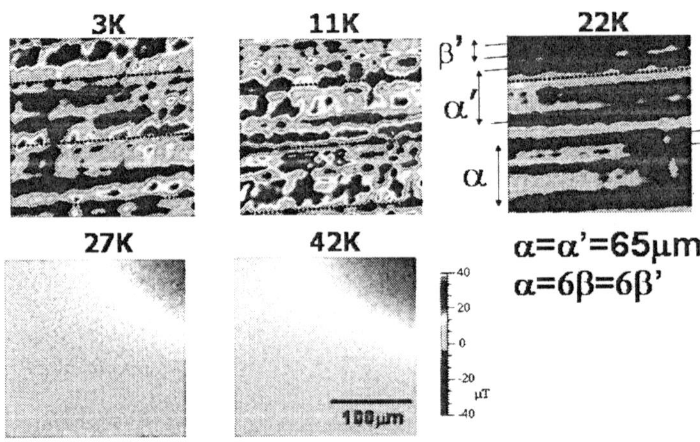


Fig. 1-12. Temperature dependence of SSM images for NCCO (x=0.15) crystals under an applied external field *Bex* using a magnetic sample stage. Based on the results at 42K, the crystal was field-cooled at *Bex*>20μT applied perpendicular to the ab plane. The detected magnetic patterns depended on the geometrical configuration between the pick-up coil and crystal.

## 1.4 High-Resolution SQUID Image for Type-304L Stainless Steel

Using SQUID sensors, the stress-induced magnetism of metals including austenitic is now being studied [18-20]. The magnetic response of austenitic stainless steel to stress likely corresponds to precursive phenomena of damage in the material. So, this section highlights high-resolution SSM measurements for type-304L stainless steel.

Based on preliminary measurement of the stress-strain curve (see Fig.1-13), various mechanical tensile loads were applied to sample of type-304L by a hydraulic testing machine. In other words, each sample was artificially fatigued only once by the tensile load. Eight samples of type-304L were prepared with different loading: (i) 0%, namely, no-loading (reference), (ii) 20% loading, (iii) 50%, (iv) 70%, (v) 90%, (vi) maximum stress (plastic deformation region), (vii) between maximum stress and rupture, and (viii) rupture. The percentages indicate the applied loading level corresponding to the yielding point, namely, 100% is the level at yielding, i.e., samples (i) to (v) are categorized as being in the elastic region. Prior to the tensile testing, all the samples were demagnetized by electrochemical polishing. For the SSM measurement, after the mechanical tensile loading, the samples were cut into rectangular-shapes of 10 × 7mm by using a wire-cutter, and only the sides (~0.14mm thick) were demagnetized again by electrochemical polishing.

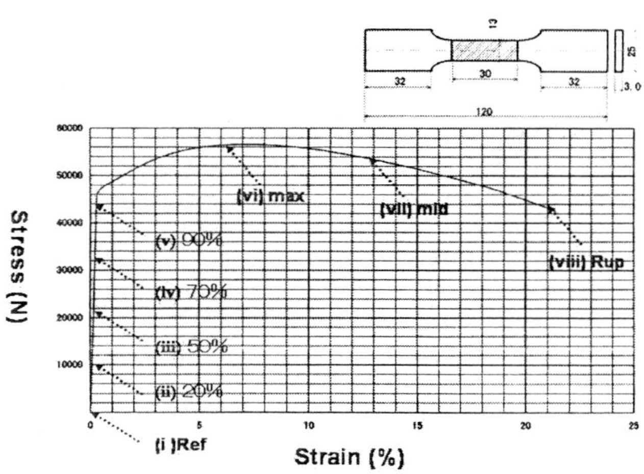


Fig.1-13. Typical strain-stress curve for type-304L used for SSM measurements. By changing the tensile loading level applied to the bar-shaped specimen (upper schematic), eight samples, from (i) to (viii), were prepared. Percentages mean the applied loading level with respect to the loading level at yielding.

Figure 1-14 shows that stress-induced magnetism in type-304L austenitic stainless steel was successfully detected by using SSM at 3K. Extremely strong stress-induced magnetism was detected for the plastic deformation level samples, i.e., samples (vi) to (viii). The spontaneous magnetism was probably induced by residual-strain. The observed spontaneous magnetization is ferromagnetism, when the nonmagnetic type-304L austenitic is transformed into stress-induced ferromagnetic martensitic. This stress-induced magnetism agrees well with the stress-induced magnetic susceptibility of austenitic described in Ref. 21. Note that the detected SQUID signals from the ferromagnets were actually higher than the dynamic range (+/- 325μTesla) of the SSM system, and therefore, the dynamic range is extended automatically using an advanced function equipped with SSM system called 'auto reset'. More importantly, even in the elastic region (samples (i) to (v)), the magnitude of the detected SQUID signals tended to intensify with increasing mechanical tensile load, suggesting that precursive phenomena of the damage in type-304L possibly detected by high-resolution measurement using SSM in the elastic region. That is, based on numerical averaging of the detected SQUID signals for samples (i) to (v), the average value of the signals in both the blue and red areas, which correspond to the position where intensive magnetic signals were detected, increased with increasing mechanical tensile load [22]. In metallography, such an indication (precursive phenomena) was not observed in samples in the elastic region, as shown in Fig. 1-15.

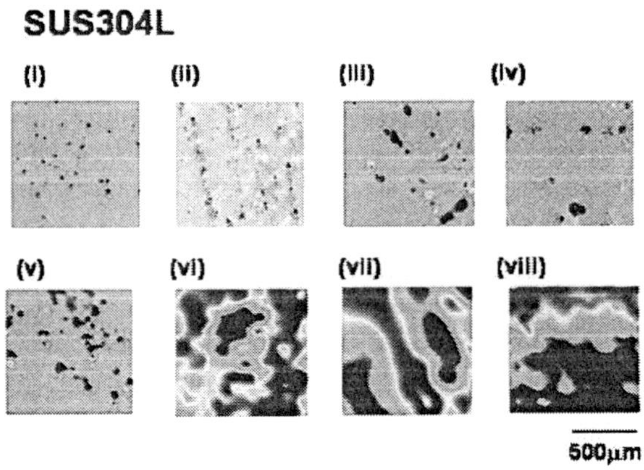


Fig. 1-14. SSM image of type-304L with different tensile loading levels (see Fig. 1-13) measured at 3K. Stress-induced magnetism is clearly evident in the plastic deformation region samples (vi), (vii) and (viii). Precursive phenomena of stress-induced damage was detectable by these high-resolution measurements even in the elastic region samples, namely, (i) to (v) (see text).

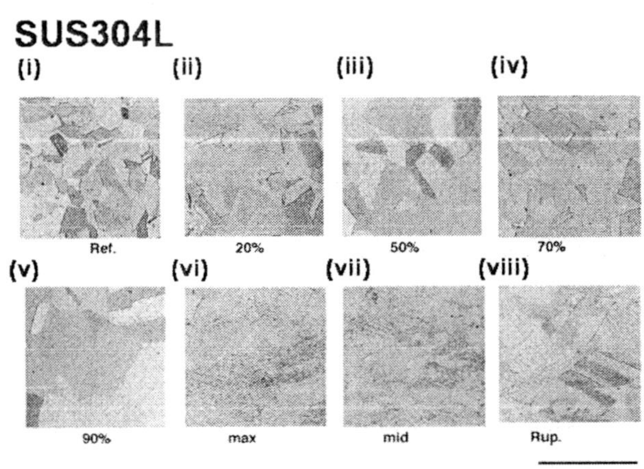


Fig.1-15. Optical microscopy images for type-304L with different tensile loading levels (see Fig. 1-13). A metallographic change was confirmed only for the plastic deformation region samples (vi), (vii), and (viii).

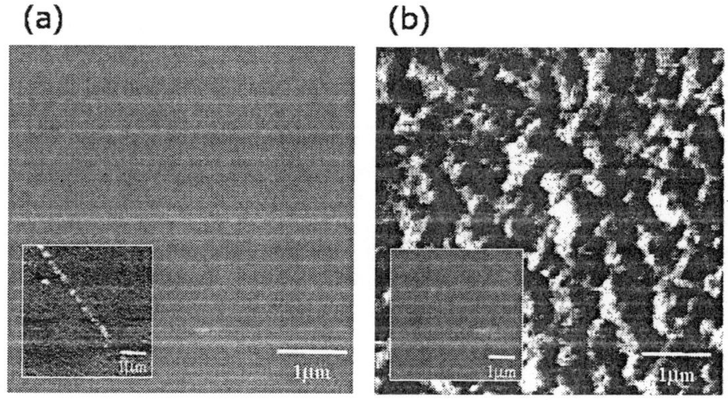


Fig.1-16. (a) MFM images for elastic region sample, (v) 90%. Inset: the full scale is enlarged. (b) MFM images for plastic deformation sample, (viii) rupture [Courtesy of SIINT]. Inset shows MFM images for plastic deformation sample, (vii) between maximum stress and rupture.

In Fig.1-15, metallographic changes were confirmed only when the stress-induced ferromagnetic martensitic phase was produced by applying a loading level above the yielding point. This tendency is confirmed by MFM (magnetic force microscope) images, as well. Figure 1-16 shows typical MFM images for samples (a) elastic region sample (v) and (b) plastic deformation sample (viii). Although MFM is also a sensitive magnetic measurement that has high spatial resolution, strong magnetic signals were detected in the plastic deformation level samples ( i. e., rupture sample), but not in the elastic region samples within their inspected areas. This indicates that the SSM level magnetic sensitivity as well as spatial resolution is crucial for observing precursive phenomena in the elastic region (see Fig. 1-14) for type-304L specimens. However, several aspects of measuring precursive phenomena in material dynamics need clarification. First, the effect of the temperature at which SSM images were obtained on precursive phenomena needs further study. SSM images were measured at 3K, assuming that cooling effects were negligible. Measurements at room temperature are necessary to investigate how phase transitions of nonmagnetic type-304L austenitic affect stress-induced magnetism in the elastic region. Presently, the available temperature range of an SSM system is 3K to 100K. Second, the dependence of time on precursive phenomena depend (i.e., relaxation effect) requires further study. Third, in-situ measurements on magnetic distribution with increasing tensile loading need further clarification. Fourth, measurements for different materials, such as type-304 and type-316, need to be done. In conclusion, observed precursive phenomena in the elastic region for type-304L might be caused by a 'local' magnetism induced by the stress.

# 2 Scanning Squid Imaging for Nondestructive Evaluation (NDE) Applications

Section 1 discussed the effectiveness of scanning SQUID imaging and the high magnetic sensitivity in high-resolution experiments achieved using SSM. Figure 2-1 shows SSM data to demonstrate the observation of quantum flux [24]. Each peak in Fig.2-1 is the quantum flux, because each intensity corresponds to $2.07 \times 10^{-15}$ Wb, namely, $\phi_0$, determined by numerical calculations [25]. The advantage of SQUID is that it is the most sensitive magnetic field sensing element known that has superior spatial resolution. Due to this advantage, its application to nondestructive evaluation (NDE) is highly anticipated. Furthermore, inspection methods involving a SQUID sensor are advantageous because they do not require contact between the sensor and sample or object being tested. Because SQUID-NDE is a magnetic measurement, the following points should be considered. As shown in Fig. 2-2, magnetic signals strongly depend on the distance (r) between the SQUID sensor and the surface of the test object. Because a magnetic signal is inversely proportional to $r^3$ [26], quadratic SQUID signals for r=15mm decay faster than those for r=10mm (see Fig. 2-2) [27]. The magnetic image of r=15mm is much less clear than that of r=10mm. Furthermore, as shown in Fig. 2-3, a SQUID magnetic imaging system senses the sum of magnetization and permeability effect:

$$B=M+\mu_0 H, \quad (1)$$

where $M$ is the magnetization, $\mu$ is the permeability ($\mu_0$ is the permeability of a vacuum), $B$ is the magnetic flux density, and $H$ is the magnetic field strength [26].

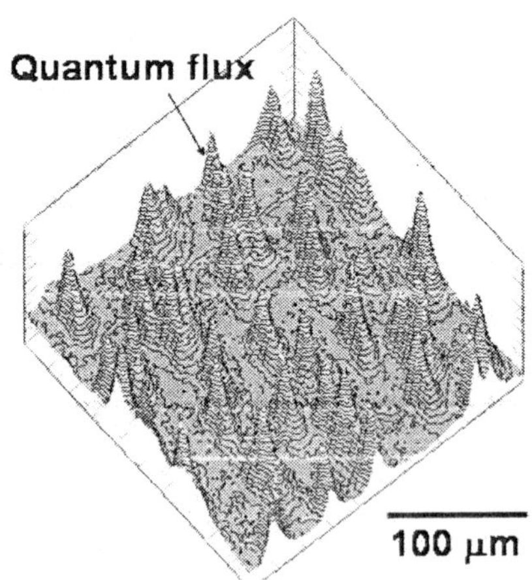


Fig. 2-1. Quantum flux measured using scanning SQUID microscope (SSM) for a $(Nd,Ce)_2CuO_4$ high-$T_c$ superconducting crystal (ab-plane). The crystal was field-cooled (remnant environmental magnetic field, ~1μT) down to 3K.

The effect of permeability would be negligible if the measurement is taken in perfect magnetic shielding room in which $H$=0. Thus, in ambient environment, the obtained SQUID-NDE image is affected by the test object's magnetization as well as its permeability. In Fig. 2-3, the so-called *edge effect* (indicated by dashed arrows in the figure) is caused by permeability, and magnetization of the object is clearly evident (indicated by solid arrow) [28].

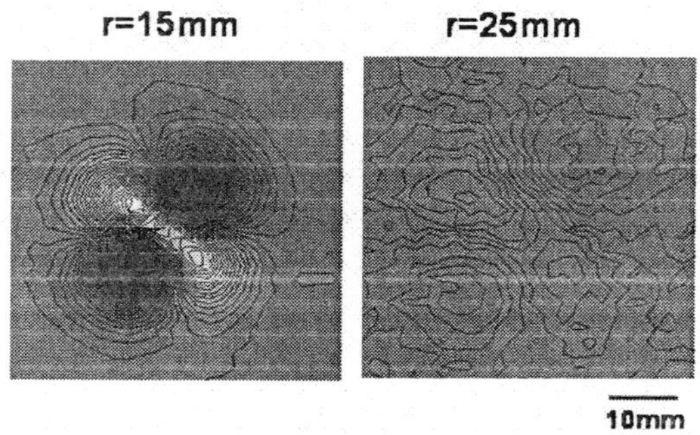


Fig.2-2. SQUID image for a small ferrite magnet (approx. 10mg) in comparison of r=15 mm and r=25mm, where, r is the distance between the ferrite magnet and SQUID sensor, to demonstrate magnetic signal are inversely proportional to $r^3$.

Historically, research and development of SQUID-NDE began in the 1980's [29]. SQUID-NDE techniques have since proved capable of detecting damage in material components or structures, such as fatigue, deformation, and cracks [30-38]. In this section, first the overall objective of research and development of SQUID-NDE is discussed, and then, current SQUID-NDE data involving laboratory-scale applications (not large-scale applications) is discussed to demonstrate that SQUID is highly promising for NDE applications.

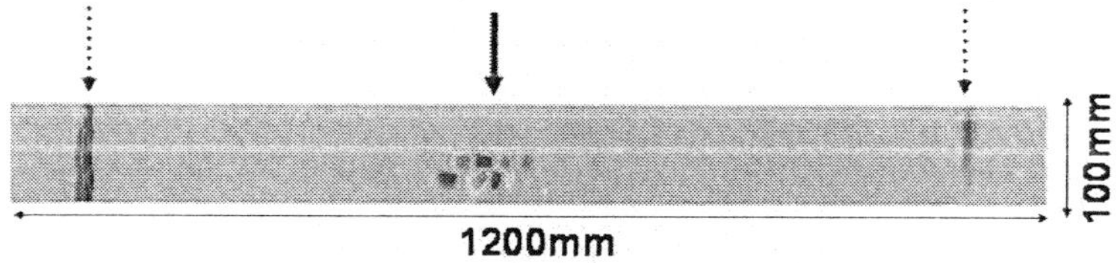


Fig.2-3. 2D magnetic images for a type-410 stainless steel plate (500mm wide × 1000mm long × 5mm thick) measured for a large scanning area (1200mm × 100mm). A ferrite magnet (15mm in diameter) was placed as dummy damage under the plate. Strong magnetic signals were detected at the position corresponding to that of the magnet as marked with the solid arrow. Also evident is the so-called *edge effect*, which is caused by the differences in permeability between type-410 and air, as indicated by dashed arrows

## 2.1 Objective of SQUID-NDE Research and Development

NDE, or nondestructive testing (NDT), are techniques used to evaluate materials, components, and structures for discontinuities or fatigue damage, to assess integrity, and to optimize life cycle management and life extension, in such a way that does not impair the future usefulness and serviceability of the test object [39,40]. Various approaches for NDE have been proposed and developed. Among them, ultrasonic testing (UT) is used to detect interior damage, whereas eddy current testing (ET) is conventionally used to inspect surface-cracks on testing materials, for instance. In developed countries, safety and proper maintenance are primary concerns to owners and operators of aging facilities such as infrastructures, transportation vehicles, buildings, and manufacturing plants. Due to limited maintenance costs, NDE is crucial to prevent further deterioration or degradation of such facilities. Significant effort has therefore been exerted to develop a cost-effective NDE method that is highly accurate for such practical applications. A sophisticated measurement method that is simple with clearly superior sensitivity is desired. Because a SQUID sensor is a candidate for such a NDE method, it is now being developed for practical applications.

## 2.2 Laboratory-Scale SQUID-NDE

SQUID-NDE techniques recently proved capable of detecting damage such as fatigue, deformation, and cracks in material components and structures [30-38]. Figure 2-4 shows a commercially available SQUID imaging system (SII Nanotechnology, SQM-3000) used for laboratory-scale experiments [41]. The notable feature of the system is that the SQUID magnetometer can be close to ~800μm from the surface of the test object by using sapphire for thermal isolation. In this section, current experimental data are described to demonstrate the possibility of SQUID for NDE applications, such as surface cracks (in type-304), stress-induced magnetism (in type-304), corrosion (in SS400), damage in the current rod (in copper), internal cracks artificially embedded in austenitic (in type-304) weld, and artificial creep damage (in type-304L).

Fig. 2-4. Photo of commercially available SQUID imaging system (SII Nanotechnology, SQM-3000) for laboratory-scale R&D applications to NDE.

### 2.2.1 Artificial Surface Crack Evaluation (Type-304)

The applicability of the SQUID-NDE was tested by evaluating different sized artificial cracks on the surface of a type-304 plate. Figure 2-5 shows the SQUID image obtained by surface scanning. The artificial cracks were 2mm wide and 1mm deep, and the lengths (CL) were 0.5cm, 1cm, 2cm, 3cm, and 4cm. The magnetic distribution detected by SQUID clearly increased with increasing CL, indicating that SQUID-NDE can detect surface cracks in objects such as type-304. Tiny surface cracks called 'hairline cracks' can also be detected by the SQUID imaging system (data not shown) [42]. Furthermore, if the test object is non-magnetic metal, applying an AC current in the measurement to evaluate surface cracks can improve the visualization of the magnetic distribution caused by the applied current.

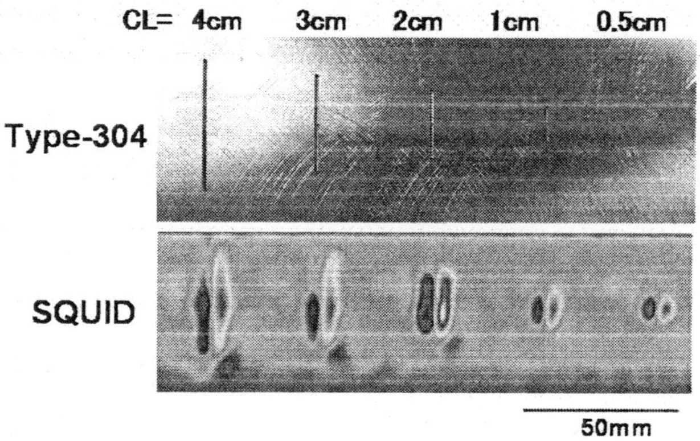


Fig.2-5. SQUID image of type-304 plate with artificially produced surface cracks of various lengths (CL). The lengths (CL) are 0.5cm, 1cm, 2cm, 3cm, and 4cm from right to the left (see upper panel).

### 2.2.2 Stress-Induced Magnetism Evaluation (Type-304)

Compared to the high-resolution experiment described in Section 1.4, stress-induced magnetism has been measured for centimeter-sized specimens at room temperature. The magnetic response of type-304 austenitic stainless steel bars to various stress levels applied by tensile testing has been examined. Recent data indicates that the magnetic response to stress likely corresponds to precursive phenomena of damage in austenitic [18-20]. Figure 2-6 shows a schematic of a bar-shaped sample (120mm × 20mm, and 3mm-thick) used in those measurements. Prior to the tensile testing, all the samples were demagnetized by annealing and by chemical etching using aqueous $HF/HNO_3$ within 10min of immersion. According to preliminary measurements of the stress-strain curve (see Fig.2-6), various mechanical tensile loads were applied to each sample by using a hydraulic testing machine: (A) no loading (reference), (B) up to yielding point, (C) 30% residual strain, (D) maximum stress, and (E) rupture. Each sample was artificially fatigued only once in correspondence with the tensile load. Figure 2-7 shows magnetic images obtained from a surface scan of the sample bars [43]. With the exception of the reference (no loading) sample, the residual magnetization induced by stress was clearly evident for each sample. In comparison, the magnitude of the detected spontaneous magnetization tended to increase with increasing mechanical tensile load, indicating that nonmagnetic austenitic is transformed into stress-induced ferromagnetic martensitic. This increase in detected spontaneous magnetization corresponds to the stress-induced magnetic susceptibilities [21] as well as to stress-induced magnetism detected by SQUID sensors [18-20]. Note that, the data shown here focuses on the plastic deformation region, i.e., martensitic phase, whereas the data discussed in Section 1.4 focused on stress-induced magnesium in the elastic region, where martensitic phase is not present. Comparison between these results (i.e., those in Sections 1.4 and 2.2.2) suggest that the local stress-induced magnetism of austenitic can be detected by SSM even in the elastic region because of the high sensitivity and spatial resolution.

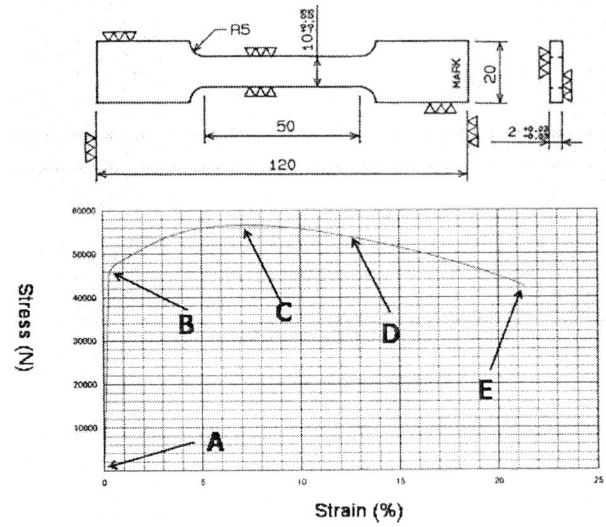


Fig.2-6. Typical strain-stress curve obtained by SQUID measurements for type-304 austenitic stainless steel at different loading levels: (A) Reference (no loading), (B)Yielding point, (C) 30% residual strain, (D) Maximum stress, and (E) Rupture. Dimensions of bar-shaped sample used for tensile testing are shown in the upper schematic.

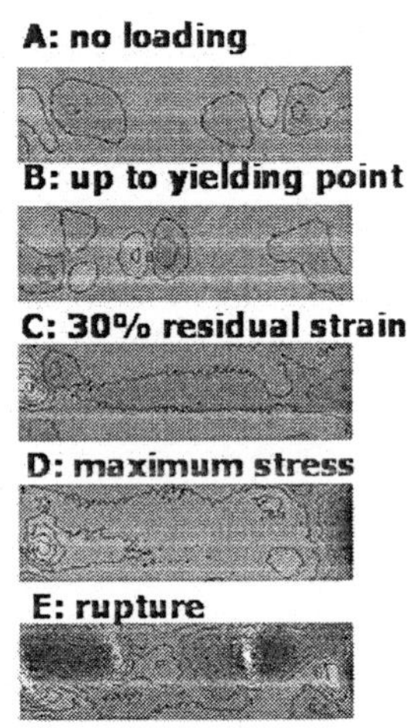


Fig.2-7. 2D magnetic image of the tensile tested sample bars obtained by SQUID imaging system measured at room temperature. Magnitude of the detected spontaneous magnetization tended to increase with increasing mechanical tensile load, indicating that nonmagnetic austenitic was transformed into stress-induced ferromagnetic martensitic.

### 2.2.3 Corrosion Evaluation (SS400)

A SQUID imaging system was used to evaluate corrosion in corroded SS400 steel samples prepared by using a conventional cyclic-corrosion-testing (CCT). CCT is an accelerated laboratory corrosion test method that can be used to determine the corrosion resistance of metals and material components. Test samples are first placed in an enclosed chamber, then exposed to a continuous indirect spray of a neutral salt water solution, and finally air-dried. The number of cycle repeats and therefore the test duration is variable. Figure 2-8 shows the

SQUID image for corroded SS400 steel samples exposed to CCT for 1 day and 9 days [24]. Only the left-sides were corroded by CCT. For the 1-day sample, the difference in the SQUID image between the corrosion area and non-corrosion area was negligible. For the 9-day sample, however, very strong SQUID signals were detected from the corrosion area only, probably due to oxidation of iron (Fe), such as $Fe_3O_4$ and $Fe_2O_3$, in other words, corrosion of iron. Thus, the SQUID-NDE system can be used to evaluate corrosion of steel, including SS4000.

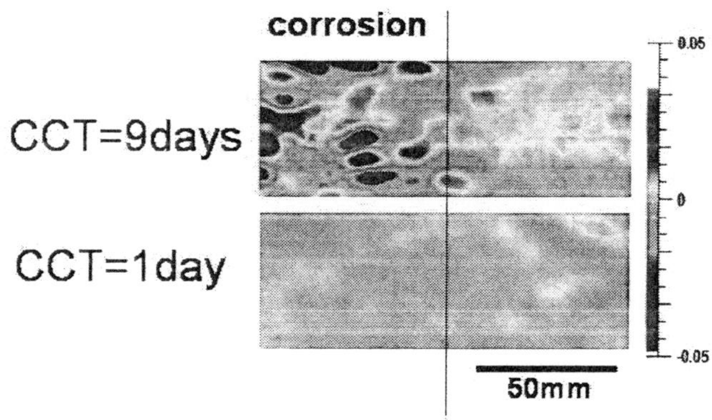


Fig. 2-8. SQUID image for SS400 steel plates corroded artificially with accelerated conditions by cyclic-corrosion-testing (CCT). Corrosion area is on the left half of the image.

### 2.2.4 Current Rod Evaluation (Copper)

SQUID-NDE techniques are likely capable of detecting damage (artificial flaws) in current rod also. The objective here was to verify the detection of a current dipole and to demonstrate the applicability of using a SQUID-NDE technique to evaluate damage in a current rod under an applied current [42]. Because the applied current must detour at the position of a flaw such as a slit and/or other resistive damage, the magnetic distribution is not uniform. Figure 2-9 shows the experimental setup for a current rod whose applied maximum capacity was 100V and 100A [42]. The applied current was controlled by changing the loading level using a combination of 0.3kW lights and an 8.4kW heater. Three artificial flaws (1mm in width, and 0.5mm, 1.5mm, and 1.0mm in depth) were made 90 degrees in circumferential direction from each other on the top of the surface of the copper current rod (see Fig.2-10). In the measurement, an AC current of 50Hz and 22.5Arms was applied. Figure 2-10 (upper panel) clearly reveals three quadruple magnetic signals directly above the positions of the artificial flaws. Such a measurement can be achieved even at much higher voltage up to 6kV (see inset) [42]. For the 6kV test, only a voltage was applied to the sample rod using two transformers, although the applied current was approximately 1Arms. These results indicate that the SQUID-NDE can successfully evaluate damage in a current rod.

Fig. 2-9. Photo of a facility used for current rod evaluation. Capacity was 100V-100A (by using two transformers, 6000V can be applied to the sample rods).

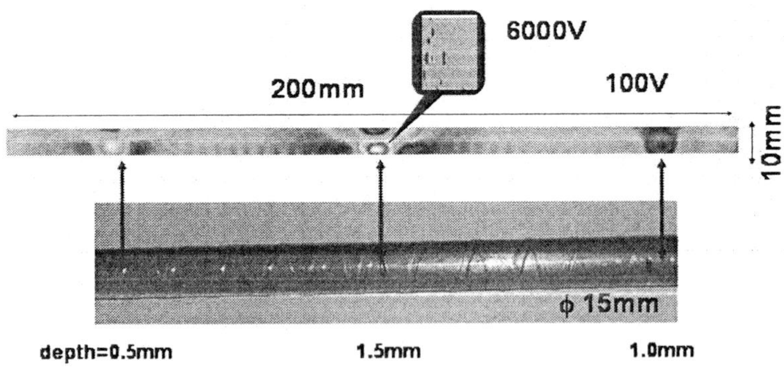


Fig. 2-10. 2D SQUID image (middle panel) obtained for copper rod with three artificial flaws shown in photo (bottom panel). Inset shows data obtained for an applied voltage of 6000V. The depth of the three flaws was 0.5mm, 1.5mm, and 1.0mm, from left to right, respectively.

### 2.2.5 Evaluation of Internal Cracks Artificially Embedded in Austenitic Weld (Type-304)

Historically, the inspection of austenitic welds has long been considered difficult by conventional ultrasonic testing (UT) methods. The difficulties have been attributed to the high levels of scatter generated by the coarse grain structure and by the consequent attenuation of ultrasonic present in the weldment [44]. Such difficulties might be overcome by using SQUID-NDE to evaluate internal cracks in austenitic welds.

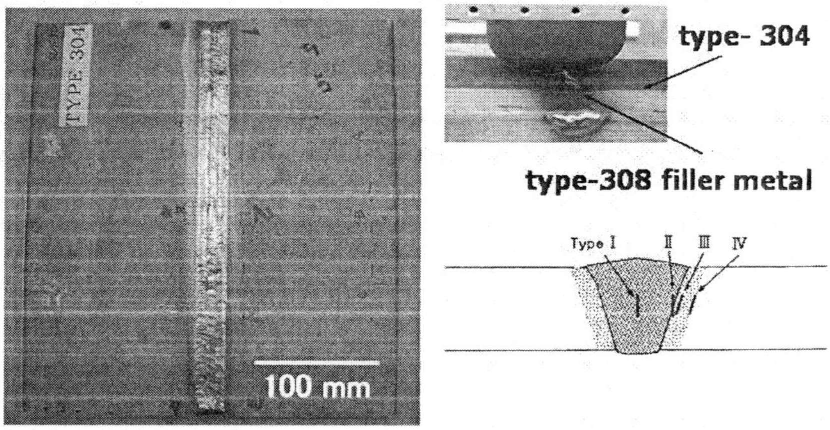


Fig.2-11. Photo of a type-304 austenitic stainless steel weld (16 kg). The filler metal was type-308. Three types of internal cracks were artificially embedded: Type I, which was a crack within the weld itself; Type II, which was a crack growing through the weld into the heat-affected zone (HAZ); and Type III, which is a coarse-grain-sized crack in the HAZ. Each type of cracks was made in two different sizes, i.e., ~10mm and ~2mm.

Figure 2-11 shows a photo of the type-304 austenitic stainless steel test sample. The mill sheet indicates that the test sample was manufactured in accordance with the Japanese industrial standard. The test sample (300mm wide, 320mm long, and 20mm thick) was welded using a TIG (Tungsten Inert Gas) welding tool, and the buttered face was welded to a type-304 sample plate using an austenitic type-308 filler metal completed joint. Three different types of internal cracks of different sizes were artificially constructed at different intervals along the test sample, and represented time-dependent degradation cracking at the pipework of boilers, for example. The cracks were graded into three categories [45] based on their location with respect to the weld: type I was a crack within the weld itself, type II was a crack growing through the weld into a heat affected zone (HAZ), and type III was a coarse-grain-sized crack in HAZ [46]. The test sample was made by KOBELCO Research Institute, Inc.(Japan) using the following procedures, which were essentially the same as those used in the preparation of standard samples for UT methods; (i) edge preparation of the test sample, (ii) TIG multi-layer butt welding up to about 13mm in thickness, (iii) groove surface profiling for embedding cracks, (iv) embedding cracks of Type I, II, and III by using an electric discharge process, (v) TIG welding up to the initial thickness, and (vi) removing the excess metal by using a grinder to smooth the surface of the test piece.

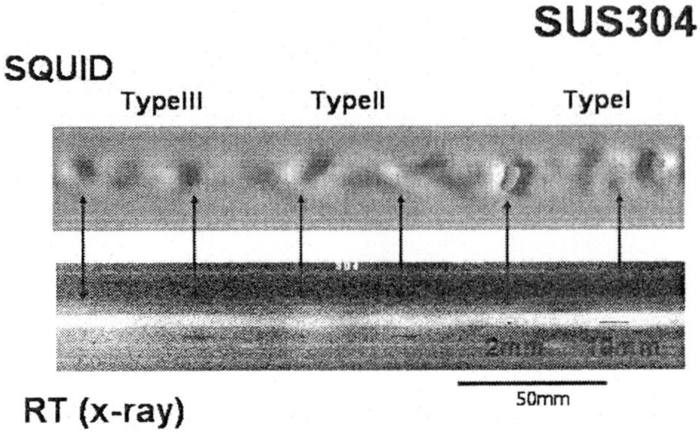


Fig.2-12. Differentiated 2-dimensional (2D) SQUID magnetic image of type-304 austenitic stainless steel weld (upper panel). The SQUID gradiometer was scanned parallel to the longitudinal axis of the weld. The inspection was completed within 14 minutes. Lower panel shows photo obtained by RT (x-ray transmission testing) for this same weldment. Three different types of internal cracks of two different lengths were clearly evident in this photo. Each arrow corresponds to the center position of a crack. The detected positions agree well with those seen in the photo obtained by the RT.

Figure 2-12 clearly shows that all three types of internal cracks were successfully detected by the SQUID imaging system [47]. The strong intensities corresponded to the larger cracks (~10mm in length) where the magnetic gradient due to spontaneous magnetism was strong. Note that all of the tiny small cracks that were ~ 2mm in length were clearly detected as well. Figure 2-12 (bottom panel) also confirmed the position of the internal cracks revealed by using radiographic testing (RT). Although in the industrial sector, UT-TOFD (time of flight diffraction) is widely used for evaluation, such a method is inefficient for use on this test sample. The inefficiency is probably due to the broad scatter of the ultrasonic wave generated by the coarse grain structure of the weld, as aforementioned. For each crack, the position detected by using RT agrees well with that detected by using differentiated SQUID signal obtained numerically magnetic image (indicated by arrows in the bottom panel). Thus, SQUID-NDE can successfully detect the positions of internal cracks in austenitic welds [47]. For practical applications, small internal cracks (e. g., 1~2 mm in length) in an object thicker than 50mm must also be detected. For practical applications, the goal is to evaluate cracks longer than 1mm and to reveal the relationship between the SQUID signal and the size of the cracks, i.e., "sizing", however, evaluating thick objects is difficult in reality. In addition, HAZ in austenitic welds can be visualized at the micron-scale by using SSM to obtain high-resolution measurements [48]. Further study is needed to determine if HAZ in austenitic welds can be characterized by a SQUID-NDE system on a much larger scale, such as manufacturing plants or buildings.

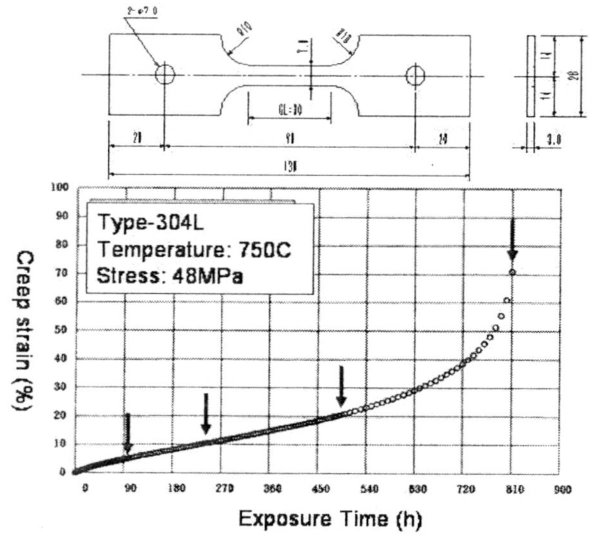


Fig.2-13. Typical creep curve (time vs. creep strain) obtained with accelerated conditions for a bar-shaped type-304L austenitic stainless steel sample (dimensions shown in upper panel). For the SQUID measurements, samples with different exposure times were prepared, as marked with arrows.

### 2.2.6 Artificial Creep Damage Evaluation (Type-304L)

Creep damage can be classified into five categories based on its severity: no damage, initial formation of cavities, oriented cavities, cavity linking and microcracks, formation of microcracks, and fracture (rupture). The damage level depends on temperature, time, and stress. Magnetic hysteresis measurement can be used to evaluate creep damage [45], and preliminary SQUID-NDE data has been obtained for artificial creep damage in type-304L stainless steel bars. Generally, test samples must be exposed to high temperatures (e.g., 600 Celsius ) for a long time (more than a year) is required for obtaining systematical creep damage data. As a preliminarily experiment to verify the SQUID measurement, creep damage samples were prepared at accelerated conditions using type-304L stainless steel bars (see Fig. 2-13). Figure 2-13 shows a typical creep curve, i.e., exposure time (ET) vs. creep strain for a sample bar fractured at 34 days (ET=814h). Three other samples were prepared for ET=96h, 222h, and 531h (marked with arrows in the figure). The creep strain (measured by a gauge) was 4.7%, 10%, 19.5%, and 71% (fracture) for ET=96h, 222h, 531h, and 814h, respectively. Figure 2-14 reveals that the SQUID image system can evaluate the creep damage in austenitic, such as type-304L. Additionally, the detected peak-to-peak SQUID signal increases with increasing ET, as clearly seen in Fig. 2-15. Metallographic analysis is necessary to check that the SQUID image corresponds well to the creep damage level.

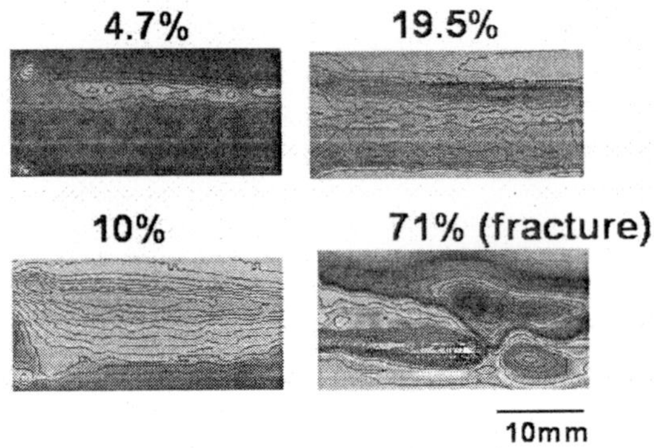


Fig. 2-14. SQUID images for creep-damaged type-304L samples with different creep strains of 4.7%, 10%, 19.5%, and 71% (fracture). Note that, the creep strain (measured by a gauge) was 4.7%, 10%, 19.5%, and 71% (fracture) for ET (exposure time) =96h, 222h, 531h, and 814h, respectively.

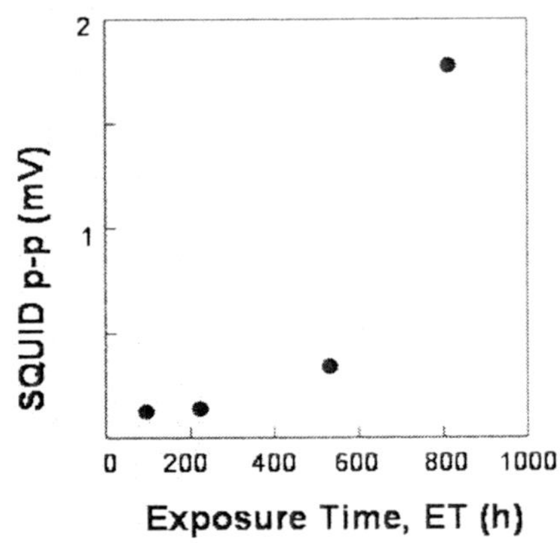


Fig.2-15. Detected peak-to-peak SQUID signal for creep-damaged type-304L samples with creep strains of 4.7%, 10%, 19.5%, and 71% (fracture) shown in Fig. 2-14. The creep strain (measured by a gauge) was 4.7%, 10%, 19.5%, and 71% (fracture) for ET=96h, 222h, 531h, and 814h, respectively.

# 3 Development of a "Traveling Squid-Nde"

As discussed in section 2, SQUID-NDE can successfully evaluate surface/internal cracks, stress-induced damage, corrosion, and creep, for instance. However, in those evaluations, a conventional SQUID-NDE was used in which the SQUID sensor was stationary while a small piece of the sample was moved by using a scanning xy stage (see Fig. 2-4). The stationary SQUID sensor makes the SQUID-NDE useless for practical applications because the available scanning area is restricted. To apply SQUID-NDE in industry, a SQUID-NDE imaging system in which the SQUID sensor travels along the surface of the test object was recently developed. This advanced system, called a "traveling" SQUID-NDE" [49], technically allows a large-scale scanning area for in-place evaluation of large fixed objects, such as buildings and infrastructures. In this section, advances in SQUID-NDE in which the SQUID gradiometer mechanically moves against the fixed object are reviewed. This unique

SQUID-NDE required key technologies such as (i) traveling SQUID gradiometer, (ii) portable cryostat, and (iii) noise cancellation technique.

## 3.1 Traveling SQUID Gradiometer

The definition of the 'traveling SQUID gradiometer' is a SQUID gradiometer in which the SQUID sensor itself moves during NDE without the need for any magnetic shielding. The key technology in the traveling SQUID gradiometer is the fabrication of a DC-SQUID gradiometer that can minimize environmental noise and magnetic noise generated from 'mechatronics', e.g., from servomotors. This section first briefly summarizes aspects of the traveling SQUID gradiometer, such as its structure, simulations of its output signals, and its flux noise spectra, and then describes future development of a traveling 'high-Tc' SQUID gradiometer.

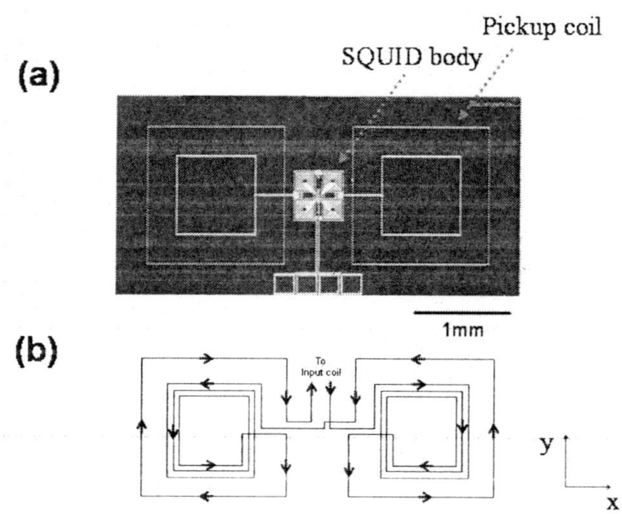


Fig.3-1. (a) Photo of traveling SQUID gradiometer consisting of a DC-SQUID made of Nb/Al-AlOx/Nb Josephson tunneling junctions combined with two coplanar concentric second-order derivative coils connected in series and positioned counterclockwise to each other. (b) Structure of the pickup coil (after Nakayama et al. [52]).

### 3.1.1 Structure of the Traveling SQUID Gradiometer

When superconducting differential pickup circuitry is properly operated, an integrated SQUID sensor can detect faint nearby signals in the presence of much stronger background interference from more distant sources [1]. This type of SQUID sensor is categorized as a gradiometer, and generally higher-order gradiometers are expected to perform better against stronger magnetic disturbances. Axial second-order gradiometers were reported in the 1980s [50,51]. The traveling SQUID gradiometer consists of two pairs of off-diagonal second-order pickup coils to suppress undesirable magnetic disturbances [52]. As schematically shown in Fig.3-1, the traveling SQUID gradiometer contains a DC-SQUID made of Nb/Al-AlOx/Nb Josephson tunneling junctions combined with two coplanar concentric second-order derivative coils connected in series and positioned counterclockwise to each other [52]. That is, a multilayer fabrication process is used to achieve a series connection of figure-8-shaped pickup coil loops (see Fig.3-1(b)) with geometrically one stroke and magnetic coupling by using a flux transformer[27]. Such a unique structure using a multilayer fabrication process is the key technology in fabricating traveling SQUID gradiometers, as discussed in section 3.1.4. The baseline of this traveling SQUID gradiometer (i.e., distance between coil centers) is 3mm. The inner and outer pick-up coils of this second-order derivative coil are $1.1547 \times 1.1547\ mm^2$ (three-turns) and $2 \times 2 mm^2$ (one-turn), respectively. The gradiometer is not perfectly symmetric in shape, and therefore the signal depends on the scanning direction. The magnetic flux resolution is about $9 \times 10^{-6}\ \phi_0\ /Hz^{1/2}$, where $\phi_0$ ($2.07 \times 10^{-15}$ Wb) is the magnetic flux quantum in magnetic shielding. This noise level is higher than that of a standard concentric double-washer-type DC-SQUID gradiometer ($4 \times 10^{-6}\ \phi_0\ /Hz^{1/2}$) developed by SII Nanotechnology Inc. (Japan) earlier [53]. The processes used to fabricate the traveling SQUID gradiometer are the same as those established by Morooka et al. [54].

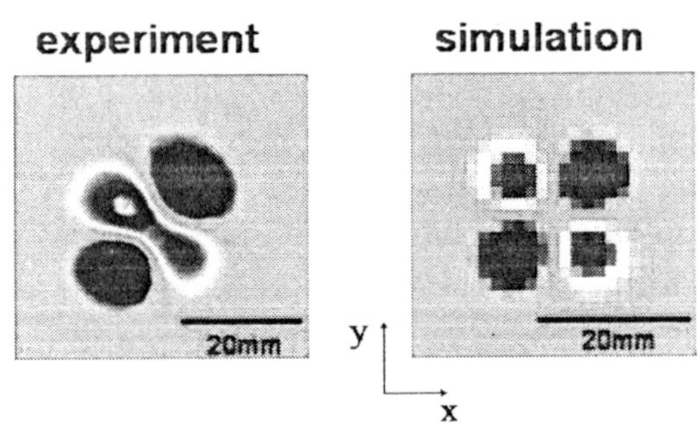


Fig.3-2. Experimental data (left image) and simulated (right image) SQUID images for magnetic dipole. In the magnetic dipole measurements, an unsymmetrical-shaped ferrite magnet (~10mg) was used. In the simulation, the gradiometer was assumed to act as a third-order gradiometer. The simulated SQUID images strongly depended on the direction of the dipole, although experimentally it was difficult to clearly determine the direction of the dipole. Therefore, the direction of the dipole in the simulation was defined based on the experimental results.

### 3.1.2 Output Signals of the Traveling SQUID Gradiometer

For the traveling SQUID gradiometer, a new type of pickup coil is connected to the SQUID body to resist environmental magnetic disturbances. In this pickup coil, two coplanar concentric second-order derivative coils are connected in series and positioned counterclockwise to each other (see Fig.3-1 bottom panel). Comparison between simulations and experimental data of the output signals [27] revealed that this traveling SQUID gradiometer can be categorized as a "quasi" third-order gradiometer. In general, off-diagonal first-order gradiometers sense the z-component of magnetic flux density $B_z(r1)$ with one loop of the gradiometer coil at r1, and sense $B_z(r2)$ with a second loop wound in the opposite direction at r2. Then, the SQUID output voltage is proportional to $\Delta B_z/\Delta r$, where $\Delta B_z = B_z(r1)-B_z(r1)$ and $\Delta r = r1-r2$ [2]. In second-order gradiometers, moreover, the SQUID voltage is proportional to $\Delta^2 B_z/\Delta r^2$. In this case, the output voltage is proportional to $\Delta (\Delta^2 B_z/\Delta r^2)/\Delta x$, indicating that the traveling SQUID gradiometer can be considered a third-order gradiometer. The pickup coil of the traveling SQUID gradiometer is rectangular-shaped, and thus the output voltage is proportional to $\Delta\{\Delta^2 B_z/\Delta x^2)+(\Delta^2 B_z/\Delta y^2)\}/\Delta x$. To verify that the traveling SQUID gradiometer can be considered a third-order gradiometer, the output SQUID voltage for a magnetic dipole was simulated [27]. Figure 3-2 shows that the simulated SQUID image for a small ferrite magnet (approx. 10mg) at a liftoff distance L of 15mm (L is defined as the distance between the surface of the object and the SQUID gradiometer) agrees well with the experimental images, indicating that the traveling SQUID gradiometer can be categorized as a third-order gradiometer. In this simulation, the output SQUID voltage of a magnetic dipole was pointed to the minus y-direction when the SQUID baseline is parallel to x direction. Furthermore, the simulated SQUID image agreed well with experimental data for the current dipole, as shown elsewhere [27]. Therefore, comparison between simulations and experimental data confirm that the traveling SQUID gradiometer is a "quasi" third-order gradiometer.

Fig.3-3. Photo of electric power substation field where the SQUID noise spectra were measured. GCB is the 275kV electric gas circuit breaker, and trans is the 450MVA transformer.

### 3.1.3 Flux Noise Spectra of the Traveling SQUID Gradiometer

Flux noise spectrum of the traveling SQUID gradiometer measured inside a magnetic shielding was reported by Nakayama et al. [52]. The magnetic flux resolution was in the white noise range and the 1/f corner frequency was lower than 1Hz. To ensure that the traveling SQUID gradiometer can be used for in-place NDE applications, the SQUID noise spectra were measured at an electric power substation [27]. The cryostat installed in the SQUID gradiometer was completely filled with liquid He, and all the SQUID electronics were transported by taxi (in about 30 minutes) from our laboratory to the substation. Before transport, the traveling SQUID gradiometer was cooled to liquid-He temperature at the laboratory. Figure 3-3 shows the 450MVA transformer and 275kV electric gas circuit breaker (GCB) where the SQUID noise spectra were measured. The traveling SQUID gradiometer operated properly in the immediate vicinity of the transformer. Figure 3-4 shows the SQUID noise spectra measured without any magnetic shielding. The noise level at 10Hz was about $2000\mu\phi_0/\mathrm{Hz}^{1/2}$, which is approximately 200 times higher than that measured with magnetic shielding. Figure 3-4 also shows that the traveling SQUID gradiometer operated normally directly under a 275kV transmission line adjacent to a 275kV GCB as well as inside a constant-voltage constant-frequency (CVCF) room at the substation building [27].

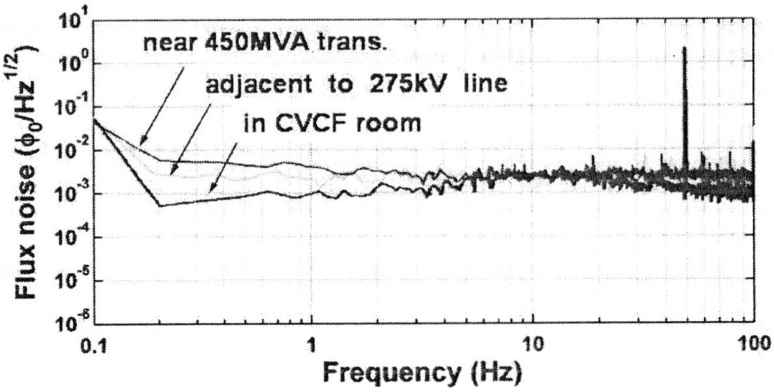


Fig.3-4. SQUID noise spectra measured at the substation without magnetic shielding. Noise spectra were measured near the 450MVA transformer (see Fig.3-3), under a 275kV transmission line close to the 275kV electric gas circuit breaker (GCB), and in a constant voltage constant frequency (CVCF) room inside the substation building.

### 3.1.4 Future Development of a Traveling 'High-$T_c$' SQUID Gradiometer

For practical applications, a high-$T_c$ SQUID is preferable to low-$T_c$ SQUID because it does not require liquid He, and is therefore easier to handle. However, the traveling SQUID gradiometer is fabricated by using low-$T_c$ SQUIDs technology. This is because low-$T_c$ SQUIDs also have potentially lower flux noise, easier fabrication processes, and batch-to-batch higher stability. To develop a traveling 'high-$T_c$' SQUID gradiometer in the future, both the advantages and disadvantages of high-$T_c$ and low-$T_c$ SQUIDs listed in Table 1 must be compared [55-57]. The main disadvantage of high-$T_c$ SQUIDs is the lack of reliable multilayered fabrication processes. In high-Tc SQUID gradiometers, the pickup coil loops are currently connected in parallel to each other and are directly connected to the SQUID ring due to limitations from the monolayer fabrication process. Such a configuration might induce magnetic flux trapping due to the shielding current running through the surrounding coil line.

In low-Tc SQUID gradiometers, on the other hand, a multilayer fabrication process [52,54] can realize a series connection of figure-8-shaped pickup coil loops with geometrically single-stroke and with magnetic coupling by using a flux transformer. (An example of "geometrically single-stroke" is writing a 'figure-8' by hand, namely, the figure can be drawn in a single stroke without having to add a line or curve to finish the figure). This unique structure using a multilayer fabrication process must be the key technology in fabricating traveling SQUID gradiometers. The series circuit of the pickup loops efficiently decreases the current density of the pickup loop lines. As a result, magnetic flux trapping is prevented, thus increasing the resistance to magnetic disturbances. Compared with direct coupling, transformer type coupling enables higher coupling efficiency. Consequently, the traveling SQUID gradiometer was based on low-Tc SQUID technologies, which have the advantage of a multilayer fabrication process, for achieving flux transformer type coupling and for achieving a series connection of the pick up coil loop.

Table 1. Comparison between high-$T_c$ SQUIDs and low-$T_c$ SQUIDs for possible future development. Although high- $T_c$ SQUIDs are promising for use in NDE applications, low- $T_c$ SQUIDs are currently superior in tolerability to magnetic disturbances due to easier fabrication process. A multilayer fabrication process is crucial for developing a SQUID gradiometer that is resistant to environmental magnetic noise.

| | **High- $T_c$ SQUID** | **Low- $T_c$ SQUID** |
|---|---|---|
| Cooling medium | liquid $N_2$ / cryocooler | liquid He |
| Operating cost | much cheaper than low-$T_c$ | expensive |
| Superconducting materials (plasticity) | ceramics (poor) | metal (not poor) |
| Reported Maximum Sensitivity | $<100fT/Hz^{1/2}$ | $\sim10fT/Hz^{1/2}$ |
| SQUID fabrication process | monolayer-type | multilayer-type |
| Coupling of pickup coil | direct coupling | transformer |
| Circuit of pickup loop | parallel | series |
| Flux trapping occurrence | infrequently (more than low-$T_c$ SQUID) | rarely |

## 3.2 Portable Cryostat

To construct the SQUID-NDE system, a portable cryostat was developed that to minimize vibration noise and to have user-friendly operation. Figure 3-5 shows a photo of this portable cryostat, which is 300mm in height, 100mm in outer diameter, and has a liquid-He capacity of 320ml [52]. The loss rate of the liquid He is about 80ml/h. The traveling SQUID gradiometer is immersed in liquid He and mounted inside the bottom cap (fiberglass) of the cryostat surrounded by the insulating vacuum of the cryostat. A spatial resolution of the SQUID-NDE system is generally determined by the dimensions of the pickup-coil and the lift-off distance between the pickup-coil and the object surface. The lift-off distance is 6 mm, which is 2 times longer than the length of the gradiometer baseline and thus, the spatial resolution is approximately 6mm. Details of the structure of this cryostat were reported by Nakayama et al. [52]. Generally, in comparison downsizing a cryocooler-based cryostat [58], downsizing of a liquid-He cryostat for portability is generally more difficult.

Fig. 3-5. Photograph of portable cryostat [52]. The cryostat is 300mm in height and 100mm in outer diameter, has a liquid-He capacity of 320ml, and has liquid-He loss rate of about 80ml/h.

## 3.3 Noise Cancellation Technique

A noise cancellation technique is another important aspect in diminishing or minimizing ambient noise. Figure 3-6 schematically shows a noise cancellation procedure developed to eliminate undesirable magnetic noise signals as follows [47]. First, a sampling interval is selected, e.g., 20ms, and then the SQUID sensor travels via the SQUID manipulator (see section 4) so that it is synchronized with the interval. For each sampling point, the movement of the SQUID sensor and the sensing action for magnetic measurement start simultaneously at the selected interval, assuming that the time-constant of the electronics is negligibly small. This procedure also shortens the time needed for NDE by roughly a factor of 5 or more, in comparison with the time needed by conventional methods, such as the data shown in Section 2.2.6 by using SQM-3000. Based on preliminary results, the optimal sampling interval is determined by a trade-off between inspection time and mapping pitch. Note that the noise cancellation procedure differs from a SQUID-based scanning eddy-current imaging system [59]. In actuality, the SQUID sensor proved useful as a detector using the Eddy current method to evaluate cracks in C-FRP (carbon fiber reinforced plastic) [60,61]. For the developed noise cancellation procedure, if the evaluated object is nonmagnetic, then an AC current or field can be applied to obtain an magnetic image. Such a data acquisition technique called "lock-in averaging" [52] can effectively shorten the sampling time. In this case, prior to acquiring data, a phase difference between the applied signal and the detected signal must be compensated for in advance.

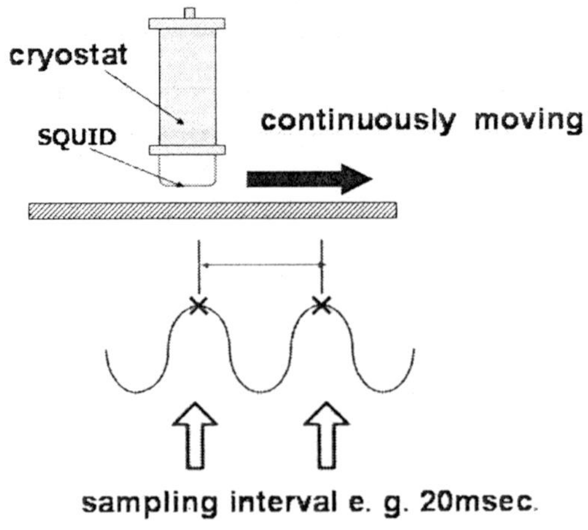


Fig. 3.6. Schematic of noise cancellation procedure to eliminate undesirable magnetic noise signals. First, a sampling interval is selected, e.g., 20ms, and then the SQUID sensor travels via the SQUID manipulator so that it is synchronized with the interval. For each sampling point, the movement of the SQUID sensor and the SQUID's sensing action for magnetic measurement start simultaneously at the selected interval.

# 4 Robotic Traveling Squid Imaging System for NDE

To manipulate the traveling SQUID gradiometer to travel across an object, a traveling SQUID-NDE system was constructed by combining it with a manipulator used by conventional industrial robots for large-scale applications. Two systems were combined with different robotic systems to extend the variation of the shape of the object: one with a rectangular-type (3-axes) robotic system and one with an articulated-type (6-axes) robotic system [28]. This section discusses the development of two robotic SQUID imaging systems for large-scale NDE applications.

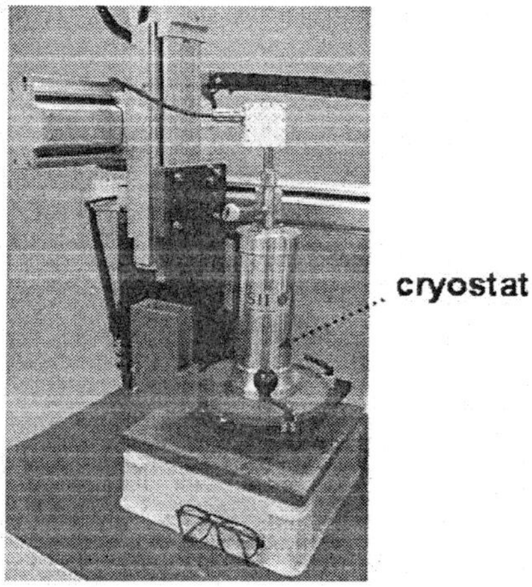


Fig.4-1. Photo of traveling SQUID-NDE system with a rectangular-type (3-axes arms) robot applicable for flat-surface test objects.

## 4.1 Rectangular-Type Robotic Traveling SQUID-NED System (3-Axes Arms System)

Figure 4-1 shows a photo of the robotic traveling SQUID-NDE system that incorporates a rectangular-type (3-axes) of robotic system applicable for relatively flat-surface objects [43,47]. In robotics terminology, this system is simply described as a Cartesian robot (X-sel, IAI, Japan) in which the end-effecter is a DC-SQUID gradiometer installed in a cryostat. As shown in Fig. 4-2, the 3-axes arms (see Fig.4-1) system consists of a traveling SQUID gradiometer, electronics for operation of the SQUID, a portable liquid helium cryostat, an xyz manipulator, and a PC for controlling the SQUID electronics and the system motion. The system noise level during robot operation is reported elsewhere [52]. The usefulness of the rectangular-type robotic traveling SQUID-NED system was discussed above for evaluation of surface cracks (section 2.2.1), stress-induced magnesium (2.2.2), corrosion (2.2.3), current rod damage (2.2.4), and austenitic weld (2.2.5). Technically the robotic traveling SQUID-NED system is useful for enlarging an available scanning area without limitation.

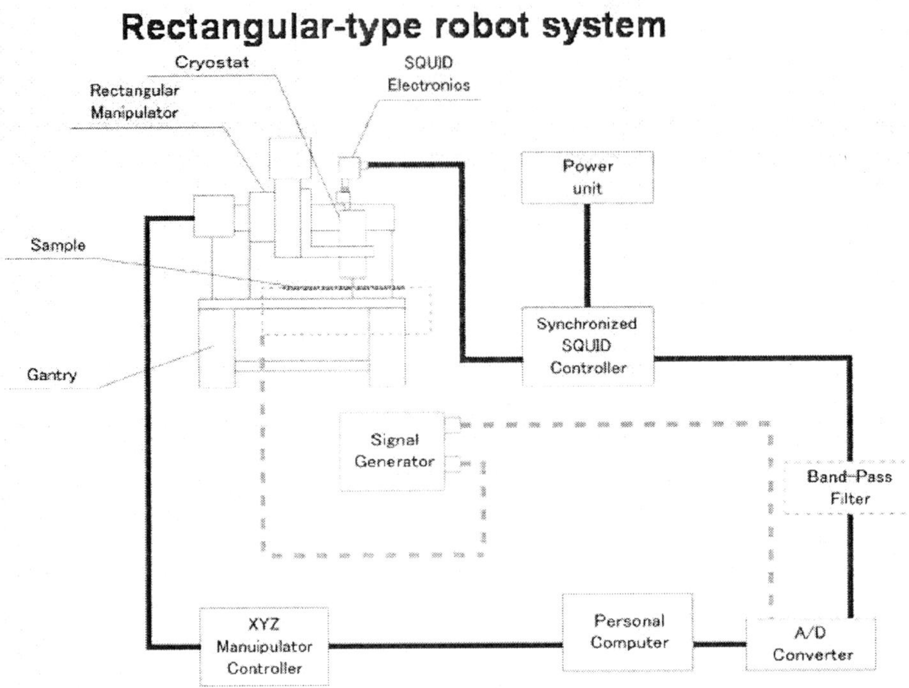


Fig.4-2. Schematic of SQUID-NDE system composed of a DC-SQUID sensor, electronics for operation, liquid-He cryostat, xyz rectangular manipulator, and laptop PC [43,47].

## 4.2 Articulated-type robot system (6-axes arms system)

To manipulate the SQUID sensor in three-dimensionally (3D) to travel across an object, the 3-axes arms system was improved by using a conventional articulated-type (6-axes) industrial robot [27,62]. The robotic 3D SQUID-NDE system described here is a state-of-the-art system, although a prototype SQUID-NDE system in which a SQUID sensor mounted on a 5-axes robot system was reported in 1993 [63]. Freely movable operation of the SQUID sensor as well as 3D NDE experimental data was not reported in Ref.63,though. For 5-axes robot systems, moreover, one degree-of-freedom in the robot movement is restricted, namely, lack of one rotational movement, and consequently, the SQUID sensor cannot be tilted with respect to the vertical direction of the surface of the object. As a result, it is impossible to maintain a constant distance between the SQUID sensor and the surface of the object. Furthermore, the conventional hand-operated teaching method is impractical for measuring a vast number of points. Thus, the following are articulated-type robot system (6-axes arms system) developed to cope with 3D measurements on curved-surface areas.

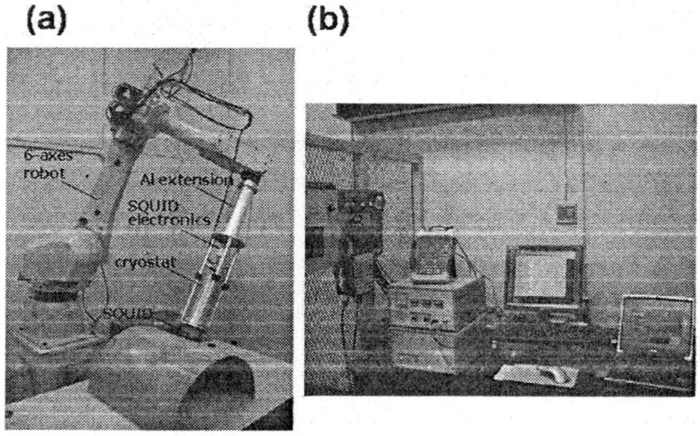


Fig. 4-3. (a) Photo of robotic 3D SQUID-NDE imaging system that incorporates an articulated-type (6-axes) robotic system applicable for large-scale objects with curved surfaces [27,62]. (b) Photo of electronics for SQUID operation and PC for robot motion control and SQUID measurement control.

Figure 4-3(a) shows the robotic 3D SQUID-NDE imaging system that incorporates 6-axes arms industrial robot system [27,62]. The robotic 3D SQUID-NDE system consists of a traveling SQUID gradiometer, electronics for SQUID operation, liquid-He cryostat, articulated-type (6-axes) robot (M-16iB, FANUC, Japan), laser CCD displacement sensor (LK-502, Keyence, Japan), and PC. Figure 4-3(b) shows a photo of the electronics for SQUID operation and PC for the robot motion control as well as for SQUID measurement control. The components of the robotic 3D SQUID-NDE system are basically the same as those for the rectangular-type (3 axes) robotic system [43,47], except that a laser CCD displacement sensor to measure the 3D coordinates of points on the surface of the object has been added.

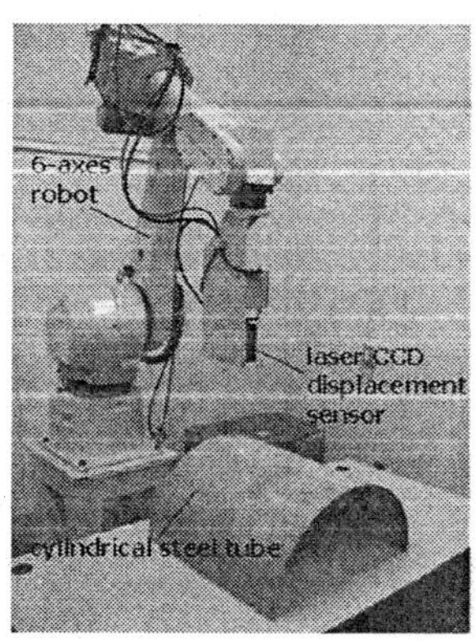


Fig.4-4. Photo of laser CCD displacement sensor equipped with an articulated-type robot. This sensor and a computer-aided geometrical surface reconstruction scheme robot were used to teach the robot the 3D coordinates (x, y, and z). In the SQUID-NDE, the laser CCD displacement sensor replaced the SQUID sensor [27,62].

The laser CCD displacement sensor (shown in Fig. 4-4) measures the shape of the object in 3D. This measured shape data is then input into a numerical surface reconstruction procedure to "teach" the robotic system about the 3D coordinates (x-, y-, and z- coordinates) automatically. The laser CCD displacement sensor is operated by using a triangulation technique to achieve high accuracy regardless of light quantity distribution that is strongly affected by the surface condition of the object. Using the following procedure also shown in Fig. 4-5, the robotic system learned the 3D coordinates (i.e., x, y, and z) without using CAD (computer-aided design) data of the evaluated object. Initially, while maintaining a constant height level above the surface of an object, a displacement sensor scans the shape of the object at each point within an x-y lattice (Fig. 4-6(a)). The sampling pitch of this x-y lattice affects the calculation of the vertical axes at each point; errors in the solid-angle (i.e., the angle in 3D) increase with decreasing sampling pitch as a function of the arctangent between the sampling pitch and the error in the laser CCD displacement sensor. Based on a preliminary experiment, error in the displacement sensor used in this robotic 3D SQUID system was 0.1-0.2mm. Although the optimal sampling pitch of the x-y lattice usually depends on the shape of the object, empirically it was 5mm. After this initial scan, the displacement sensor is removed and the SQUID sensor is connected as the end-effecter of the articulated-type robot. Prior to this replacement, a numerical surface reconstruction procedure is used to calculate the vertical vector perpendicular to the surface of the object. For SQUID measurements, the distance between the SQUID sensor and the surface of the object must be kept constant (see Fig. 4-6(b)). Based on the data acquired by the displacement sensor, the following geometrical methods are used [27,62]. As computational-geometry oriented techniques for reconstructing the surface of an object using data points acquired from an x-y lattice, geometrical expressions called Bezier surfaces as well as B-spline surfaces are adapted [64,65]. The rigorous numerical expressions are theoretically described elsewhere [66,67]. In short, B-spline interpolation is categorized as a global approach, whereas Bezier interpolation is a local approach. Figure 4-7 shows the reconstructed surface image of the evaluated object by B-spline interpolation. The bottom figure indicates the difference in vertical distance between the numerically approximated distance and experimentally measured distance by laser CCD displacement sensor. Here, this difference in vertical distance is defined as $\gamma$. The $\gamma$ is less than 0.1mm (including the error), indicating this surface reconstruction procedure is sufficient for the automatic recognition of the shape of an evaluated object and for handling a vast number of data points. Thus, experiments (e.g., Fig. 4-7) confirmed that such a surface reconstruction scheme is useful for the automatic recognition of the shape of an evaluated object and for handling a vast number of data points.

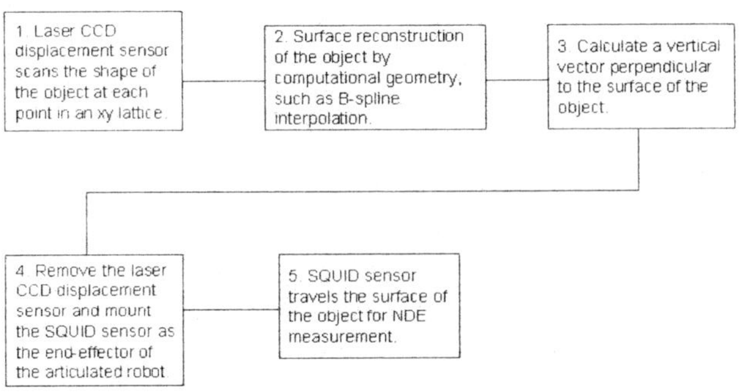


Fig.4-5. Flowchart of the five-step process to evaluate an object by robotic 3D SQUID imaging system for NDE applications. To "teach" the robot system the 3D (i.e., xyz coordinates) of the object at the inspection area, the acquired data utilized by the laser CCD displacement sensor is conveyed to the computer-aided numerical interpolation procedure to reconstruct the surface of the object in 3D.

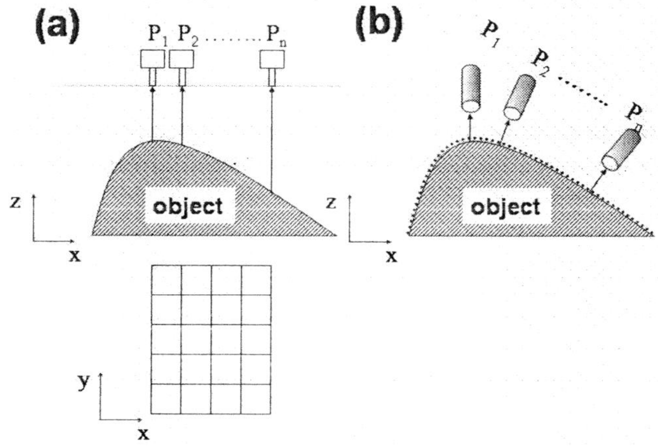


Fig. 4-6. (a) Measurement of the shape of an object at each point in an x-y lattice by a laser CCD displacement sensor while maintaining a constant height level above the surface. A constant interval, i.e., linearity, is requisite to the surface reconstruction. (b) Prior to the SQUID measurements, the vertical direction, which is perpendicular to the surface of the object, is continuously calculated by using the surface reconstruction program, e.g., dotted line (see Section 4.2).

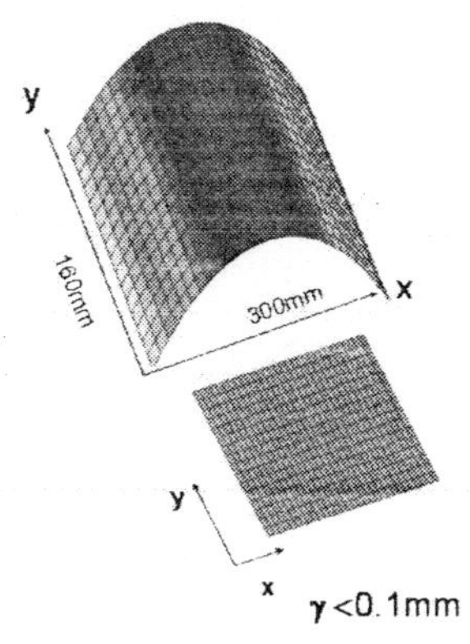


Fig. 4-7. Surface reconstruction of a cylindrical-shaped steel tube (see Fig. 4-4) using B-spline interpolation. Scan area was 300 × 160mm.

## 4.3 Applicability of the Robotic 3D SQUID Imaging System to NDE

Applicability of the robotic 3D SQUID imaging system to NDE was examined by using the system to detect artificial damage in cylindrical-shaped steel tubes [27,62]. A flowchart of the procedures in the measurements is shown in Fig. 4-5. The test samples were commercially available type-304 stainless steel tubes with 508mm outer diameter, 500mm length, and 6mm thickness. A small ferrite magnet (~10mg, 5mm-long) attached under the cylindrical-shaped tube was used as dummy damage. In the 3D SQUID image (Fig. 4-8) the detected signals (arrows) corresponded to the position of the dummy magnet, demonstrating the success of the developed robotic SQUID imaging system in 3D operation [27,62]. These results also confirm that the laser CCD displacement sensor and the numerical interpolation 3D surface reconstruction procedure used in the developed imaging system were sufficient for 3D NDE measurements. This imaging system enables large-scale, curved-surface areas to be scanned and possibly evaluated "in-place". Various curved-surface samples used at high temperature in manufacturing plants, for example, can be eventually evaluated by using this robotic 3D SQUID imaging system. So, the core-design of the SQUID imaging system, namely, traveling SQUID-NDE system described here is useful for large-scale applications since it technically allows enlarging an available scanning area without limitation.

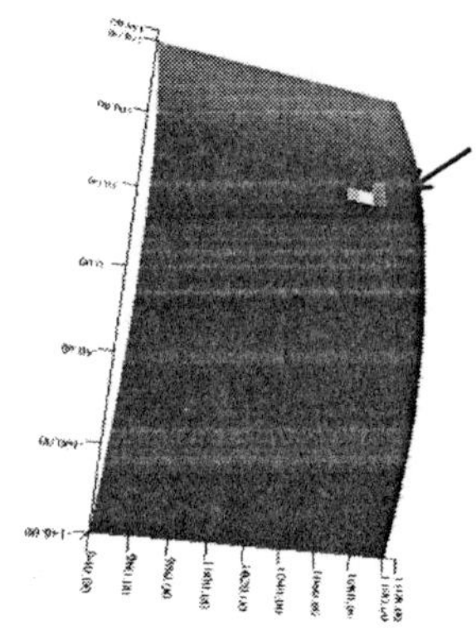

Fig. 4-8. 3D SQUID image of artificial damage in a cylindrical-shaped SUS304 stainless tube to demonstrate capability of robotic 3D SQUID-NDE system. Small ferrite magnet (approx. 10mg in weight and 2mm in length) was attached as artificial damage to the reverse side of the tube. The scanning area was 300mm × 160mm. Strong signals were detected at the position of the magnet (indicated by arrow).

## 5 Summary

In this paper, current progress in scanning SQUID imaging systems are reviewed mainly from an application viewpoint. Scanning SQUID imaging systems are clearly successful not only for academic research but also for practical applications. These systems can visualize the magnetic distribution two-dimensionally with unparalleled magnetic sensitivity from micron-scale to meter-scale by the surface scanning of a sample. The traveling SQUID gradiometry (one that is not stationary, but in which the recently developed SQUID sensor itself moves or scans along the surface of the test object) technically allows the "in-place" evaluation of large-scale, curved-surface areas to be scanned using the SQUID-NDE without magnetic shielding. Basic experimental SQUID-NDE data were reviewed here to demonstrate the capability of this system for practical usage, including evaluation of surface cracks, stress-induced magnetism, corrosion, current rod damage, internal cracks artificially embedded in austenitic weld, and creep damage. Recently, SQUID-NDE data proved the capability to successfully detect damage such as fatigue, deformation, and cracks in material components and structures.

Finally, the outlook and prospective development of scanning SQUID imaging were discussed as a concluding remark. In academic research, spatial resolution of ~10μm is not sufficient, especially in semiconductor research and quantitative analyses in material science. Although the lower limit of spatial resolution depends on the diameter of the pickup coil integrated into the SQUID body, submicron or nano-order resolution is expected. For practical SQUID-NDE applications, on the other hand, the need for a comprehensive and pinpoint database (i.e., for example, just select one evaluation issue among fatigue, deformation, cracks, etc.) for SQUID-NDE is urgent. Comparisons between conventional NDE techniques and the robotic SQUID-NDE system used to evaluate the same large-scale object must be done. Opinions from end-users will be helpful for further improvements of SQUID-NDE. If the SQUID can operate relatively easy and user-friendly manner in applications involving superconductivity, cooling, or magnetic shielding, then SQUID-NDE will be convenient for practical use. Due to the rapid progress in the research and development of SQUID-NDE during the last few years, widespread use of SQUID-NDE is expected in the near future.

## Collaborators

This paper is written on the basis of current experimental results by the following collaborators.

S. Nakayama, M. Ikeda, T. Morooka and K. Chinone
SII NanoTechnology Inc.
1-8 Nakase, Mihama-ku, Chiba 261-8507, JAPAN

S. Takagi, and S. Tosaka
R & D Center, Tohoku Electric Power Co., Inc.
2-1 Nakayama 7-chome, Aoba-ku, Sendai, 981-0952 JAPAN

N. Kasai
National Institute of Advanced Industrial Science and Technology,
Tsukuba Central 2, Tsukuba, Ibaraki 305-8568, JAPAN

K. Ikeuchi, T. Uefuji, and K. Yamada
Institute for Materials Research, Tohoku University
Katahira 2-1-1, Aoba-ku, Sendai, 980-8577, JAPAN

## Acknowledgement

The author is grateful to Prof. Kokichi Sugihara and Dr. Kohei Murotani of the University of Tokyo for their coding of the surface reconstruction program for Bezier and B-spline interpolation based on geometrical theory. The author also indebted to the staff at the electric power substation for their kind help in the SQUID noise spectra measurements.

## References


[1] For example, Weinstok, H. *SQUID Sensors: Fundamental, Fabrication and Application*; Kluwer Academic: Dordrecht, 1996, pp 82-84.

[2] Wikswo, J. P. Jr. *IEEE Trans. Appl. Supercond.* 1995 , vol.5 no.2, 74-120.

[3] For example, Kirtley, J. R.; Tsuei, C. C.; Sun, J. Z.; Chi, C. C. *Nature.* 1995, 373, 225-228.

[4] Morooka, T.; Nakayama, S.; Odawara, A.; Chinonne, A. *Jpn. J. Appl. Phys.* 1999, 38, L119-L122.

[5] Hasegawa, T.; Kageyama, T.; Fukumura, T.; Okazaki, N.; Kawasaki, M.; Koinima, H.; Yoo, Y. K.; Duewer, F.; Xiang, X. D. *Appl. Surf. Sci.* 2002, 189, 210-215.; Murakami, M.; Masumoto, Y.; Nagano, M.; Hasegawa, T.; Kawasaki, M.; Koinuma, H. *Appl. Surf. Sci.* 2004, 223, 245-248.

[6] Iguchi, I.; Yamaguchi, T.; Sugimoto, A. *Nature.* 2001, 412, 420.; Sugimoto, A.; Iguchi, I.; Miyake, T.; Sato, H. *Jpn. J. Appl. Phys.* 2002, 41, L497-L499.

[7] Isawa, K.; Nagano, M.; Ikeuchi, K.; Yamada, K. *Physica.* 2003, B329-333, 783-784.

[8] Kasai, J.; Togawa, Y.; Sasagawa, T.; Shimoyama, J.; Kishio, K.; Koinuma, Y.; Hasegawa, T. *Jpn. J. Appl. Phys.* 2001, 40, 1359-1361; Kasai, J.; Okazaki, N.; Togawa, Y.; Sasagawa, T.; Shimoyama, J.; Kishio, K.; Koinuma, H.; Hasegawa, T. *Appl. Phys.* 2001, A72, S263-S265.

[9] Ikeuchi, K.; Uefuji, T.; Fujita, M.; Ymada, K. *Physica B.* 2003, 329-333, 1497-1480; Fujita, M.; Yamada, K.; Hiraka, H.; Gehring, P. M.; Lee, S.-H.; Wakimoto, S.; Shirane, G. *Phys. Rev. B.* 2002, 65, 064505,1-6.

[10] Kasai, J.; Okazaki, N.; Hasegawa, T.; Togawa, Y.; Sasagawa, T.; Shimoyama, J.; Kishio, K.; Ohtani, M.; Fukumura, T.; Kawasaki, M.; Koinuma, H. *Jpn. J. Appl. Phys.* 2004, 43, L1327-L1329.

[11] For example, Shirane, G.; Birgeneau, R. J.; Endoh, Y.; Kantner, M. A. *Proc. LT-20.* 1993. and their references in.

[12] Yamada, K.; Kurahashi, K.; Uefuji, T.; Fujita, M.; Park, S.; Lee, S. H.; Endoh, Y. *Phys. Rev. Lett.* 2003, 90, 137004,1-4.

[13] Alff, L.; Krockenberger, Y.; Welter, B.; Schonnecke, M.; Gross, R.; Manske, D.; Naito, M. *Nature.* 2003, 422, 698-701.

[14] Kurahashi, K.; Matsushita, H.; Fujita, M.; Yamada, K. J. Phys. Soc. Jpn. 2002, 71, 910-915; Uefuji, T.; Kubo, T.; Yamada, K.; Fijita, M.; Kurahashi, K.; Watanabe, I.; Nagamine, K. *Physica C.* 2001, 357-360, 208-211.

[15] Isawa, K.; Uefuji, T.; Yamada, K. *Physica C.* 2005, 426-431, 202-207.

[16] Zhao, X.; Kasai, J.; Okazaki, N.; Nakayama, Y.; Togawa, Y.; Shimoyama, J.; Kishio, K.; Kitazawa, K.; Fukumura, T.; Kawasaki, M.; Koinuma, H.; Hasegawa, T. *Supercond. Sci. Technol.* 2001, 14, 1124-1127.

[17] Mang, K.; Larochelle, S.; Mehta, A.; Vajk, O. P.; Erickson, A. S.; Lu, L.; Buyers, W. J. L.; Marshall, A. F.; Prokes, K.; Greven, M. *Phys. Rev.* 2004, B70, 094507-094522.

[18] Kasai, N.; Chinonne, K.; Nakayama, S.; Odawara, A.; Yamakawa, H.; Ishiyama, N. *Jpn. J. Appl. Phys.* 1998, 37, 5965-5970.

[19] Tomiya, T.; Matsuda, N.; Shinada, K.; Arakawa, A.; Yamada, Y.; Yoshida, Y. Simazu. 1998, 55, 53-60 (in Japanese).

[20] Nagaya, S.; Hirano, N.; Shintomi, K.; Yahara, Y.; Sakano, M.; Matsui, M.; Matsuda, H. *Advances in Superconductivity* VII; Springer-Verlag: Tokyo, 1995, pp 1279-1282.

[21] For example, Takahashi, S.; Echigoya, J.; Ueda, T.; Li, X.; Hatafuku, H. J. *Mat. Processing Tech.* 2001, 108, 213-216.

[22] Isawa, K.; Sato, F.; Takagi, S.; Tosaka, S. *Proc. of Tohoku-Section Joint Convention of Institutes of Electrical and Information Engineers*. Japan. 2004, 158 (in Japanese); Sato, F.; Isawa, K.; Takagi, S.; Tosaka, S. *Proc. of Tohoku-Section Joint Convention of Institutes of Electrical and Information Engineers*. Japan. 2004,159 (in Japanese).

[23] Isawa, K.; Takagi, S.; Tosaka, S. *Inspection Engineering*. 2005, 10, 59-65 (in Japanese).

[24] Isawa, K.; Takagi, S.; Tosaka , S. *Electrical Review*.2005, 90, 72-76 (in Japanese).

[25] Morooka, T.; Tanaka, K.; Odawara, A.; Nakayama, S.; Nagata, A.; Ikeda, M.; Chinonne, K. *Physica C*. 2000, 335, 157-163.

[26] For example, Mizoguchi, T. In Magnetic and Magnetism I; Baifukan Inc: Tokyo, 1995 (in Japanese).

[27] Isawa, K.; Nakayama, S.; Ikeda, M.; Takagi, S.; Tosaka, S.; Kasai, N. *Physica C*. 2005, 432, 182-192.

[28] Isawa, K.; Takagi, S.; Tosaka, S.; Nakayama, S.; Ikeda, M.; Chinone, K. *IEEE Trans. Appl. Supercond.* 2005, vol.15, no.2, 715-718.

[29] Weinstock, H.; Nisenoff, M. "Nondestructive evaluation of metallic structure using a SQUID magnetometer", *SQUID'85-Superconducting Interference Device and Their Applications*; Hahlbohm, H. D.; Lubbing, H.; Eds.; de Gruyer: Berlin, 1985, pp 853-858.

[30] Tralshawala, N.; Claycomb, J. R.; Miller, J. H. Jr. *Appl. Phys. Lett.* 1997, vol. 71, 1573-1575.

[31] Hatsukade, Y.; Kasai, N.; Ishiyama, A. *Jpn. J. Appl. Phys.* 2001, vol. 40, L606-L608.

[32] Lee, S.-Y.; Viswanathan, V.; Huckans, J.; Matthews, J.; Wellstood, F. C. *IEEE Trans. Appl. Supercond.* 2005, vol.15, no.2, 707-710.

[33] Krause, H.-J.; Wolf, W.; Glaas, W.; Zimmermann, E.; Faley, M. I.; Sawade, G.; Neudert, G.; Gampe, U.; Krieger, J. *Insight*. 2001, vol.43, no.7, 458-461.

[34] Krause, H.-J.; Kreutzbruck, M. V. *Physica C*. 2002, 368, 70-79.

[35] Qing, T.; and Enpuku, K. *Physica C*. 2003, 392-396, 1396-1400.

[36] Itozaki, H.; Kondo, Y.; Nagaishi, T. *Physica C*. 2003, 392-396, 1392-1395.

[37] Tanaka, S.; Matsuda, K.; Yamazaki, O.; Natume, M.; Ota, H.; Mizoguchi, T. *Jpn. J. Appl. Phys.* 2001, 40, L431-L433.

[38] Tanaka, S.; Natsume, M.; Uchida, M.; Hotta, N.; Mastuda, T.; Spanut, Z.; Hatsukade, Y. *Supercond. Sci., Technol.* 2004, 17, 620-623.

[39] Collins, R.; Dover, W. D.; Bowler, J. R.; Miya K. eds. *Nondestructive testing of materials*; IOS press: Amsterdam, 1995, pp 1-124.

[40] Blitz, J. *Electrical and magnetic method of nondestructive testing*; Adam Hilger: New York, 1991.

[41] Ono, Y.; Ishiyama, A.; Kasai, N.; Chinonne, K. *Appl. Phys. Lett.* 2004, 85, 332-344.

[42] Isawa, K.; Takagi, S.; Tosaka S. *Power Electric Technique*. 2004, 43, 57-61 (in Japanese).

[43] Isawa, K.; Nakayama, S.; Morooka, T.; Ikeda, M.; Takagi, S.; Chinone, K.; Tosaka , S. *J. Cryo. Soc. Jpn.* 2004, 39, 488-494 (in Japanese).

[44] Tomlinson, J. R.; Wagg, A. R.; and Whitte, M. J. *British Journal of NDE* . 1980, 22, 119-127.

[45] Govindaraju, M. R.; Kaminski, D. A.; Devine, M. K.; Biner, S. B.; Jiles, D. C. *NDT & E Inter.* 1997, 30, 11-17.

[46] Also, type IV is categorized as the damages or cracks in the intercritical zone very near to the base metal.

[47] Isawa, K.; Nakayama, S.; Morooka, T.; Ikeda, M.; Takagi, S.; Chinone, K.; Tosaka, S. *Physica C*. 2005, 418, 1-8.

[48] Watanabe, Y.; Kang, S. H.; Chan, J. W.; Morries, J. W. Jr.; Clark, J. *Appl. Phys. Lett.* 2003, 88, 1878-1880.

[49] Isawa, K.; Takagi, S.; Tosaka, S. IERE (International Energy Research Exchange) *Newsletter*. Oct. 1st. 2004, no.34.

[50] de Waal, V. J.; van Nieuwenhuyzen, G. J.; Klapwijk, T. M. *IEEE Trans. Magn.* 1983, vol.19, 648-651.

[51] van Nieuwenhuyzen, G. J.; de Waal, V. J. *Appl. Phys. Lett.* 1985, vol.46, 439-441.

[52] Nakayama, S.; Morooka, T.; Ikeda, M.; Chinone, K.; Isawa, K.; Takagi, S.; Tosaka, S. *IEEE Trans. Appl. Supercond.* 2005, vol.15, no.2, 699-702.

[53] Nakayama, S.; Odawara, A.; Morooka, T.; Shimizu, N.; Chinone, K. T. *IEE Jpn* .1995, 115A, 1228-1235 (in Japanese).

[54] Morooka, T.; Nakayama, S.; Odawara, A.; Shimizu, N.; Chinonne, K.; Akita, T.; Kasai, N. *Jpn. J. Appl. Phys.* 1996, vol.35, L486-L488.

[55] Enpuku, K.; Minotani, T.; Kandori, A.; Shiraishi, F.; Beyer, J.; Drung, D.; Fudwig, F. *Jpn. J. Appl. Phys.* 1998, 37, 4769-4773.

[56] Beyer, J.; Drung, D.; Ludwig, F.; Minotani, T.; Enpuku, K. *Appl. Phys. Lett.* 1998, 72, 203-205; Ludwig, F.; Drung, D. *Appl. Phys. Lett.* 1999, 75, 2821-2823.

[57] Dantsker, E.; Tanaka, S.; Clarke, J. *Appl. Phys. Lett.* 1997, 70, 2037-2039.

[58] Carr, C.; Macfarlane, J. C.; Donaldson, G. B. *IEEE Trans. Appl. Supercond.* 2003, vol.13, no.2, 245-249.

[59] Kreutzbruck, M. v.; Baby, U.; Theiss, A.; Muck, M.; Heiden, C. *IEEE Trans. Appl. Supercond.* 1999, vol.9, no.2, 3805-3808.

[60] Kasai, N.; Suzuki, D.; Hatsukade, Y.; Koyanagi, M. *Jpn. J. Appl. Phys.* 2000, 39, 1399-1404.

[61] Ruosi, A.; Valentino, M.; Lopresto, V.; Caprino, G. *Composite Structures*. 2002, 56, 141-149.

[62] Isawa, K.; Nakayama, S.; Ikeda, M.; Takagi, S.; Tosaka, S. *J. Cryo. Soc. Jpn.* 2005, 40, 327-334 (in Japanese).

[63] Otaka, M.; Enomoto, K.; Hayashi, M.; Sakata, S.; Shimizu, S. *Transactions of the Japan Society of Mechanical Engineers Series* .1993, A59, 25-30 (in Japanese).

[64] Sugihara, Kokichi. In *Mathematics in Graphics*; Kyoritsu- Publishers: Tokyo,1995 (in Japanese).

[65] Farin, G. In *NURBS From projective geometry to practical use*. AK Peters; LTD; Natick, MA, USA, 1999.

[66] For example, Murotani, Kohei.; Sugihara, Kokichi. *International Journal on Shape Modeling*. 2003, Vol.9, No.1, 21-39.

[67] Sapidis, N.; Farin, G. *Computer-aided design*. 1990, 22, 121-129.

In: New Research on Superconductivity and Magnetism ISBN: 1-60021-540-8
Editor: Lannie K. Tran, pp. 143-174 

*Chapter 5*

# CHEMICAL AND MACROSCOPIC PHYSICAL PROPERTIES OF THE SODIUM COBALT OXYHYDRATE SUPERCONDUCTOR

***Hiroya Sakurai***[a, b]***, Kazunori Takada***[c]
***and Eiji Takayama-Muromachi***[a]

[a]Superconducting Materials Center, National Institute for Materials Science, Namiki 1-1, Tsukuba, Ibaraki 305-0044, Japan
[b]International Center for Young Scientists, National Institute for Materials Science, Namiki 1-1, Tsukuba, Ibaraki 305-0044, Japan
[c]Advanced Materials Laboratory, National Institute for Materials Science, Namiki 1-1, Tsukuba, Ibaraki 305-0044, Japan

## Abstract

This report reviews chemical and macroscopic physical properties of the bilayer sodium cobalt oxyhydrate (BLH) superconductor and its related compounds, focusing on the recent development of research. The exact composition of BLH is $Na_x(H3O)_zCoO2{\cdot}y'H_2O$ and it has three adjustable compositional parameters: Na content, $x$, $H_3O$ content, $z$, and $H_2O$ content, $y'$. The former two parameters greatly affect superconductivity and related properties of the system. With adjusting them, a magnetically ordered phase appears in the immediate vicinity of the superconducting phases in a phase diagram, suggesting a certain magnetic origin for superconductivity. Regarding the band structure of BLH, hole pockets near the K points look crucially important relating to the mechanism of superconductivity. Furthermore, there is a possibility that two-gap superconductivity is realized in the present system corresponding to the hole pockets near the K points and a large cylindrical Fermi surface around the Γ point. At the end of the paper, the mother oxide of $\gamma$-$Na_xCoO_2$ is discussed. Its rich physical properties originate from strong electron-electron correlation in the $CoO_2$ layer and seem to relate to the unconventional superconductivity in BLH.

# 1 Introduction

Motivated by the discovery of high $T_c$ cuprates (HTC), much effort had been made to discover new superconductors for over 15 years in other transition metal oxide systems such as cobalt and nickel oxides. Thus, the discovery of sodium cobalt oxyhydrate superconductor, $Na_xCoO_2{\cdot}yH_2O$, has given a strong impact to the condensed matter physics [1]. Furthermore, since the so-called soft-chemical method is indispensable for synthesis, control of the Co valence, modification of the distance between the $CoO_2$ layers, etc., the compound has also attracted much attention from chemists. Thanks to enormous research carried out so far, this compound has been proved to be one of key materials for deep understanding of superconductivity. In comparison with HTC, characteristics and importance of this superconductor may be summarized in three points.

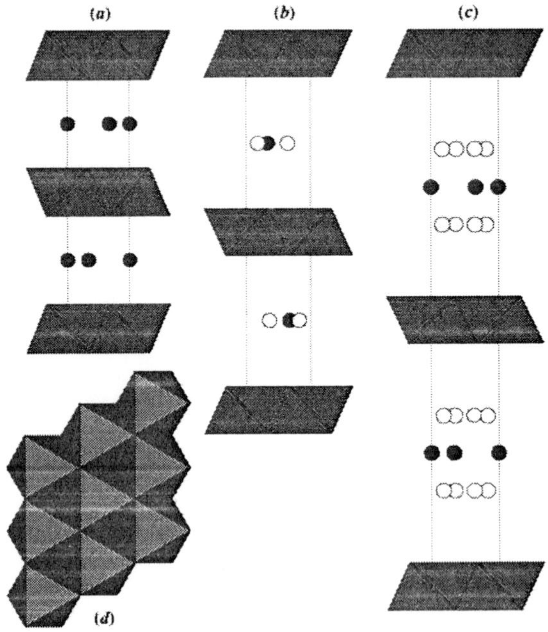


Figure 1. Structures of sodium cobalt oxides and oxyhydrates. (a) $\gamma$-$Na_xCoO_2$, (b) $Na_{0.35}CoO_2{\cdot}0.7H_2O$ (MLH), (c) $Na_{0.35}CoO_2{\cdot}1.3H_2O$ (BLH), and (d) a $CoO_2$ layer. The figure (*a*), (*b*) or (*c*) is a projection to 100 plane of each unit cell, and the figure (*d*) is a projection of the $CoO_2$ layer to 001 plane. The octahera, filled and open circles represent $CoO_6$ octahedra, Na sites, $H_2O$ sites, respectively.

First, the electronic state of the compound has a feature common to HTC. As well-known mother oxides of HTC have the antiferromagnetic ground state of the $S$ = 1/2 copper spins and superconductivity is induced by carrier (hole or electron) doping there. On the other hand, the Co ion in the superconducting oxide has an intermediate oxidation state between trivalent ($3d^6$) and tetravalent ($3d^5$) in low-spin configurations, being regarded as an electron doped $S$ = 1/2 system. In addition, the compound is unique because superconductivity, induced by chemical doping is rather rare except for HTC.

Second, the compound has a layered structure consisting of $CoO_2$ layers as in HTC with $CuO_2$ planes. The $CoO_2$ layer is separated from the others by a thick insulating triple layers of $H_2O$-Na(H3O)-$H_2O$. As shown later, the superconductivity is sensitive to the interlayer distance, which strongly suggests that two-dimensional electronic structure plays a key role in superconductivity.

Third, the Co sites in the $CoO_2$ layer form a triangular lattice as seen in Fig. 1, whereas the Cu sites in the $CuO_2$ layer of HTC is a square lattice. Since Anderson proposed a resonating-valence-bond (RVB) state related to the mechanism of the high $T_c$ superconductivity [2], superconductivity on a triangular lattice has became a central issue of condensed matter physics because the RVB state was originally proposed and investigated for a triangular lattice [3]. Though the RVB model may not be simply applicable for the cobalt system due to a heavy doping level, it played an important role in stimulating the interst among theoreticians in the compound [4-13]. Now the compound is attracting much attention due to its possible disconnected Fermi surface and multi-orbital effect [14-20].

The main concerns in this review will be the chemical aspects and macroscopic physical properties of the compound. Microscopic experiments will be reviewed in another article in this book. A review article is already available on the present compound [21], and thus, we will focus here on the recent progress which was not described in the previous review article.

## 2 Chemical Aspects Of Sodium Cobalt Oxyhydrate

Through various experiences on HTC, it has been well recognized that understanding of chemical aspects, such as structure, composition, chemical stability, etc. are very important in research of oxide superconductors. It is also the case for the sodium cobalt oxyhydrate, because superconductivity is strongly dependent on the structure and the chemical composition, and thus, on the synthesis processes.

### 2.1 Layered Cobalt Oxide

Many compounds are known to have the composition of $AMO_2$ (A: alkali metal elements, etc., and M: transition metal elements, etc.) with triangular $MO_2$ layers. The $MO_2$ layer is composed of two close-packed oxygen layers with M ions in the octahedral sites having the same structure as $CdI_2$ layer in $CdI_2$. The A ion occupies an octahedral or trigonal prismatic site depending on the size of the A ion and synthesis condition. The content of the A ion can be less than unity with a formula of $A_xMO_2$ ($x < 1$), and thereby, fast ion mobility is expected when A is an alkali metal element. Indeed, $Li_xCoO_2$ is a representative rechargeable battery material.

$Na_xCoO_2$ crystallizes in three different structures depending on $x$ and synthesis conditions: $\alpha$-, $\beta$-, and $\gamma$-type structures shown in Fig. 2. If A, B, and C are used as notations of two-dimensional relative positions of the triangular oxygen lattice while a, b, and c for the Na and Co lattices, the unit staking sequence of the $\alpha$-phase is represented as ABCABCABC for the oxygen atoms with a manner of cubic closest packing, and A-c(Co)-B-a(Na)-C-b(Co)-A-c(Na)-B-a(Co)-C-b(Na)-A-c(Na)-B-a(Co)-C, adding the Na and Co atoms (i.e., an ordered NaCl structure or $\alpha$–NaFeO2 structure), where the Na atoms occupy the octahedral site. In the same manner, the unit stacking sequences of $\beta$- and $\gamma$-phases are A-c(Co)-B-c/a(Na)-B-a(Co)-C-a/b(Na)-C-b(Co)-A and A-c(Co)-B-c/a(Na)-B-c(Co)-A, respectively ("/" means and). These structures are sometimes classified as O3 ($\alpha$), P3 ($\beta$), and P2 ($\gamma$) with the first letter standing for the coordination type of the Na site (O: octahedral, P: prismatic) and the following number standing for the number of $CoO_2$ layers in the unit cell [22].

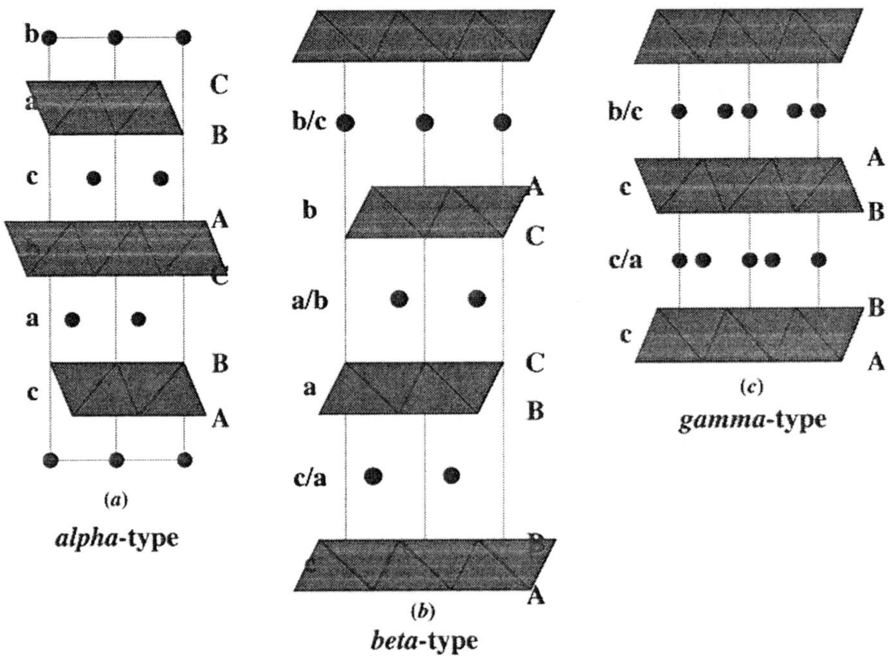


Figure 2. Structures of $\alpha$- (a), $\beta$- (b), and $\gamma$-$Na_xCoO_2$ (c). These are projections to 100 planes of $2a \times 2b \times c$. The octahedra and filled circles represent $CoO_6$ octahedra and Na sites, respectively.

## 2.2 Structure of the Sodium Cobalt Oxyhydrates

The sodium cobalt oxyhydrates are synthesized starting from the aforementioned sodium cobalt oxides through the soft-chemical processes with final structures depending on the structure of starting oxide and the water content. Crystal structures of the oxyhydrates derived from $\gamma$-$Na_{0.7}CoO_2$ are shown in Fig. 1(b) and 1(c) [1, 23]. Bilayered hydrate (BLH) is obtained when the water content is large ($y \sim 1.3$ in $Na_xCoO_2 \cdot yH_2O$), whereas monolayered hydrate (MLH) when small ($y \sim 0.7$). These $y$ values may reflect $4x$ and $2x$ [24, 25] although they include the contents of oxonium ions ($H_3O^+$) at Na site as described later. The compounds with smaller $H_2O$ contents of $y = 0.3$ and 0.1 were also reported [25-28], but they are not obtained at room temperature [29] and their detailed structures have not been elucidated yet. The superconductivity is reported only for BLH among these phases. Both BLH and MLH phases have P2-type structures inheriting them from the starting oxide.

In the BLH phase, the distance between the adjacent $CoO_2$ layers are greatly enlarged compared with the mother compound of $Na_{0.7}CoO_2$. Two-dimensional character should also be greatly enhanced in the electronic structure of BLH, which is widely believed to be important for inducing superconductivity. However, enhancement of the two-dimensional character alone does not seem to account well for the appearance of superconductivity in BLH because the structure of non-superconducting MLH also looks to be enough two-dimensional [30]. Two other possible factors which should be taken into account may be screening of random potential and squashing of the $CoO_2$ layer. The Na ions can cause a random Coulomb potential on $CoO_2$ layers. As shown in Fig. 1, the Na layer is sandwiched by the water layers in BLH, and thus, the electric polarization of the $H_2O$ molecules can effectively shield the random potential of the Na ions. On the contrary, water molecules are located on the same plane as the Na atoms in MLH with less effective shielding.

The $CoO_2$ layer becomes thinner with increasing the $y$ value. The thickness of the $CoO_2$ layers in non-hydrated $\gamma$-$Na_xCoO_2$ has been reported $d$ =1.93 Å ($x$ = 0.40) [31], 1.97 Å ($x$ = $0.67 \leq x \leq 0.75$) [32, 33], and 1.96 Å ($x$ = 0.84) [34], at room temperature. On the other hand, those reported for MLH and BLH are $d$ = 1.84 Å [23], and 1.78 Å [1] or 1.80 Å [31], respectively. The distance between two opposite triangle planes of a regular octahedron with the side length of $a$ may be calculated as $l = \sqrt{2/3}a$, and a distortion factor can be defined as $\alpha = d/l = \sqrt{3/2} \cdot d/a$ ($a$: lattice parameter). The $\alpha$ values of the non-hydrates is 0.839 ($x$ = 0.40), 0.851 to 0.852 ($x = 0.67 \leq x \leq 0.75$), and 0.846 to 0.847 ($x$ = 0.84), whereas those of MLH and BLH are 0.795 and 0.768 or 0.781, respectively. These values mean that the octahedra in the non-hydrates, MLH and BLH shrink, compared with an ideal one, to about 85%, 80% and 77% along the $c$-axes, respectively. The trigonal distortion influences the crystal field splitting of the $d$-orbitals and the electronic band structure, which will be discussed later. (It may be better to point out here that deuterated BLH phases have been reported to have thicker $CoO_2$ layers with $d$ = 1.84 Å or 1.87 Å ($\alpha$ = 0.800 or 0.813) [24, 35] in spite of the fact that they showed superconductivity.)

Arrangement of the $H_2O$ molecules in BLH is very interesting by itself related to two-dimensional ice. Jorgensen et al. proposed an arrangement of water where the oxygen atoms were located on a single plane [24], while Lynn et al. proposed step-like structure for the water layer which was similar to local structure in three-dimensional ice [35]. However, both groups did not take into account the presence of the $H_3O^+$ ions assuming wrong chemical compositions, and reexamination seems to be needed for both models. The structure refinement with the $H_3O^+$ ions taken into account has been carried out for a synchrotron X-ray powder diffraction pattern [31], and it has been proved that the water layer has a step-like structure. But, the step is very small and the local structure is quite different from that of three-dimensional ice. Consistent with this result, X-ray absorption spectroscopy suggests that hydrogen bonding network in BLH is different from that in the three-dimensional ice [36].

BLH and MLH oxyhydrates have been synthesized also from $\alpha$- and $\beta$-$Na_xCoO_2$ [37-41]. Superconductivity was observed only in BLH prepared from the $\alpha$-phase. Both BLH and MLH derived from the $\alpha$-phase have P3-type structures [37, 38], in spite of the O3-type structure of the $\alpha$-phase. This is caused by sliding of the $CoO_2$ layer relative to the adjacent $CoO_2$ layers during the hydration process. However, any type of relative sliding for an O3-type starting oxide does not bring about a P2-type hydrate, and thus, $\alpha$- and $\gamma$-phases inevitably produce different hydrates. The structures of BLH and MLH prepared from the $\beta$-phase have not been refined yet. In this article, P2-type BLH and MLH prepared from the $\gamma$-phase are mainly discussed because they have been most widely and deeply studied.

## 2.3 Synthesis of Sodium Cobalt Oxyhydrates

The P2-type BLH phase is usually synthesized from $\gamma$-$Na_{0.7}CoO_2$ by partial deintercalation of the Na ions followed by water intercalation. The Na deintercalation is performed by a chemical method or electrochemical method [42]. In the former method, $Br_2$ [1], $KMnO_4$ [43], $NaMnO_4$ [44], or $Na_2S_2O_8$ [45] has been used as an oxidizing agent. The $Br_2$ oxidation in acetonirile ($CH_3CN$) has been adopted most frequently, and in its typical procedure, $Na_{0.7}CoO_2$ is immersed in $Br_2/CH_3CN$ for several days to obtain $Na_xCoO_2$ ($x \sim 0.4$) [31]. The $x$ value of the filtrated sample depends on the concentration of $Br_2$ in $CH_3CN$ as a function of redox potential of $Br_2/Br^-$. However, it should be stressed that the $x$ value just after the oxidation process is not equal to the final value in the oxyhydrate because it decreases further in the hydration process. Moreover, as discussed later, cobalt oxidation state is not only determined by the $x$ value. Due to this complexity, relationships of $x$ vs. $Br_2$ concentration [46] or $x$ vs. $T_c$ [28, 46] reported earlier needs to be reconsidered.

For the water intercalation, three different methods have been adopted: washing of a sample by water [1], immersing of a sample in water [31], and exposing of a sample to water vapor [29, 47]. The second method is suitable to know details of the reaction occurring in water. Table 1 shows the $x$ value and Co valence, $s$, obtained by chemical analyses after each synthesis step [31]. This table clearly indicates that both the $x$ value and the Co oxidation state decrease during the hydration process. Thus, the reaction in water is not the simple intercalation of water molecules. It has been elucidated that two side reactions occur in water besides the water intercalation, (i) ion exchange between the $Na^+$ ions and the $H_3O^+$ ions and (ii) reduction by water with evolution of $O_2$ which has been known to occur in Mn and Ni oxides [48-51]. The formulas of the two side reactions are as follows:

$$Na_aCoO_2 + bH_2O \rightarrow Na_{a-b/2}(H_3O)_{b/2}CoO_2 + b/2Na^+ + b/2OH^- \quad (A)$$

$$Na_aCoO_2 + bH_2O \rightarrow Na_a(H_3O)_{2b/3}CoO_2 + b/6O_2 \quad (B)$$

Consistent with the reaction (A), pH of the water after sample immersion becomes approximately 11. Moreover, the existence of the $H_3O^+$ ions has been directly proved by Raman spectroscopy, which showed characteristic vibration modes of the $H_3O^+$ ions and qualitative correlation between the content of the $H_3O^+$ ions determined by the chemical analyses and the shift of the Raman peak [31]. Taking into account the $H_3O^+$ ions, the exact chemical formula of the hydrate should be $Na_x(H_3O)_zCoO_2 \cdot y'H_2O$, rather than $Na_xCoO_2 \cdot yH_2O$ and the overall reaction in water in the typical condition is,

$$Na_{0.406}CoO_2 + (0.384+1.10)\,H_2O \rightarrow$$
$$Na_{0.337}(H_3O)_{0.233}CoO_2 \cdot 1.10H_2O + 0.069\,NaOH + 0.041\,O_2.$$

The importance of the existence of the $H_3O^+$ ions should be stressed. First, the cobalt valence cannot be determined only from the Na content $x$, but the $H_3O^+$ content should be known for that. Second, the $H_3O^+$ content substantially influences superconductivity even when Co valence is kept constant (see below). Low Co valence was confirmed by several groups for the P2-type BLH [47, 52-54], as well as for the P3-type BLH [38]. The reactions A and B will occur, not only during the immersion in water, but also during washing by water or exposing to water vapor.

Table 1. Results of chemical analyses [31]. Na contents were determined by inductively coupled plasma atomic emission spectroscopy (ICP-AES). The chemical analyses were performed for two batches of samples for each kind. The calculated Co valences are estimated from the $x$ values assuming that the nominal compositions are true, and the analyzed Co valences were determined by redox titration.

| | **Na content, $x$** | **Co valence, $s$ (calculated)** | **Co valence, $s$ (analyzed)** |
|---|---|---|---|
| the precursor of $\gamma$-$Na_xCoO_2$ | 0.695 | +3.305 | +3.30 |
| | 0.701 | +3.299 | +3.29 |
| $Na_xCoO_2$ just after the immersion in $Br_2/CH_3CN$ | 0.406 | +3.594 | +3.56 |
| | 0.402 | +3.598 | +3.57 |
| $Na_xCoO_2 \cdot yH_2O$ (BLH) | 0.337 | +3.663 | +3.43 |
| | 0.349 | +3.651 | +3.42 |

MLH is synthesized from BLH by controlling the temperature and the humidity in the air based on a reversible reaction between BLH and MLH [23, 26, 47]. At room temperature, the lowest relative humidity where BLH is stable is about 43%, and MLH is obtained below 33% through a dehydration process [29]. From hydration experiments for a starting oxide of $Na_{0.3}CoO_2$, it was reported that BLH and MLH coexisted in a humidity condition between 52% and 75%. According to Gibbs' phase rule ($f = c - p + 2$) and assuming two components system ($c = 2$ with water and remaining part of the oxide) and three phase coexistence ($p = 3$ with BLH, MLH and the gas phase), number of freedom becomes unity ($f = 1$). Thus, if the freedom is used to fix temperature, humidity should be decided unequivocally. The aforementioned variable range of humidity for the BLH-MLH coexistence may suggest that complex side reactions occur in the hydration/dehydration process through the gas phase as well as that through the liquid phase.

Li- and K-containing oxyhydrates, $Li_x(H_3O)_zCoO_2 \cdot y'H_2O$ and $K_x(H_3O)_zCoO_2 \cdot y'H_2O$, have also been reported [55-58]. However, superconductivity is not observed in them. In the case of the Li system, BLH with $y' + z = 2$ has a distorted structure of monoclinic system. In the K system, BLH has not been prepared.

## 2.4 Control of the Composition of Sodium Cobalt Oxyhydrate

The sodium cobalt oxyhydrate has three adjustable compositional parameters: $x$, $y'$ and $z$ in $Na_x(H_3O)_zCoO_2 \cdot y'H_2O$. The $y'$ value is controlled by the humidity of the air in storage. The $x$ value can vary by the reaction A, while the $z$ value by both reactions A and B. The reaction A is reversible ion exchange, and the equilibrium state is determined, in principle, by pH. The reaction is, of course, promoted in acid environment. On the other hand, the reaction B is irreversible reduction generating oxygen gas. The reaction is affected by the redox potential of water which is a function of pH [59]. Since half reaction formula of the reaction may be represented as $4OH^- \rightarrow 2H_2O + O_2 + 4e^-$, the reaction is considered to be promoted in alkaline environment. However, the situation is not so simple because the reaction is also affected by the redox potential of the hydrate (more precisely, by the difference between the redox potentials of water and the hydrate). At present, redox potential of the hydrate has not been known as a function of $x$ and $z$, hence as a function of pH although the synthesis experiments described later suggest that there is an optimum pH or optimum process where the reaction B proceeds most thoroughly. In the following paragraph, we will discuss details of sample preparation processes and resulting compositions in three concrete cases according to recent reports [60-62].

(1) *Synthesis 1.* 1g precursor of $Na_{0.7}CoO_2$ was immersed in $Br_2/CH_3CN$ for 5 days at room temperature to deintercalate Na ions, and the sample filtrated was immersed in 400 ml distilled water for 3 days. Then, $v_{HCl}$ ml of 0.1 M HCl aqueous solution was added to the water. The samples were filtrated and dried for 3 days in the air [60]. After the immersion in distilled water, pH of the aqueous solution became approximately 11 due to the reaction A, then pH decreased by the addition of HCl depending on $v_{HCl}$ as shown in Fig. 3(a). This titration curve deviates significantly from one calculated assuming that ion exchange reaction did not proceed during the process of the HCl addition, indicating that the ion exchange is promoted by HCl. Indeed, the $x$ value determined by chemical analysis decreases with increasing $v_{HCl}$ as shown in Fig. 3(b). The Co valence, $s$, also decreases for $v_{HCl} > 10$ ml. However, in the case of $v_{HCl} > 10$ ml, a secondary phase of CoOOH with trivalent Co ions or its related phase appeared in powder X-ray diffraction (XRD) patterns, in contrast to that single-phase samples were obtained for $v_{HCl} \leq 10$ ml, as seen in Fig. 4(a). This suggests that additional reactions such as partial dissolution of the sample occurred in the condition of $v_{HCl} > 10$ ml. For $v_{HCl} \leq 10$ ml, on the contrary, the $s$ value is almost constant to be +3.40 to +3.41. This may be consistent in that the reaction B is irreversible and does not proceed under acid circumstances.

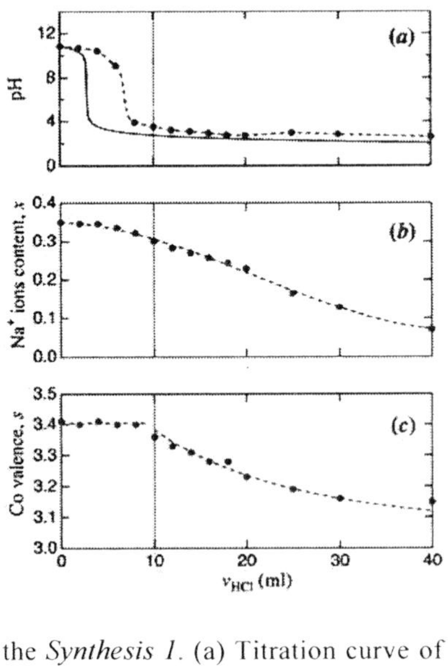


Figure 3. Chemical analyses for the *Synthesis 1.* (a) Titration curve of the solution. The solid line is a calculated curve assuming no further ion exchange. (b) $Na^+$ ions content, $x$, estimated by ICP-AES. (c) Co valence, $s$, determined by redox titration. The horizontal axis represents the volume of 0.1 M HCl aqueous solution added to the water. All the broken lines are gudes to eyes, and the dotted lines represent the border of single phase region of $v_{HCl} \leq 10$ ml.

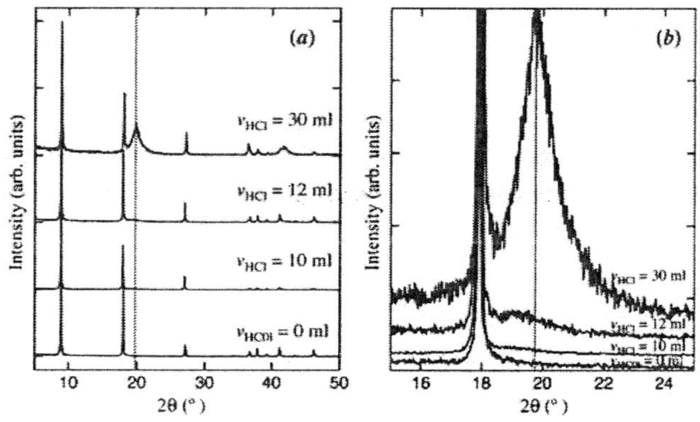


Figure 4. (a) X-ray diffraction patterns of the samples obtained by the *Synthesis 1*, and (b) their expansions. The $\nu_{HCl}$ value is the volume of 0.1 M HCl aqueous solution added to the water. The dotted line indicates the 002 peak position of the secondary phase of CoOOH.

(2) *Synthesis 2.* After the same $Br_2$ oxidation procedure for 1g precursor of $Na_{0.7}CoO_2$, the sample was immersed for 5 days in 400 ml HCl or NaOH aqueous solution with a certain pH to control the redox potential of water [61]. As shown in XRD patterns in Fig. 5(a), the single-phase BLH samples were obtained for $3 \leq pH_i \leq 11$ where $pH_i$ stands for initial pH of the aqueous solution. At $pH_i = 1$ or 14, almost single-phase $Na_xCoO_2$ with $c \sim 11$ Å was synthesized and at $pH_i = 2$, 12 and 13, both BLH and the $Na_xCoO_2$ were seen in XRD patterns, as seen in Fig. 5(a). It is noteworthy that the final pH ($pH_f$) which was measured after the immersion, was almost constant (~11) independent of $pH_i$ across the single phase region of BLH (i.e., for $3 \leq pH_i \leq 11$), which indicates that single-phase BLH is obtainable only when $pH_f \approx 11$. The $x$ value increases slightly with increasing $pH_i$ in the single phase region as seen in Fig. 5(b), because the reaction A is suppressed with increasing $pH_i$. However, no clear $pH_i$ dependence of $s$ was detectable. On the other hand, the samples obtained at $pH_i = 1$ and 14 are identified as $Na_{0.4}CoO_2$ and $Na_{0.5}CoO_2$, respectively [63]. It is interesting that the ion exchange A does not occur significantly for the nonhydrate even in a strong acid environment. In addition, Co valence of the latter phase was larger than that of BLH.

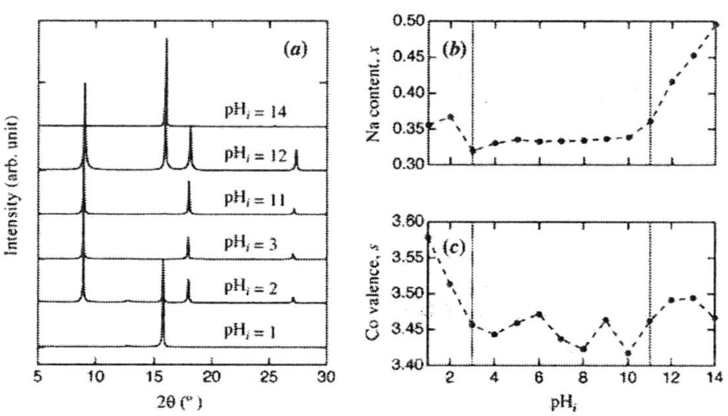


Figure 5. (a) X-ray diffraction patterns of the samples obtained by the *Synthesis 2*. (b) Na content, $x$, and (c) Co valence, $s$, of the samples. The $pH_i$ represents the pH of the solution before the immersion.

(3) *Synthesis 3.* 1g precursor of $Na_{0.7}CoO_2$ is immersed in $Br_2/CH_3CN$ for 5 days, at room temperature, and the samples filtrated were immersed in 400 ml distilled water for 3 days. Then, $v_{NaOH}$ ml of 0.1 M NaOH aqueous solution was added to the water (instead of $v_{NaOH}$ = 100 and 500 ml, 10 ml and 50 ml of 1 M NaOH solution were used, respectively), and the samples were filtrated and dried 3 more days after the addition [62]. The $v_{NaOH}$ = 500 ml sample contained a small amount of the secondary phase of $Na_xCoO_2$ as seen in Fig. 6(c), and further addition of NaOH made the impurity content larger.

The $x$ value determined by the chemical analysis increased with increasing $v_{NaOH}$ as shown in Fig. 6(a). On the other hand, Co valence was almost constant to be $s$ = +3.47 to +3.50 in spite of the addition of alkaline NaOH solution. The redox potential of BLH seems to depend on $x$ and $z$ which is changed by the reaction A, and a more alkaline environment does not always promote the reaction B. (The compositional difference between the $v_{NaOH}$ = 0 ml and $v_{HCl}$ = 0 ml samples is unclear at the present.)

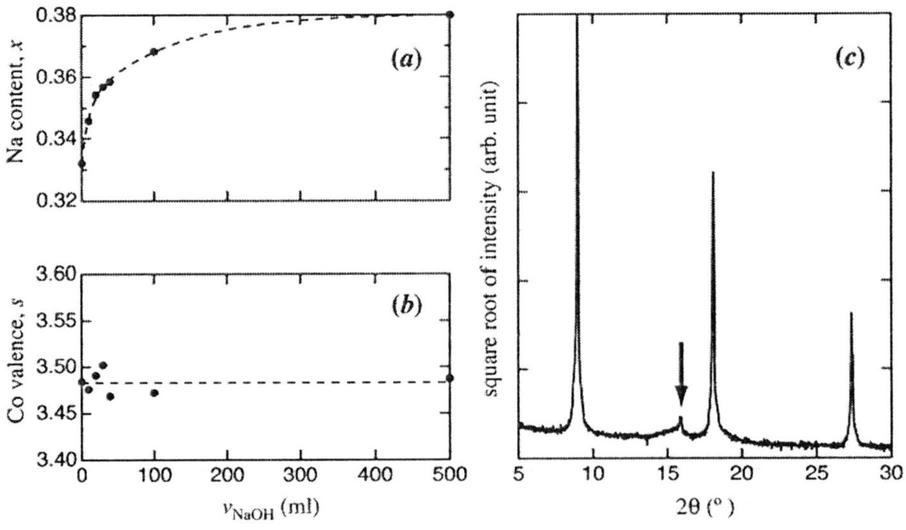


Figure 6. (a) Na content, $x$, and (b) Co valence, $s$, of the sample obtained by the *Synthesis 3*. The broken lines are guides to eyes. (c) X-ray diffraction pattern of the $v_{NaOH}$ = 500 ml sample. The arrows indicate the peak due to the secondary phase of $Na_xCoO_2$.

## 3 Physical Aspects Of Sodium Cobalt Oxyhydrate

Here, we will discuss macroscopic physical properties, such as magnetization and specific heat, as well as electronic structure and superconducting phase diagram. Microscopic tools, such as nuclear magnetic resonance (NMR), nuclear quadrople resonance (NQR), and muon spin relaxation/rotation ($\mu$SR), are sometimes quite helpful to elucidate the mechanism of superconductivity. Nevertheless, information given by the macroscopic experiments is indispensable to understand physics of superconducting and normal states of the system.

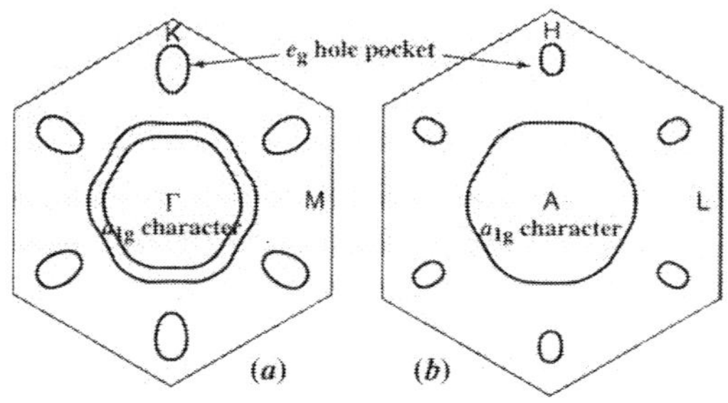


Figure 7. Fermi surface at $k_z$ =0 (*a*) and 0.5 (*b*) planes calculated by Singh [64]. The fine lines represent Brillouin zone boundary.

## 3.1 Electronic Structure

Before the discovery of the sodium cobalt oxyhydrate superconductor, Singh had calculated the band structure of $Na_{0.5}CoO_2$ by local density approximation (LDA) as shown in Fig. 7 [64]. According to it, Co $d$ – O $p$ hybridization is weak, and the Co $d$ bands are clearly split by the crystal field into $t_{2g}$ and $e_g$ manifolds, whose widths are 1.6 eV and 1.2 eV, respectively, with an energy gap of approximately 2.5 eV. Since the Fermi level is located at the former manifold, the compound can be regarded as a low-spin system. The $t_{2g}$ manifold is further split into two $a_{1g}$ and four $e_g$ bands by the trigonal distortion mentioned above ("$e_g$" is used in this sense hereafter). The large cylindrical Fermi surface around the Γ-A line has dominant $a_{1g}$ character, while the small hole-like pockets near the K points have more $e_g$ character.

Some plausible theories reported are based on the existence of the hole pockets [14-20]. The holes in the pockets are expected to have large ferromagnetic or nearly ferromagnetic fluctuations, which can be an origin of the superconductivity. In addition, the disconnected Fermi surface topology does not significantly reduce the superconducting condensation energy even in the case of higher superconducting gap symmetry such as *f*-wave. This is because line nodes can be located between the hole pockets so that the nodes affect only passive $a_{1g}$ bands. The hole pockets can play important roles for unconventional superconductivity, and thus, their existence should be checked rigorously.

As pointed out by Singh [64], the size of the hole pocket is very sensitive to the thickness of the $CoO_2$ layer. As the layer thins, the pocket becomes large, because the larger trigonal distortion pushes up the $e_g$ bands [65, 66]. Singh assumed that the thickness was 1.82 Å, which is much smaller than the real thickness in $Na_xCoO_2$ but is comparable to that in MLH and BLH as mentioned above. Since the highest energy of the band related to the hole pocket is lower than that of the band related to the cylindrical Fermi surface, the pockets are expected to appear only when sufficient number of holes are introduced: $x <\sim 0.6$ ($s >\sim +3.4$) by rough estimation from the Singh's result [64]. However, this simple rigid-band picture may be altered, especially in consideration of on-site Coulomb interaction $U$ [67-70]. The interaction greatly diminishes the size of the hole pocket and it disappears even for $x$ = 0.3 ($s$ = +3.7) when $U$ has a moderately strong value ($U \sim 4$ eV). Though dynamical effect tends to stabilize the hole pocket, its degree is limited [71]. Thus, the major factors which enlarge the size of the pocket are (i) thinner CoO2 layer, (ii) higher Co valence, and (iii) smaller $U$.

Fermi surface is directly observed by angle-resolved photoemission spectroscopy (ARPES). ARPES studies on the mother oxides have clearly shown no hole pockets at Fermi level for $Na_{0.7}CoO_2$ [72], $Na_{0.6}CoO_2$ [73], and even for "$Na_{0.3}CoO_2$" although its size of the Fermi surface suggested $x \sim 0.4$ rather than $x = 0.3$ [74, 75]. The $U$ is estimated to be ~ 4 to 5 eV from the ARPES experiments [72, 75], which is consistent with soft X-ray absorption experiment [76]. These results seem to be consistent with the LDA + $U$ calculations, rather than the LDA calculation. However, based on a tight-binding investigation, Johannes et al. have pointed out that the hole pocket should be preserved even for $x = 0.7$ in consideration of three-dimensionality, and that the absence of the hole pocket in ARPES could be due to surface effects [77]. ARPES experiments and optical property measurements using $Na_xCoO_2$ samples with various $x$ values suggest another reason for the absence of the hole pockets [74, 78]. At present, it is difficult to conclude only from the ARPES results that the hole pockets are absent near the K-points. Thus, it seems worth noting as another evidence for the absence of the hole pocket that small temperature dependence of magnetic susceptibility for $x = 0.35$ [79] is inconsistent with the existence of the hole pocket [66].

The hole pockets are likely to be absent in non-hydrated $Na_xCoO_2$, but it does not mean that BLH and MLH are also free from them because their $CoO_2$ layers are much thinner than those of the non-hydrates ($\alpha_{BLH}/\alpha_{non\text{-}hydrate} \sim 0.9$). Indeed, the presence of the hole pocket has been suggested for BLH by photoemission spectroscopy (PES) [36] and other experiments. Relating to this, it is to be recalled that superconductivity in BLH is very sensitive to electrical field gradient at the Co site which can reflect the thickness of the $CoO_2$ layer [80]. Experimental and theoretical studies for BLH at low temperature are highly desirable to elucidate the Fermi surface topology. However, the experimental data are very limited due to difficulties in sample preparation, especially of single crystals and chemical instability of BLH under vacuum condition. From PES, Mott-Hubbard type electronic structure has been proposed for BLH [81], in contrast to Charge-Transfer type one for the non-hydrate [76], although this contrast may depend on the interpretation of the data. In addition, pseudo-gap behavior was observed only in BLH [82]. These suggest that the band structure of BLH is substantially different from that of the non-hydrate.

The enlargement of the distance between the $CoO_2$ layers is a conspicuous effect of the hydration. However, according to band calculations [83, 84], this itself does not influence the band structure so substantially, although the calculations were performed assuming $s = +3.67$, which is much higher than the real value for BLH. The effects of the enlargement for the band structure are less marked dispersion along the $k_z$-direction and degeneration of the bands around the Γ-point. Recently, it has been suggested that the dispersion along $k_z$ regarding the central Fermi surface is more pronounced in MLH than BLH, which may be related to the absence of superconductivity in MLH [85].

The presence of the $e_g$ hole pockets may enhance the superconducting pairing comparing to the situation with only the central Fermi surface because the pockets have a much higher density of state (DOS) [84] and cause strong magnetic fluctuations [17, 65, 86]. That is the reason why many theoretical works have been done presupposing the hole pockets [14-20]. Of course, there are theories focusing on the central Fermi surface, though some of them have assumed a positive sign of hopping $t$ in spite of the negative one expected experimentally [72, 75]. In order to elucidate the Fermi surface topology, it seems necessary to refer not only to ARPES, but other various indirect tools, some of which are discussed below.

## 3.2 Phase Diagram

Several phase diagrams have been proposed for BLH [28, 46, 52, 60-62] but they are inconsistent with each other. Among three adjustable compositional parameters of $x$, $y'$ and $z$ in BLH, the water content $y'$ does not vary so greatly depending on the synthesis condition and its effect on the physical properties has not been clarified yet. Thus, we focus here on $x$ and $z$ dependences of the properties. Schaak et al. proposed the dome-shape phase diagram regarding $T_c$ and the Na content $x$ with the superconducting range of $0.25 \leq x \leq 0.33$, while Chen et al. reported almost constant $T_c$ in this range of $x$. A crucial problem is that both Schaak et al. and Chen et al. did not take into account the oxonium ion. Milne et al. considered the oxonium ion and reported that superconductivity appeared regarding the Co valence $s$ in the range of $\sim +3.2 \leq s \leq +3.45$ with the optimal region of $+3.24 \leq s \leq +3.35$ which gave $T_c$ ranging from 4.3 to 4.8 K. However, their optimal $s$ value is significantly lower than that reported by others [31, 47, 53, 54]. Furthermore, their two-dimensional phase diagram is insufficient because $T_c$ depends not only on $s$ but also $x$. Hereafter, we will discuss the phase diagram concretely based on the recent reports [60-62].

(1) *Phase diagram 1.* This part corresponds to *synthesis 1* in the preceding section with samples synthesized by the addition of $v_{HCl}$ ml 0.1 M HCl after the immersion in distilled water [60]. As seen in Fig. 8(a), so long as the single phase region ($v_{HCl} \leq 10$ ml) was concerned, superconductivity was observed for $v_{HCl} = 0$, 8 and 10 ml, but suppressed completely for intermediate range of $v_{HCl} = 2$, 4 and 6 ml. Including the multi-phase region, superconductivity appeared in three separated parts of $SC1$, $SC2$ and $SC3$ (see Fig.8(b)), which had the lattice parameters of $c < \sim 19.72$ Å ($SC1$ and $SC3$) and $c > \sim 19.79$ Å. Non-superconducting samples derived from $v_{HCl} = 2$, 4, or 6 ml showed a magnetic anomaly around 6 K as seen in temperature dependence of magnetic susceptibility and its derivative (Fig. 9), indicating a certain magnetic ordering [80]. For the single phase region of $v_{HCl} \leq 10$ ml, the cobalt valence was almost constant to be $s = +3.40$ to $+3.41$ while $x$ varies as shown in Fig. 3(b), and thus, an $x$–$T$ two-dimensional section of the $x$–$s$–$T$ phase diagram is depicted as shown in Fig. 10(a).

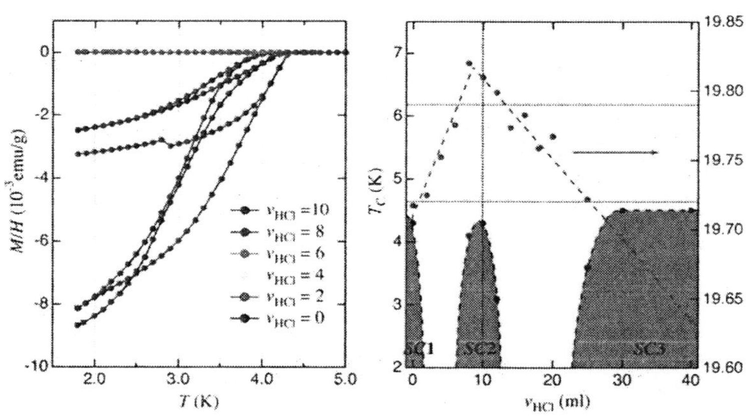


Figure 8. (a) Magnetic susceptibility measured under 0.001 T for the $v_{HCl} = 0$ to 10 ml samples and (b) the transition temperature, $T_C$, and the lattice parameter, $c$ of the samples. The broken and the black dotted lines are guides to eyes and the border of single phase region, respectively, and the red dotted lines indicate 19.72Å and 19.79Å.

This phase relation looks to be consistent with the idea that the superconductivity is induced in the region with a moderately strong magnetic interaction [80]. The interaction seems to be the strongest at $x \sim 0.34$, and superconductivity appears with the reduction of the magnetic interaction in both sides of the magnetic phase *M*. This remarkable behavior of superconductivity and magnetism cannot be explained by the variation in the carrier concentration because *s* is almost constant and only the $Na^+/H_3O^+$ ratio changes in the region in question. Although the *c*-axis is enlarged with increasing $\nu_{HCl}$, the difference between $c$ = 1.972 Å (at the border between *SC*1 and *M* phases) and 19.79 Å (at the border between *M* and *SC*2 phases) is only 0.35%. As mentioned previously, the band structure is not influenced significantly by an increase of the interlayer distance in this extent. Moreover, the change in the *c*-axis length may be converted to pressure change of −0.2 GPa [45] which seems negligible considering that $T_c$ is hardly influenced by positive pressure of 0.2 GPa [87]. A possible explanation for the remarkable changes in physical properties with $\nu_{HCl}$ (or *x*) is given by assuming a correlation between the thickness of the $CoO_2$ layer and $\nu_{HCl}$. As stated before, the size and shape of the hole pockets near the K-points is quite sensitive to the thickness. With increasing $\nu_{HCl}$ (decreasing *x*), the interlayer distance becomes larger by the substitution of smaller $Na^+$ ions by the larger $H_3O^+$ ions, and the $CoO_2$ layer can become thinner because the oxygen ions are less attracted by the $Na^+$ and $H_3O^+$ ions. It is necessary to determine the thickness of the $CoO_2$ layer as a function of $\nu_{HCl}$ (or *x*) for a more quantitative discussion. In Fig. 10(b), *T*–*H* phase diagram is shown for the $\nu_{HCl}$ = 0 ml sample ($x$ = 0.350). Details of this diagram will be discussed later.

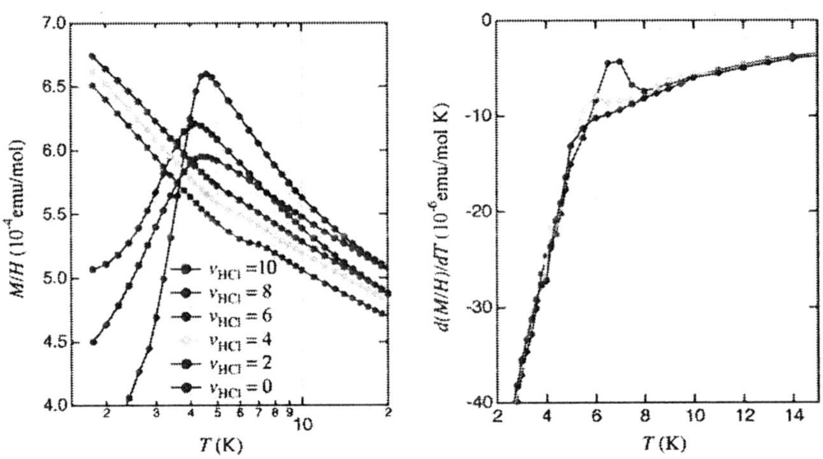


Figure 9. (a) Magnetic susceptibility measured under 1T for the $\nu_{HCl}$ = 0 to 10 ml samples and (b) the differentiated susceptibilities of the $\nu_{HCl}$ = 2, 4, and 6 ml samples.

(2) *Phase diagram 2.* This part corresponds to *synthesis 3* in the preceding section with samples synthesized by the addition of $\nu_{NaOH}$ ml 0.1 M (or 1 M) NaOH after immersion in distilled water [62]. The superconductivity was observed in all the samples as shown in Fig. 11(a) but $T_c$ was suppressed substantially for $\nu_{NaOH}$ = 100 and 500 ml. The *x*–$T_c$ two-dimensional section of the phase diagram is depicted as shown in Fig. 11(b) where the Co valence is almost constant to be $s$ = +3.47 to +3.50 (see Fig. 6(b)). The *x* values of the $\nu_{NaOH}$ = 0 and 10 ml samples were located in the magnetically ordered region in Fig. 10(a); they were not magnetic, but superconducting was caused by the different values of the Co valence. The

*SC*1 phase of Fig. 10(a) seems to extend slightly to lower $x$ direction with increasing $s$. The $s$ dependence of $T_c$ seems to be relatively small compared with HTC. If the Fermi surface is composed of only the $a_{1g}$ bands around Γ-point, it seems reasonable to assume that $T_c$ and other physical properties are more sensitive to the change in $s$ than $x$. The pronounced $x$ dependencies of the physical properties may support the presence of the hole-like pockets near the K points.

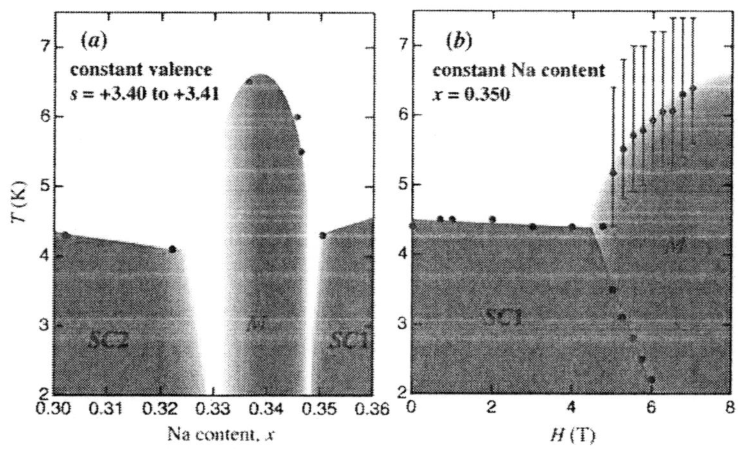


Figure 10. (a) Superconducting $x$-$T$ phase diagram of the sample obtained by the *Synthesis 1*. *SC* and *M* represent superconducting phases and magnetic ordering phase, respectively. (b) The $T$-$H$ phase diagram for the $v_{HCl} = 0$ sample.

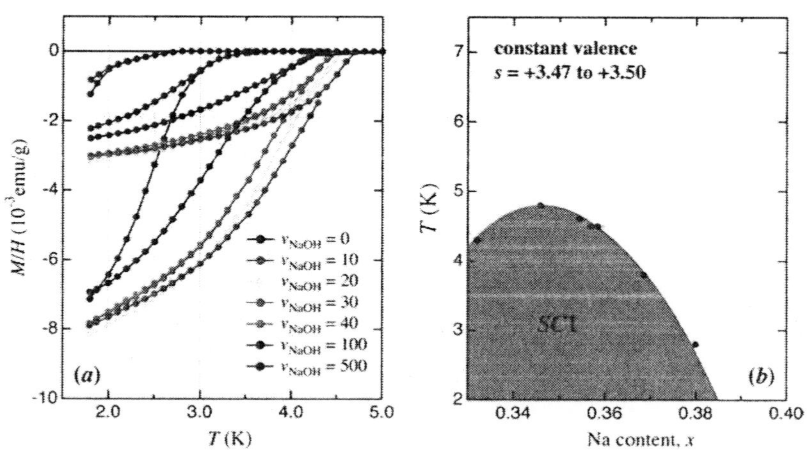


Figure 11. (a) Magnetic susceptibility of the samples obtained by the *Synthessis 3* measured under 0.001 T. (b) Superconducting $x$-$T$ phase diagram.

## 3.4 Superconducting State

It seems quite certain that BLH is an extreme type-II superconductor with an unconventional mechanism [21]. This picture is supported by the presence of the magnetically ordered phase next to the superconducting phase. Specific heat is a useful tool to elucidate the superconducting mechanism because it gives information not only on phenomenological properties such as superconducting volume fraction, but also on more microscopic properties related to thermal excitations of quasi-particles. After the discovery of the BLH superconductor, a great amount of specific heat data have been reported, to give clear evidence for the bulk superconductivity. However, they are not always fully reliable [21]. For instance, some of the analyses to determine the gap function were not appropriate because entropy balance regarding the superconducting transition was not taken into account. Moreover, in all cases, samples measured had not been identified in a strict sense, i.e., their $x$, $s$ and $y'$ values were not known separately. Thus, we have to be careful to compare two sets of specific data because their difference may come from differences in the sample.

Typical specific heat data of BLH are shown in Fig. 12 [88]. As seen in the figure, the specific heat has two distinguished features: large tail of the transition above $T_c$ and $T^2$ dependence below ~2 K, which have been confirmed by several groups [21, 89-91]. The former is probably caused by strong superconducting fluctuations [91]. Two possibilities have been proposed for the latter feature: an existence of line nodes in the superconducting gap structure [88-91] or superimposition of contributions from two superconducting gaps [91]. Of course, the second interpretation is based on the existence of the hole pockets.

There is a discrepancy among the reports in the specific heat lower than ~0.8 K. Jin et al. reported a sharp decrease [90], while Yang et al. and Oeschler et al. did not observe such a decrease down to 0.6 K and 0.34 K, respectively [88, 91]. The discrepancy may be caused by differences in the samples. Jin et al. suggested that the decrease was not due to a second superconducting gap because of its small field dependence.

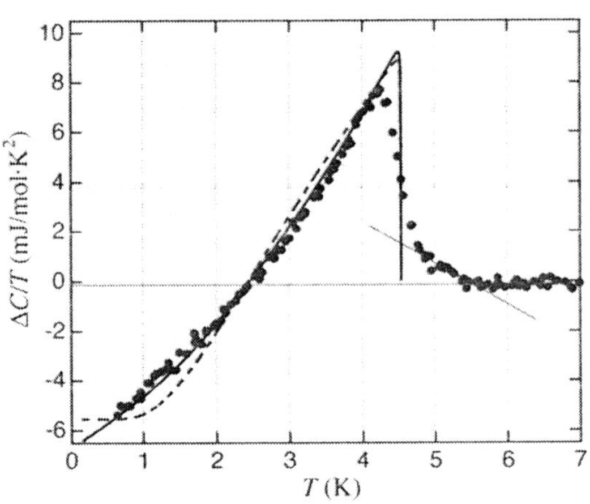


Figure 12. Specific heat anomaly caused by the superconducting transition. $\Delta C = C_{obs} - (C_n + C_{ph})$ ($C_{obs}$: observed spcific heat, $C_n$: electron contribution above $T_C$, and $C_{ph}$: phonon contribution) [88]. The solid and broken lines are expectation curves for line-nodal and isotropic superconducting gaps, respectively. The fine lines are guides to eyes.

The field dependence of the residual $\gamma_r$ ($\gamma$: Sommerfeld constant) implied the existence of line nodes but with a certain confusion. Yang, et al. and Jin et al. observed approximate $H^{0.5}$ dependence of $\gamma_r$ as expected for unconventional (non $s$-wave) superconducting gap symmetry with line nodes [88, 90]. However, Oeschler et al. pointed out that the ranges of magnetic field and temperature did not satisfy the condition necessary for the $H^{0.5}$ dependence [91].

Instead, they observed approximate $H$ linear dependence of $\gamma_r$ and suggested that it was evidence for the line nodes [89]. At present, it still seems hard to conclude from the specific heat data alone how the gap structure is.

A large sample dependence of the specific heat reported by Oeschler et al. [91] is worth noting because sample dependence seems to be one of keys to understand apparent discrepancies in specific heat data or other physical properties. They tested two samples with almost the same $T_c$ of ~4.6 K and obtained quite different $\gamma_r$ values under zero magnetic field: $\gamma_r$ = 6.67 mJ/mol·K$^2$ and 11.0 mJ/mol·K$^2$, which were about 41% and 71% of $\gamma$ at normal state (16.1 mJ/mol·K$^2$ and 15.7 mJ/mol·K$^2$, respectively). The temperature variation of the specific heat of the former sample was similar to that in Fig. 12 and was well reproduced by two-gap models; one having a smaller gap with line nodes and a larger node-less gap, and the other having two node-less gaps. On the other hand, the specific heat of the latter sample had exponential dependence rather than $T^2$ dependence below 2 K and was reproducible assuming two node-less gaps. $T_c$ was a little higher for the latter sample in spite of larger $\gamma_r$, which is inconsistent with the simple pair breaking theory and these results may suggest that superconducting gap symmetry changes with the composition of BLH. Change in superconducting symmetry with change in composition is seen in $AOs_2O_6$ or $AOs_4Sb_{12}$. In the former system, anisotropic superconducting gap with nodes is changed into isotropic one with changing A = K to A = Rb [92-97]. In the latter case, a Cooper pair seems to have $s$-wave symmetry for A= La while chiral symmetry for A = Pr [98, 99].

For the BLH samples, the two-gap model gives almost constant gap values of $\Delta_1(0)/k_BT_c$ = 2.15 to 2.30 and $\Delta_2(0)/k_BT_c$ = 0.70 to 1.10, independent of the samples, while the weight of $\gamma$ concerning the larger and the smaller gaps are largely different to be 55% and 45%, respectively in the sample with a smaller $\gamma_r$, whereas 80% and 20% in the sample with a larger $\gamma_r$. This may be consistent with theories based on the hole pockets [14-20], because, the more contribution there is from the hole pockets, $T_c$ increases and the effect of the line nodes is suppressed, although the origin of the large residual $\gamma_r$ is unclear at present.

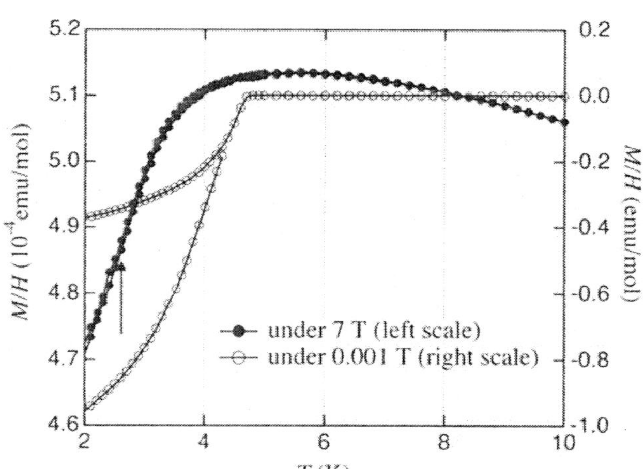


Figure 13. Magnetic susceptibility of the $v_{NaOH}$ = 10 ml sample. The arrow indicates 2.6 K.

### 3.5 Vortex State

As reviewed previously [21], upper-critical fields, $H_{c2}$, reported thus far are scattered quite largely. With magnetic field along the $c$-axis, $H_{c2}(T=0)$ was determined to be ~1.7 T for a sample with $T_c$ = 4.5 K [100], ~ 3 T for $T_c$ = 4.2 K [42] by electrical resistivity measurements, while 3.6 T for $T_c$ = 3.7 K [101] by magnetization measurement. On the other hand, with magnetic field along the $ab$-plane, $H_{c2}(T=0)$ was determined to be ~8 T for $T_c$ = 4.2 K [42] and 4.5 K [100] by electrical resistivity measurements. The large anisotropy is understandable from the layered structure, and $H_{c2}$ with a magnetic field along the $ab$-plane is consistent with the Pauli limit based on a weak coupling theory ($1.84T_c$ ~ 8 T) [102]. However, in magnetic susceptibility curve of Fig. 13, the superconducting transition is clearly observed for a polycrystalline sample at around 4 K even under 7 T, which is much higher than corresponding $T_c$ of ~2.6 K [42] and ~1.7 K [100] in the resistivity measurements with magnetic field along the $ab$-plane. Indeed, much larger values of $H_{c2}(T=0)$ of 15.6 T for $T_c$ = 3.7 K [101], 17.1 T for $T_c$ = 4.7 K [90], 28 T for $T_c$ = 4.3 K [103], or 61 T for $T_c$ = 4.6 K [104] have been observed by magnetization or specific heat measurements, based on WHH model [105].

Three factors are worth considering regarding the discrepancies in $H_{c2}$. First, the electric resistiviy data will strongly reflect the flux flow resistivity because of weak pinning force. Indeed, irreversibility field of BLH was estimated as low as that of Bi cuprates [106]. Thus, $H_{c2}$ may be underestimated by the resistivity measurement. Second, magnetic and specific heat measurements may be affected by strong superconducting fluctuations. As seen in Fig. 14(a), the tails of the superconducting transition seems to extend to a higher temperature with an increasing field [107]. The large tail is ascribed to the fluctuations by Oeschler et al. [89, 91], although the on-set of the tail of their sample was independent of the field and they estimated $H_{c2}$ lower than 9 T. The superconducting fluctuations may cause overestimation of $H_{c2}$. (It may be worth describing here that the tail extending to higher temperature with increasing magnetic field suggests spin triplet pairing of Cooper pair as expected by the theories based on the hole pockets [14-20]).

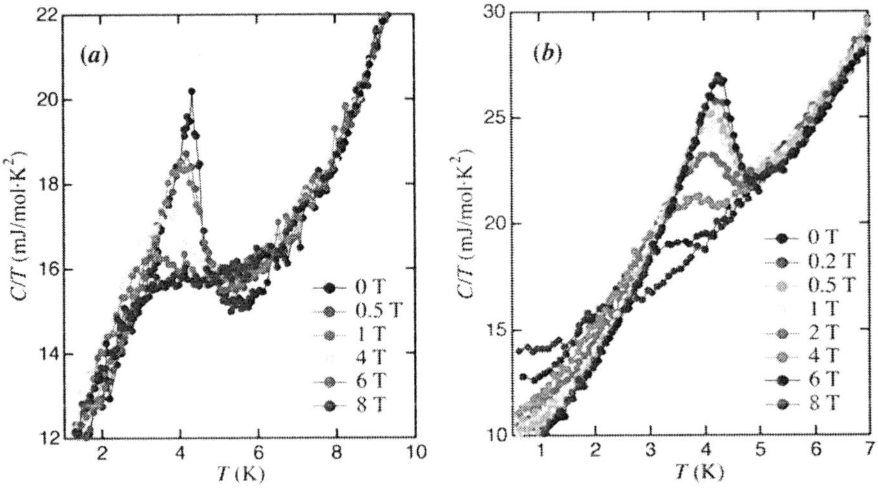


Figure 14. Specific heat of samples with higher (a) and lower (b) $H_{c2}$ under magnetic fields.

Although the flux flow resistivity and the superconducting fluctuations have caused the discrepancies to some extent, the third factor, i.e., the sample dependence seems to be more essential. Figure 14(b) shows the specific heat for a different sample measured at various fields. The superconducting transition looks to disappear under ~8 T in spite that $T_c$ at 0 T is almost same as that of the sample in Fig. 14(a). It is unclear what caused this difference because the detailed chemical analyses had not been carried out for the samples. Recently, it has been confirmed that $H_{c2}(T=0)$ greatly depends on the composition of the BLH phase, at least in a certain case. $^{23}$Na NMR measurements suggest that $H_{c2}(T=0)$ depends on surroundings at Na sites [108]. It is notable that a broad bending was observed at ~ 4 K in magnetization vs. temperature curve for the $v_{NaOH}$ = 10 ml sample ($s$ = +3.48, $x$ = 0.346) under 30 T [62], which suggested a very large value of $H_{c2}(T=0)$. The opposite case was the $v_{HCl}$ = 0 ml sample ($s$ = +3.40, $x$ = 0.350) in Fig. 15. Figure 15 shows temperature and magnetic field dependences of magnetic susceptibility for this sample. The superconducting transition is seen at ~ 4.5 K under 4.75 T, but two broad peaks appear under 5 T at about 3.5 K and 5 K. With an increasing field, these peaks shift to lower and higher temperatures, respectively. For instance, they are well distinguished at about 3 K and 5.7 K under 5.5 T, respectively. These peaks that appear at lower and higher temperatures have been assigned to superconducting and magnetic transitions, respectively [60]. Thus, the $T$–$H$ phase diagram is depicted for this particular sample as shown in Fig. 10(b) where the magnetically ordered phase appears under higher magnetic field and $H_{c2}(T=0)$ is estimated to be ~8 T. Since the superconducting transition of the $v_{HCl}$ = 8 ml sample under 7 T is as broad as that of the $v_{HCl}$ = 0 ml sample under 5 T, the magnetically ordered phase seen in Fig. 10(a) probably extends to both lower and higher $x$ directions under a high magnetic field.

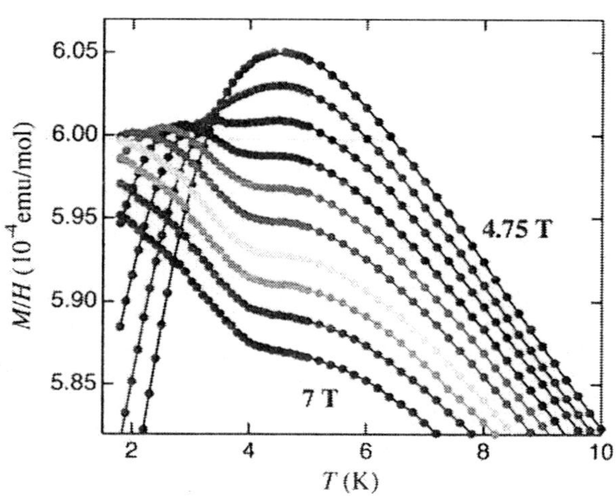


Figure 15. Magnetic susceptibility of the $v_{HCl}$ = 0 ml sample measured under magnetic fields from 4.75 T to 7 T with an increment of 0.25 T. The data of 5 T and 5.5 T are colored purple and blue, respectively.

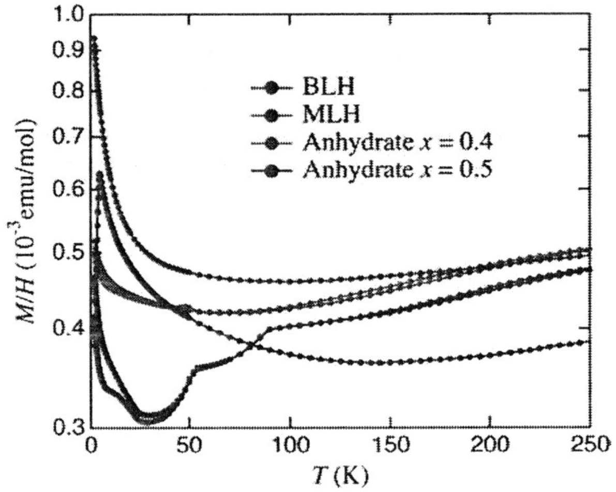


Figure 16. Magnetic susceptibility of BLH, MLH, $Na_{0.4}CoO_2$ (the $pH_i$ = 1 sample) and $Na_{0.5}CoO_2$ (the $pH_i$ = 14 sample). Logarithm scale is adopted for the vertical axis. The data were collected under 1 T for the hydrates and under 0.1 T for the non-hydrates.

## 3.6 Normal State

The normal state magnetic susceptibility of BLH shows Curie-Weiss type of upturn below 130 K and weak temperature dependence above 130 K, as seen in Fig. 16. It is an important problem relating to the mechanism of superconductivity whether the upturn is intrinsic or not. Temperature dependence of Knight shift, $K$, usually gives a clear answer to this kind of problem because $K$ reflects the intrinsic dynamical susceptibility $\chi(q = 0)$. However, there are discrepancies in $K$ for BLH reported thus far. Ning et al. obtained almost temperature independent $^{59}K$ and $^{17}K$ by $^{59}Co$ and $^{17}O$ NMR in spite of strong temperature dependence of $^{59}(1/T_1T)$ and $^{17}(1/T_1T)$ [109, 110], which means that the upturn of the bulk susceptibility had extrinsic origin such as magnetic impurities, and that dominant magnetic fluctuations had antiferromagnetic features. A similar result of $^{59}K$ has been reported by Mukhamedshin et al [111]. On the other hand, other studies of $^{59}Co$, $^{23}Na$, $^{2}D$ and $^{17}O$ NMR, as well as muon Knight shift, showed a clear upturn of Knight shifts and good linear relation of $K$-$\chi$ plot, at least, down to ~20 K [108, 112] indicating that the upturn of the bulk susceptibility reflected the intrinsic nature. The discrepancies are probably caused by Van-Vleck part of $^{59}K$ much larger than spin part and by small hyperfine coupling constant at the $^{17}O$ site, which makes it difficult to detect weak temperature dependence of $K$. Related to this, the deviation of $^{17}(1/T_1T)$ from the bulk susceptibility at lower temperatures can be explained by the change of magnetic fluctuations from ferromagnetic to nearly ferromagnetic [112] as expected by theoretical calculations assuming the hole pockets [18]. Together with other experimental results shown below, it is likely that the upturn has intrinsic origin and the magnetic fluctuations are of ferromagnetic (or nearly ferromagnetic) nature.

The magnetic susceptibilities of BLH, MLH and the non-hydrates are compared in Fig. 16. It should be noted here that MLH was prepared from BLH by removing water in a less humid atmosphere without any detectable change in the Co valence. Nevertheless, the upturn near 130 K was greatly suppressed and seems to move to near 90 K in MLH which strongly suggests the intrinsic nature of the upturn because it is unlikely that extrinsic sources such as magnetic impurities or lattice defects disappeared during the soft process of the MLH formation. It is interesting to note that the upturn disappeared completely in the non-hydrate with $x = 0.4$. Regarding the non-hydrate, a susceptibility curve similar to that for $x = 0.4$ has been reported for $x = 0.6$ [113] suggesting an apparent systematic tendency that the upturn is more pronounced in a phase with a larger water content. The hydrate with $x = 0.5$ has a complicated variation of the susceptibility due to charge ordering (see below).

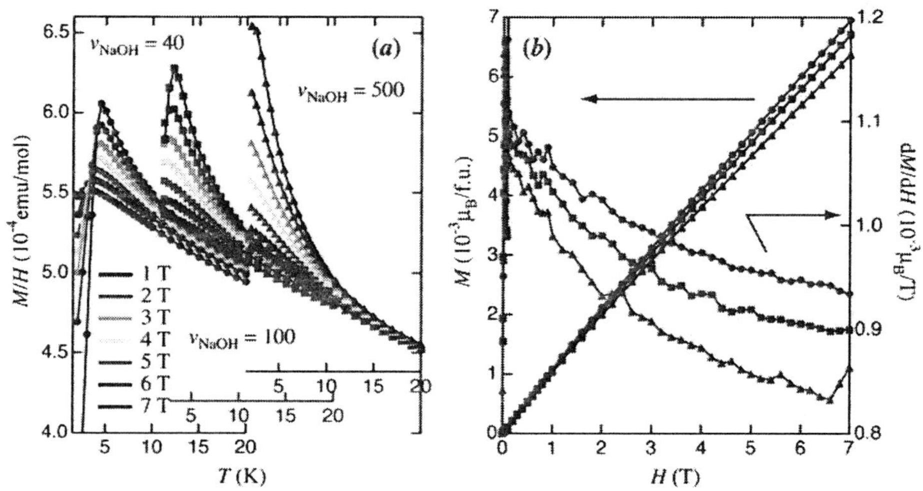


Figure 17. (a) Magnetic susceptibility of the $\nu_{NaOH}$ = 40, 100, and 500 ml samples measured under various fields. (b) Magnetization curves of them at 5 K and its differentiated curves.

PES results showed pseudo-gap behavior for BLH and MLH [82], which also supports the intrinsic origin of the upturn. According to PES, strong reduction of density of states at Fermi level (DOS) started just near 130 K in BLH and 90 K in MLH, i.e., at the upturn temperatures, suggesting that the upturn reflects the change of the band structure with temperature. A possible scenario may be considered assuming that the hole pockets appear around the upturn temperature and grow significantly with decreasing temperatures caused by thermal shrinkage. According to the suggestion that the hole pockets are hard to be detected by PES [77], the apparent reduction of DOS may reflect only a change in the cylindrical Fermi surface around Γ-A line as a result of transferring holes to the indetectable hole pockets. This idea may solve the problem that the magnetic susceptibility is enhanced in spite of the reduction of DOS, because holes in the hole pockets would cause much stronger magnetic fluctuations [17, 65, 86].

The rapid upturn below ~10 K in BLH was not well reproduced by Curie-Weiss type function plus a constant term. The muon Knight shift measured under 6 T was apparently inconsistent with the bulk susceptibility measured under 1 T [114], which may suggest that this low temperature upturn is extrinsic. Recently, however, it has been proved that the magnetic susceptibility in that temperature region depends on the applied field [61]. Thus, the $K$-$\chi$ plot in the region should be performed by the data obtained under the same field, and indeed, muon Knight shift looks to reflect the bulk susceptibility if the magnetic data under 6 T were compared. In addition, it is notable that the field dependence of the susceptibility is varied with the chemical composition as shown in Fig. 17. If the upturn was caused by magnetic impurities, it is hard to explain why the field dependence was changed depending on the composition [61]. These results suggest that the upturn in the low temperature range has also an intrinsic nature. The composition variation of the field dependence may be caused by the size and the shape of the hole pockets, because $\chi(q)$ in this temperature range is expected to have a peak at $q \neq 0$ and the ferromagnetic fluctuation corresponding to $\chi(q=0)$ reflects the tail of the peak [18]. This shift of the peak may be the reason why Curie-Weiss fit is unavailable in the lower temperature region.

Relatively small temperature dependence of the magnetic susceptibility of BLH ($s = +3.4$ to $+3.5$) may be due to the absence of the hole pockets, as suggested above. Magnetic susceptibility of $Na_{0.35}CoO_2$ ($s = +3.65$) was calculated to have a large (small) temperature dependence when the hole pockets appear (do not appear) [66]. However, the calculation seems very sensitive to the Co valence, because $Na_{0.7}CoO_2$ is likely to have no hole pocket but shows Curie-Weiss behavior of the magnetic susceptibility. Thus, the calculation for $s = +3.65$ is not applicable for compounds with other Co valence, such as BLH. The Sommerfeld constant of BLH estimated from the specific data at low temperature is 13.3 to 16.1 $mJ/mol{\cdot}K^2$ [88, 89, 90, 91], which agrees well with 10.5 $mJ/mol{\cdot}K^2$ of the calculation for $Na_{0.5}CoO_2$ with the hole pockets [64].

# 4 Sodium Cobalt Oxide

Here, the mother oxide of $\gamma$-$Na_xCoO_2$ will be discussed. The $\gamma$-$Na_xCoO_2$ shows a variety of physical properties originated from strong correlation between electrons in the $CoO_2$ layer. Although its band structure does not necessarily closely resemble that of BLH or MLH, its rich physical features seem to relate to the unconventional superconductivity in BLH. As proposed by Foo et al. [79], $\gamma$-$Na_xCoO_2$ phase can be separated into three parts depending on the $x$ value: 0.6 (or 0.5) $< x$, $x = 0.5$ and $x < 0.6$ (or 0.5) (except for $x = 0.5$). A difference in the way of ordering of the Na atoms [115, 116] seems to cause this separation. The ordering can affect the physical properties, because interlayer magnetic interaction of non-hydrate $Na_xCoO_2$ is comparable to intralayer one [117, 118] due to multiple $c$-axis exchange paths of direct O-O hopping and intermediate Na $sp^2$ hybridization coupling nearest and next-nearest Co ions [119].

The upper limit of the $x$ value is $x = 0.78$ if a sample is prepared by conventional solid state reaction starting from $Na_2CO_3$ and cobalt oxide ($Co_3O_4$ or CoO) [120]. Although a metastable phase with a higher $x$ value may be prepared at a lower synthesis temperature, samples reported as $\gamma$-$Na_xCoO_2$ with $x > 0.78$ were not usually well characterized. The lower limit of the $x$ value attained by the solid state reaction is $x = 0.7$, and it can extend to $x \sim 0.4$ by chemical oxidation or electrochemical oxidation. A phase separation, such as the spinodal decomposition, can occur introducing inhomogeneity in a sample, because Na ordering would take place in various manners depending on the $x$ value [115, 116]. A quasi-open-circuit potential vs. time curve showed a step-like structure in electrochemical deintercalation of the $Na^+$ ions [42] suggesting that homogeneous phases exist only for discrete $x$ values.

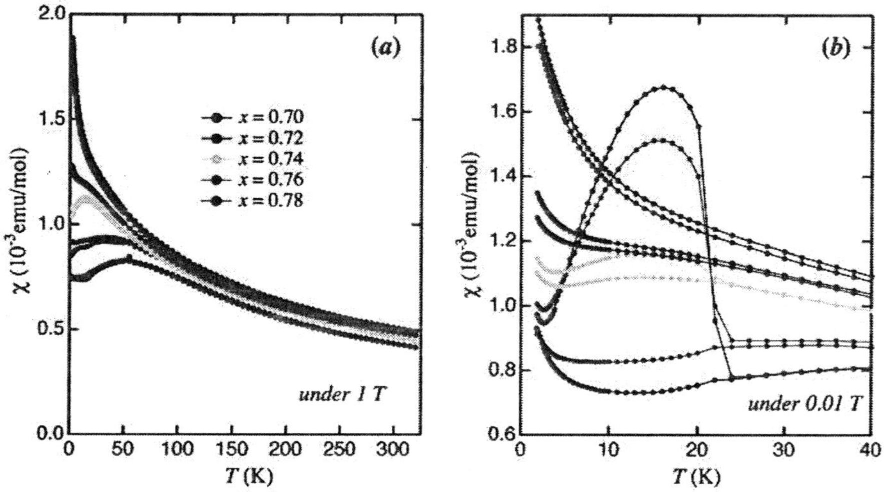


Figure 18. Magnetic susceptibility of $Na_xCoO_2$ ($x$ = 0.70 to 0.78) measured under 1 T (a) and 0.01 T (b).

The compound with a higher $x$ value ($x > 0.6$ or 0.5) is classified as Curie-Weiss metal [79], because its electrical resistivity showed metallic behavior independent of the composition and temperature and its magnetic susceptibility obeyed Curie-Weiss law in a higher temperature range as shown in Fig. 18 [120]. However, physical properties in a lower temperature range are not so clearly understood in spite of intensive works. The broad maximum of the magnetic susceptibility below 50 K appearing for higher $x$ values was explained by an enhancement of the effective mass [121]. However, $\gamma$( ~ 30 mJ/Co mol·K$^2$ ) is not so large, and moreover, the temperature of the broad maximum decreased rapidly with slightly increasing $\gamma$ when $x$ became smaller [122].

A magnetic transition has been observed at 22 K for $x$ values near the higher limit of 0.78 and considered to be caused by spin-density-wave (SDW) formation [123, 124]. DOS became smaller at the transition, consistent with the SDW formation [122]. Magnetic structure below 22 K has been solved by neutron diffraction to be A-type antiferromagnet where the ferromagnetic $CoO_2$ layer coupled antiferromagnetically with adjacent layers [117, 118]. The transition temperature was independent of the $x$ value between $x = 0.74$ and 0.78 and only the degree of magnetic anomaly depended on $x$ as seen in Fig. 18(b), which strongly suggests occurrence of a phase separation [120]. Namely, the phase with $0.74 \leq x \leq 0.78$ seems to undergo a spinodal-like decomposition separated to Na rich domains, Na poor domains and domains surviving the decomposition having the nominal composition, and the magnetic transition is likely to take place in the Na rich domains. Indeed, $\mu$SR measurement indicates that the magnetic phase in $Na_{0.75}CoO_2$ is only ~21% [124], and it is unusual for homogeneous sample that electrical resistivity drops at the transition temperature [123], in spite of partial reduction of DOS. The magnetic transition was found to undergo in ~100% volume fraction at 20 K for $Na_{0.82}CoO_2$ [125] and at 27 K for $Na_{0.85}CoO_2$ [126], which suggests that the phase separation occurs below $x = 0.78$ if the nominal compositions were true. However, ~100% volume fraction was reported also for $Na_{0.75}CoO_2$ by other $\mu$SR measurement [126].

$Na_{0.7}CoO_2$ has been considered to be located at the vicinity of a quantum critical point [122, 127-129]. The critical point is believed to be just below $x = 0.70$ [122, 127, 129], rather than above $x = 0.70$ [128]. By $^{59}$Co NMR measurement, ferromagnetic fluctuations were observed for the compound with $x = 0.7$, and it was regarded as an itinerant electron system [130], consistent with the metallic behavior and very sensitive $x$ dependence of the physical properties [120, 122]. It is noteworthy that upward deviation from the Curie-Weiss behavior below 20 K observed for $Na_{0.70}CoO_2$ [130] is intrinsic, which may be something to do with the upturn of the susceptibility of BLH below 10 K. Inconsistent with aforementioned results, charge ordering or tendency of charge ordering has been reported according to ESR and $^{59}$Co NMR measurements [131-133]. The discrepancy may be caused by differences in the composition of the sample (see below).

Cautions are needed on sample quality because sample dependencies seem to often cause the inconsistencies in physical properties. $Na_xCoO_2$, especially when the $x$ value is large, easily absorbs water molecules (and $CO_2$) in the air changing its composition into $Na_x(H_3O)_zCoO_2$ through ion exchange of $Na^+$ and $H_3O^+$ ions (sub-product of NaOH or $NaHCO_3$ would probably remain on surface of particles). The position of the broad maximum in the magnetic susceptibility of $Na_{0.78}CoO_2$ shifted from 50 K to 40 K after being left in the air for one week, and disappeared completely 6 months later. The ion exchange has been confirmed for $Na_{0.70}CoO_2$ as well; its magnetic susceptibility changed from Curie-Weiss type to Pauli paramagnetic one after being left in the air [134]. Aging effect of surface degradation observed by Raman spectroscopy [135] may also be due to the ion exchange. If a sample contains $H_3O^+$ ions, the $x$ value is not a measure of the Co valence any more. In addition, since the $c$-axis length depends both on the $x$ and $z$ values [134], it cannot be a measure of the $x$ value and the Co valence. Unless the sample is stored very carefully in dry atmosphere, the Co valence or $x$ value estimated from the initial composition or the $c$-axis length is not reliable. Furthermore, special attention is needed on single crystals grown by flux method, i.e., water cannot be used to remove flux.

The $\gamma$-$Na_xCoO_2$ with a lower $x$ value ($x < 0.6$ or 0.5) is considered to be Pauli paramagnetic metal (PPM), whose magnetic susceptibility shows only a weak temperature dependence. The typical magnetic susceptibility of a sample in this region is shown in Fig. 16 [61, 63]. A pseudo-gap is proposed for the origin of the temperature dependence of the magnetic susceptibility of PPM, based on analysis of $^{59}$Co NMR [113]. Furthermore, non-Fermi liquid behavior is seen in $x = 0.36$ sample [127].

A higher limit of $x$ of the PPM region is reported to be $x = 0.5$ [79] or 0.6 [113]. The former value may look plausible because it corresponds exactly to a singularity at $x = 0.5$ described later. However, sampling widths regarding $x$ were too discrete in Foo et al.'s experiment to determine the limit with enough accuracy. On the other hand, the latter value was supported by a sufficient number of samplings regarding $x$, although it has not been discussed why the limit was located at $x = 0.6$. For both experiments, sample characterization may not be sufficient in particular on the $x$ value and Co valence by the previously discussed reason.

$Na_xCoO_2$ with $x$ exactly equal to 0.5 ($s = +3.5$) shows a metal-insulator transition at $T_{c2} =$ 52 K accompanying with a magnetic anomaly [79]. In addition, two other magnetic transitions have been observed at $T_{c1} = 88$ K and $T_{c3} = 24$ K [61, 63, 79], as shown in Fig. 16. These anomalies have been confirmed to have bulk nature by specific heat, $\mu$SR, $^{59}$Co NMR, and neutron diffraction measurements [113, 126, 136, 137] in addition to the bulk magnetic susceptibility and electrical resistivity measurement [63, 79]. The metal-insulator transition at

$T_{c2}$ had been considered to be caused by charge ordering of $Co^{3.5+} \rightarrow Co^{3+}$ (non-magnetic) + $Co^{4+}$ (magnetic) [79, 115, 137], but Co sites with a small magnetic moment were observed by recent $^{59}$Co NMR and neutron diffraction measurements [113] instead of the non-magnetic $Co^{3+}$ sites. Thus, charge disproportionation, rather than charge ordering, seems to occur as suggested by the calculation of Lee et al. [138].

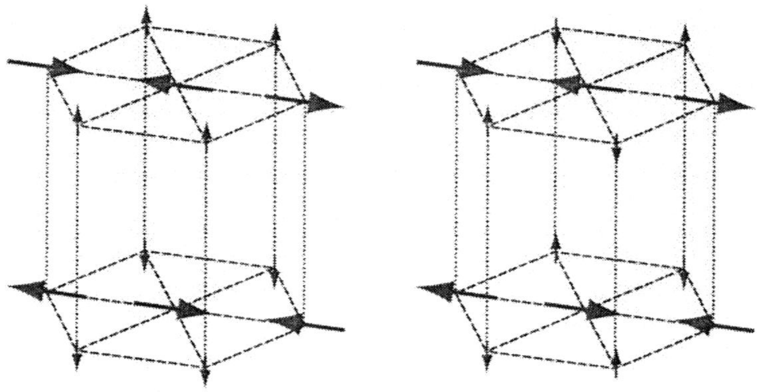


Figure 19. Proposed magnetic structures of $Na_{0.5}CoO_2$. $Co^{3.5+\delta}$ ions with a large magnetic moment and $Co^{3.5-\delta}$ ions with a small magnetic moment form lines. The large moments orders antiferromagnetically parallel to the line, while the small ones ferromagnetically (a) or antiferromagnetically (b) in the $CoO_2$ layer and antiferromagnetically between the layers parallel to the *c*-axis.

Huang et al. suggested that the anomaly at $T_{c1}$ corresponded to a structural transition [137]. On the other hand, a study of $\mu$SR argued that two kinds of internal field appeared below $T_{c2}$ probably caused by an antiferromagnetic transition and one of them changed at $T_{c3}$ implying a fist-order transition there [136]. On the contrary, other studies of $\mu$SR, neutron diffraction and $^{59}$Co NMR showed that magnetic ordering took place at $T_{c1}$ [126, 113], and two possible magnetic structures shown in Fig. 19 have been proposed [113]. The neutron diffraction and $^{59}$Co NMR measurements detected no anomaly at $T_{c2}$ and $T_{c3}$, while two of three internal fields changed at $T_{c2}$ and $T_{c3}$ according to the $\mu$SR measurement. Details of the transition at $T_{c3}$ has not been known, but canting of the antiferromagneticaly ordered moments is likely to occur at this temperature because of a small spontaneous magnetization in spite of the localized electron magnetism in the electric insulator. These transitions have also been observed for a sample with $x < 0.5$ but it was confirmed to have $s = +3.5$ due to the presence of the $H_3O^+$ ions [139].

Before the discovery of the hydrate superconductor, the compound of "$Na_{0.5}CoO_2$" or "$NaCo_2O_4$" had been intensively studied from the viewpoint of high thermoelectrical performance [140]. However, in almost all cases, samples did not really have the composition of $Na_{0.5}CoO_2$ but were mixtures of $Na_{0.7}CoO_2$ and an impurity of $Co_3O_4$ as detected by XRD pattern [141]. Indeed, it has been proved by $^{59}$Co NMR measurement that a sample with an apparent composition of $Na_xCoO_2$ with $x < 0.7$ synthesized by conventional solid state reaction is always a mixture of $Na_{0.7}CoO_2$ and impurity of $Co_3O_4$ or CoO [142]. In addition, a possible formation of $Na_x(H_3O)_zCoO_2$ ($x + z \sim 0.7$) needs to be considered as already discussed.

# 5 Conclusion

We reviewed chemical and macroscopic physical properties of the BLH superconductor and its related compounds, focusing on the recent development of the research. BLH has three adjustable compositional parameters: Na content, $x$, $H_3O$ content, $z$, and $H_2O$ content, $y'$. The former two parameters greatly affect superconductivity and other physical properties of the system. They can be controlled through a soft chemical process in sample preparation. Three different superconducting phases *SC*1, *SC*2 and *SC*3, and a magnetically ordered phase *M* appear in the $x$–$T$ phase diagram with fixing $s$ at $s = +3.40$. *SC*1 and *SC*2 are located on both sides of the magnetically ordered phase. This strongly suggests that superconductivity of the system has a certain magnetic origin with unconventional nature, consistent with the line-nodal superconducting gap structure and the large superconducting fluctuations observed by specific heat measurements.

Regarding the band structure of BLH, the hole pockets near the K-points look crucially important relating to the mechanism of superconductivity. They are not present in non-hydrate $Na_xCoO_2$ according to ARPES, but it does not seem to be the case for BLH because of its thinner $CoO_2$ layer and thus much larger splitting of the $t_{2g}$ bands. Indeed, the phase diagram, the normal-state magnetic susceptibility, and specific heat support the existence of the hole pockets in BLH. There is a possibility that two-gap superconductivity is realized in the present system corresponding to the large cylindrical Fermi surface around the Γ-A line and the hole pockets near the K-points.

Finally, the mother oxide of $\gamma$-$Na_xCoO_2$ was discussed. Its rich physical properties originate from a strong correlation between electrons in the $CoO_2$ layer and seem to relate to the unconventional superconductivity in BLH.

## Acknowledgements

We would like to thank T. Sasaki, N. Tsujii (NIMS), Y. Yanase (Tokyo University), H. Kontani (Nagoya University), and all other collaborators for useful discussion and suggestions. H. D. Yang (Natl. Sun Yat Sen University) and J.-Y. Lin (Natl. Chiao Tung University) kindly offered us their specific heat data shown in the figures. This work is partially supported by CREST, Japan Science and Technology Agency and by Grants-in-Aid for Scientific Research (1634011, 16076209) from Japan Society for the Promotion of Science and MEXT.

# References

[1] Takada, K.; Sakurai, H.; Takayama-Muromachi, E.; Izumi, F.; Dilanian, D. A.; Sasaki, T. *Nature* 2003, 422. 53-55.

[2] Anderson, P. W. *Science* 1987, 235. 1196-1198.

[3] Anderson, P. W. *Mater. Res. Bull.* 1973, 8. 153-160.

[4] Baskaran, G. *Phys. Rev. Lett.* 2003, 91. 097003.

[5] Honerkamp, C. *Phys. Rev. B* 2003, 68. 104510.

[6] Kumar, B.; Shastry, B. S. *Phys. Rev. B* 2003, 68. 104508.

[7] Motrunich, O. I.; Lee, P. A. *Phys. Rev. B* 2004, 69. 214516.
[8] Ogata, M. *J. Phys. Soc. Jpn.* 2003, 72. 1839-1842.
[9] Tanaka, Y.; Yanase, Y.; Ogata, M. *J. Phys. Soc. Jpn.* 2004, 73. 319-322.
[10] Tanaka, A.; Hu, X. *Phys. Rev. Lett.* 2003, 91. 257006.
[11] Wang, Q.-H. *Phys. Rev. Lett.* 2004, 92..057003.
[12] Wang, Q.-H.; Lee, D.-H.; Lee, P. A. *Phys. Rev. B* 2004, 69. 092504.
[13] Zheng, W.; Oitmaa, J.; Hamer, C. J.; Singh, R. R. P. *Phys. Rev. B* 2004, 70. 020504.
[14] Kuroki, K.; Tanaka, Y.; Arita, R. *Phys. Rev. Lett.* 2004, 93. 077001.
[15] Johannes, M. D.; Mazin, I. I.; Singh, D. J.; Papaconstantopoulos, D. A. *Phys. Rev. Lett.* 2004, 93. 097005.
[16] Kuroki, K.; Tanaka, Y.; Arita, R. *Phys. Rev. B* 2005, 71. 024506.
[17] Mochizuki, M.; Yanase, Y.; Ogata, M. *Phys. Rev. Lett.* 2005, 94. 147005.
[18] Yanase, Y; Mochizuki, M.; Ogata, M. *J. Phys. Soc. Jpn.* 2005, 74. 430-440.
[19] Yanase, Y.; Mochizuki, M.; Ogata, M. *J. Phys. Soc. Jpn.* 2005, 74. 2568-2578.
[20] Yanase, Y.; Mochizuki, M.; Ogata, M. *J. Phys. Soc. Jpn.* 2005, 74. 3351-3364.
[21] Takada, K.; Sakurai, H.; Takayama-Muromachi, E. *Frontiers in Superconducting Materials*; Narlikar, A. V.; Ed.; Springer: Berlin Heidelberg New York, 2005; pp 651-682.
[22] Delmas, C.; Fouassier, C.; Hagenmuller P. *Physica B* 1980, 99. 81-85.
[23] Takada, K.; Sakurai, H.; Takayama-Muromachi, E.; Izumi, F.; Dilanian, R. A.; Sasaki, T. *J. Solid State Chem.* 2004, 177. 372-376.
[24] Jorgensen, J. D.; Avdeev, M.; Hinks, D. G.; Burley, J. C.; Short, S. *Phys. Rev. B* 2003, 68. 214517.
[25] Lin, C. T.; Chen, D. P.; Lemmens, P.; Zhang, X. N.; Maljuk, A.; Zhang, P. X. *J. Cryst. Growth* 2005, 275. 606-616.
[26] Foo, M. L.; Schaak, R. E.; Miller V. L.; Klimczuk, T.; Rogado, N. S.; Wang, Y.; Lau, G. C.; Craley, C.; Zandbergen, H. W.; Ong, N. P.; Cava, R. J. *Solid State Comm.* 2003, 127. 33-37.
[27] Cmaidalka, J.; Baikalov, A.; Xue, Y. Y.; Meng, R. L.; Chu, C. W. *Physica C* 2004, 403. 125-131.
[28] Chen, D. P.; Chen, H. C.; Maljuk, A.; Kulakov, A.; Zhang, H.; Lemmens, P.; Lin, C. T. *Phys. Rev. B* 2004, 70. 024506.
[29] Foo, M. L.; Klimczuk, T.; Cava, R. J. *Mater. Res. Bull.* 2005, 40. 665-670.
[30] Sakurai, H.; Takada, K.; Izumi, F.; Dilanian, R. A.; Sasaki, T.; Takayama-Muromachi, E. *Physica C* 2004, 412-414. 182-186.
[31] Takada, K.; Fukuda, K.; Osada, M.; Nakai, I.; Izumi, F.; Dilanian, R. A.; Kato, K.; Takata, M.; Sakurai, H.; Takayama-Muromachi, E.; Sasaki, T. *J. Mater. Chem.* 2004, 14. 1448-1453.
[32] Balsys, R. J.; Davis, R. L. *Solid State Ionics* 1997, 93. 279-282.
[33] Ono, Y.; Kato, N.; Ishii, Y.; Miyazaki, Y.; Kajitani, T. *J. Jpn. Soc. Powder and Powder Metallurgy* 2003, 50. 469-474.
[34] Nakatsugawa, H.; Nagasawa, K. *J. Solid State Chem.* 2004, 177. 1137-1145.
[35] Lynn, J. W.; Huang, Q.; Brown, C. M.; Miller, V. L.; Foo, M. L.; Schaak, R. E.; Jones, C. Y.; Mackey, E. A.; Cava, R. J. *Phys. Rev. B* 2003, 68. 214516.

[36] Kubota, M.; Takada, K.; Sasaki, T.; Kumigashira, H.; Okabayashi, J.; Oshimra, M.; Suzuki, M.; Kawamura, N.; Takagaki, M.; Fukuda, K.; Ono, K. *Phys. Rev. B* 2004, 70. 012508.
[37] Takada, K.; Sakurai, H.; Takayama-Muromachi, E.; Izumi, F.; Dilanian, R. A.; Sasaki, T. *Adv. Mater.* 2004, 16. 1901-1905.
[38] Takada, K.; Osada, M.; Izumi, F.; Sakurai, H.; Takayama-Muromachi, E.; Sasaki, T. *Chem. Mater.* 2005, 17. 2034-2040.
[39] Sakurai, H.; Takada, K.; Sasaki, T.; Izumi, F.; Dilanian, R. A.; Takayama-Muromachi, E. *J. Phys. Soc. Jpn.* 2004, 73. 2590-2591.
[40] Foo, M. L.; Klimczuk, T.; Li, L.; Ong, N. P.; Cava, R. J. *Solid State Comm.* 2005, 133. 407-410.
[41] Mistry, S.; Arnold, D. C.; Nuttall, C. J.; Lappas, A.; Green M. A. *Chem. Comm.* 2004, 21. 2440-2441.
[42] Chou, F. C.; Cho, J. H.; Lee, P. A.; Abel, E. T.; Matan, K.; Lee, Y. S. *Phys. Rev. Lett.* 2004, 92. 157004.
[43] Liu, C.-J.; Liao, C.-Y.; Huang, L.-C.; Su, C.-H.; Neeleshwar, S.; Chen, Y.-Y. *Physica C* 2004, 416. 43-46.
[44] Liu, C.-J.; Hung, W.-C.; Wang, J.-S.; Liu, C.-J. C. *J. Am. Chem. Soc.* 2005, 127. 830-831.
[45] Park, S.; Lee, Y.; Moodenbaugh, A.; Vogt, T. *Phys. Rev. B* 2003, 68. 180505.
[46] Schaark, R. E.; Klimczuk, T.; Foo, M. L.; Cava, R. J. *Nature* 2003, 424. 527-529.
[47] Karppinen, M.; Asako, I.; Motohashi, T.; Yamauchi, H. *Chem. Mater.* 2004, 16. 1693-1696.
[48] Ooi, K.; Miyai, Y.; Katoh, S.; Maeda, H.; Abe, M. *Langmuir* 1989, 5. 150-157.
[49] Kanoh, H.; Feng, Q.; Miyai, Y.; Ooi, K. *J. Electrochem. Soc.* 1993, 140 3162-3166.
[50] Li, W.; McKinnon, W. R.; Dahn, J. R. *J. Electrochem. Soc.* 1994, 141. 2310-2316.
[51] Arai, H.; Sakurai, Y. *Denki Kagaku* 1998, 66. 1206-1208.
[52] Milne, C. J.; Argyriou, D. N.; Chemseddine, A.; Aliouane, N.; Veira, J.; Landsgesell, S.; Alber, D. *Phys. Rev. Lett.* 2004, 93. 247007.
[53] Bañobre-López, M.; Rivadulla, F.; Caudillo, R.; López-Quintela, M. A.; Rivas, J.; Goodenough, J. B. *Chem. Mater.* 2005, 17. 1965-1968.
[54] Yang, H. X.; Li, J. Q.; Xiao, R. J.; Shi, Y. G.; Zhang, H. R. *Phys. Rev. B* 2005, 72. 075106.
[55] Liu, C.-J.; Liao, C.-Y.; Neeleshwar, S.; Chen, Y.-Y. *cond-mat/0407420.*
[56] Park, S.; Lee, Y.; Vogt, T. *cond-mat/0407778.*
[57] Tang, H.-Y.; Lin, H.-Y.; Wang, M.-J.; Liao, M.-Y.; Liu, J.-L.; Hsu, F.-C.; Wu, M.-K. *Chem. Mater.* 2005, 17. 2162-2164.
[58] Park, S.; Lee, Y.; Vogt, T. *Solid State Comm.* 2005, 134. 607-611.
[59] O'M Bockris, J.; Reddy, A. K. N. *Modern Electrochemistry* vol. 2; Plenum Press: New York, 1970.
[60] Sakurai, H.; Takada, K.; Sasaki, T.; Takayama-Muromachi, E. *J. Phys. Soc. Jpn.* 2005, 74. 2909-2912.
[61] Sakurai, H.; Takada, K.; Sasaki, T.; Takayama-Muromachi, E. unpublished.
[62] Sakurai, H.; Tsujii, N.; Suzuki, O.; Kitazawa, H.; Kido, G.; Takada, K.; Sasaki, T.; Takayama-Muromachi, E. *cond-mat/0511416.*

[63] Sakurai, H.; Takada, K.; Sasaki, T.; Takayama-Muromachi, E. *Proceedings of LT24, to be published in the AIP Conference Proceedings Series*.
[64] Singh, D. J. *Phys. Rev. B* 2000, 61. 13397.
[65] Mochizuki, M.; Yanase, Y.; Ogata, M. *J. Phys. Soc. Jpn.* 2005, 74. 1670-1673.
[66] Yada, K.; Kontani, H. *J. Phys. Soc. Jpn.* 2005, 74. 2161-2164.
[67] Zou, L.-J.; Wang, J.-L.; Zeng, Z. *Phys. Rev. B* 2004, 69. 132505.
[68] Zhang, P.; Luo, W.; Crespi, V. H.; Cohen, M. L.; Louie, S. G. *Phys. Rev. B* 2004, 70. 085108.
[69] Zhang, P.; Luo, W.; Cohen, M. L.; Louie, S. G. *Phys. Rev. Lett.* 2004, 93. 236402.
[70] Zhou, S.; Gao, M.; Ding, H.; Lee, P. A.; Wang, Z. *Phys. Rev. Lett.* 2005, 94. 206401.
[71] Ishida, H.; Johannes, M. D.; Liebsch, A. *Phys. Rev. Lett.* 2005, 94. 196401.
[72] Hasan, M. Z.; Chuang, Y.-D.; Qian, D.; Li, Y.W.; Kong, Y.; Kuprin, A.; Fedorov, A. V.; Kimmerling, R.; Rotenberg, E.; Rossnagel, K.; Hussain, Z.; Koh, H.; Rogado, N. S.; Foo, M. L.; Cava, R. J. *Phys. Rev. Lett.* 2004, 92. 246402.
[73] Yang, H.-B.; Wang, S.-C.; Sekharan, A. K. P.; Matsui, H.; Souma, S.; Sato, T.; Takahashi, T.; Takeuchi, T.; Campuzano, J. C.; Jin, R.; Sales, B. C.; Mandrus, D.; Wang, Z.; Ding, H. *Phys. Rev. Lett.* 2004, 92. 246403.
[74] Yang, H.-B.; Pan, Z.-H.; Sekharan, A. K. P.; Sato, T.; Souma, S.; Takahashi, T.; Jin, R.; Sales, B. C.; Mandrus, D.; Fedorov, A. V.; Wang, Z.; Ding, H. *Phys. Rev. Lett.* 2005, 95. 146401.
[75] Hasan, M. Z.; Qian, D.; Li, Y.; Fedorov, A. V.; Chuang, Y.-D.; Kuprin, A. P.; Foo, M. L.; Cava, R. J. *cond-mat/0501530.*
[76] Wu, W. B.; Huang, D. J.; Okamoto, J.; Tanaka, A.; Lin, H.-J.; Chou, F. C.; Fujimori, A.; Chen, C. T. *Phys. Rev. Lett.* 2005, 94. 146402.
[77] Johannes, M. D.; Papaconstantopoulos, D. A.; Singh, D. J.; Mehl, M. J. *Europhys. Lett.* 2004, 68. 433-439.
[78] Johannes, M. D.; Mazin, I. I.; Singh, D. J. *Phys. Rev. B* 2005, 71. 205103.
[79] Foo, M. L.; Wang, Y.; Watauchi, S.; Zandbergen, H. W.; He, T.; Cava, R. J.; Ong, N. P. *Phys. Rev. Lett.* 2004, 92. 247001.
[80] Ihara, Y.; Ishida, K.; Michioka, C.; Kato, M.; Yoshimura, K.; Takada, K.; Sasaki, T.; Sakurai, H.; Takayama-Muromachi, E. *J. Phys. Soc. Jpn.* 2005, 74. 867-870.
[81] Chainani, A.; Yokoya, T.; Takata, Y.; Tamasaku, K.; Taguchi, M.; Shimojima, T.; Kamakura, N.; Horiba, K.; Tsuda, S.; Shin, S.; Miwa, D.; Nishino, Y.; Ishikawa, T.; Yabashi, M.; Kobayashi, K.; Namateme, H.; Taniguchi, M.; Takada, K.; Sasaki, T.; Sakurai, H.; Takayama-Muromachi, E. *Phys. Rev. B* 2004, 69. 180508.
[82] Shimojima, T.; Yokoya, T.; Kiss, T.; Chainani, A.; Shin, S.; Togashi, T.; Watanabe, S.; Zhang, C.; Chen, C. T.; Takada, K.; Sasaki, T.; Sakurai, H.; Takayama-Muromachi, E. *Phys. Rev. B* 2005, 71. 020505.
[83] Marianetti, C. A.; Kotliar, G.; Ceder, G. *Phys. Rev. Lett.* 2004, 92. 196405.
[84] Johannes, M. D.; Singh, D. J. *Phys, Rev. B* 2004, 70. 014507.
[85] Arita, R. *Phys. Rev B* 2005, 71. 132503.
[86] Mochizuki, M.; Yanase, Y.; Ogata, M. *Physica B* 2005, 359-361. 594-596.
[87] Lorenz, B.; Cmaidalka, J.; Meng, R. L.; Chu, C. W. *Phys. Rev. B* 2003, 68. 132504.
[88] Yang, H. D.; Lin, J.-Y.; Sun, C. P.; Kang, Y. C.; Huang, C. L.; Takada, K.; Sasaki, T.; Sakurai, H.; Takayama-Muromachi, E. *Phys. Rev. B* 2005, 71. 020504.

[89] Oeschler, N.; Fisher, R. A.; Phillips, N. E.; Gordon, J. E.; Foo, M. L.; Cava, R. J. *Chinese J. Phys.* 2005, 43. 574-582.

[90] Jin, R.; Sales, B. C.; Li, S.; Mandrus, D. *Phys, Rev. B* 2005, 72. 060512.

[91] Oeschler, N.; Fisher, R. A.; Phillips, N. E.; Gordon, J. E.; Foo, M.-L.; Cava, R. J. *cond-mat/0503690.*

[92] Koda, A.; Higemoto, W.; Ohishi, K.; Saha, S. R.; Kadono, R.; Yonezawa, S.; Muraoka, Y.; Hiroi, Z. *J. Phys. Soc. Jpn.* 2005, 74. 1678-1681.

[93] Arai, K.; Kukuchi, J.; Kodama, K.; Takigawa, M.; Yonezawa, S.; Muraoka, Y.; Hiroi, Z. *J. Phys. Soc. Jpn.* 2005, 74. 1678-1681.

[94] Arai, K.; Kikuchi, J.; Kodama, K.; Takigawa, M.; Yonezawa, S.; Muraoka, Y.; Hiroi, Z. *Physica B* 2005, 359-361.

[95] Brühwiler, M.; Kazakov, S. M.; Zhigadlo, N. D.; Karpinski, J.; Batlogg, B. *Phys. Rev. B* 2004, 70. 020503.

[96] Magishi, K.; Gavilano, J. L.; Pedrini, B.; Hinderer, J.; Weller, M.; Ott, H. R.; Kazakov, S. M.; Karpinski, J. *Phys. Rev. B* 2005, 71. 024524.

[97] Khasanov, R.; Eshchenko, D. G.; Karpinski, J.; Kazakov, S. M.; Zhigadlo, N. D.; Brütsch, R.; Gavillet, D.; Castro, D. D.; Shengelaya, A.; Mattina, F. L.; Maisuradze, A.; Baines, C.; Keller, H. *Phys. Rev. Lett.* 2004, 93. 1570004.

[98] Kotegawa, H.; Yogi, M.; Imamura, Y.; Kawasaki, Y.; Zheng, G.-q.; Kitaoka, Y.; Ohsaki, S.; Sugawara, H.; Aoki, Y.; Sato, H. *Phys. Rev. Lett.* 2003, 90. 027001.

[99] Aoki, Y.; Tsuchiya, A.; Kanayama, T.; Saha, S. R.; Sugawara, H.; Sato, H.; Higemoto, W.; Koda, A.; Ohishi, K.; Nishiyama, K.; Kadono, R. *Phys. Rev. Lett.* 2003, 91. 067003.

[100] Sasaki, T.; Badica, P.; Yoneyama, N.; Yamada, K.; Togano, K.; Kobayashi, N. *J. Phys. Soc. Jpn.* 2004, 73. 1131-1134.

[101] Badica, P.; Kondo, T.; Togano, K.; Yamada, K. *cond-mat/0402235.*

[102] Clogston, A. M. *Phys. Rev. Lett.* 1962, 9. 266-267.

[103] Maśka, M. M.; Mierzejewski, M.; Andrzejewski, B.; Foo, M. L.; Cava, R. J.; Klimczuk, T. *Phys. Rev. B* 2004, 70. 144516.

[104] Sakurai, H.; Takada, K.; Yoshii, S.; Sasaki, T.; Kindo, K.; Takayama-Muromachi, E. *Phys. Rev. B* 2003, 68. 132507.

[105] Werthamer, N. R.; Helfand, E.; Hohenberg, P. C. *Phys. Rev.* 1966, 147. 295-302.

[106] Badica, P.; Kondo, T.; Togano, K. *Appl. Phys. Lett.* 2004, 84. 559-561.

[107] Yang, H. D.; Lin, J.-Y.; Sun, C. P.; Kang, Y. C.; Takada, K.; Sasaki, T.; Sakurai, H.; Takayama-Muromachi, E. *cond-mat/0308031.*

[108] Michioka, C.; Ohta, H.; Itoh, Y.; Waki, T.; Yoshimura, K.; Kato, M.; Ishida, K.; Ihara, Y.; Sakurai, H.; Takayama-Muromachi, E.; Takada, K.; Sasaki, T. unpublished/ Kato, M.; Michioka, M.; Taki, T.; Itoh, Y.; Yoshimura, K.; Ishida, K.; Sakurai, H.; Takayama-Muromachi, E.; Takada, K.; Sasaki, T. *J. Phys.: Condens. Matter* 2006, 18 669-682/ Michioka, C.; Kato, M.; Waki, T.; Yoshimura, K.; Ishida, K.; Ihara, Y.; Sakurai, H.; Takayama-Muromachi, E.; Takada, K.; Sasaki, T. *Physica B* 2005, 359-361. 485-487.

[109] Ning, F. L.; Imai, T.; Statt, B. W.; Chou, F. C. *Phys. Rev. Lett.* 2004, 93. 237201.

[110] Ning, F. L.; Imai, T. *Phys. Rev. Lett.* 2005, 94. 227004.

[111] Mukhamedshin, I.R.; Alloul, H.; Collin, G.; Blanchard, N. *Phys. Rev. Lett.* 2005, 94. 247602.

[112] Ihara, Y.; Ishida, K.; Yoshimura, K.; Takada, K.; Sasaki, T.; Sakurai, H.; Takayama-Muromachi, E. *J. Phys. Soc. Jpn.* 2005, 74. 2177-2180.

[113] Yokoi, M.; Moyoshi, T.; Kobayashi, Y.; Soda, M.; Yasui, Y.; Sato, M. *J. Phys. Soc. Jpn.* 2005, 74. 3046-3056.

[114] Higemoto, W.; Ohishi, K.; Koda, A.; Saha, S. R.; Kadono, R.; Ishida, K.; Takada, K.; Sakurai, H.; Takayama-Muromachi, E.; Sasaki, T. *Phys. Rev. B* 2004, 70. 134508.

[115] Zandbergen, H. W.; Foo, M.; Xu, Q.; Kumar, V.; Cava, R. J. *Phys. Rev. B* 2004, 70. 024101.

[116] Huang, Q.; Foo, M. L.; Prscal Jr., R. A.; Lynn, J. W.; Toby, B. H.; He, T.; Zandbergen, H. W.; Cava, R. J. *Phys. Rev. B* 2004, 70. 184110.

[117] Boothroyd, A. T.; Coldea, R.; Tennant, D. A.; Prabhakaran, D.; Helme, L. M.; Frost, C. D. *Phys. Rev. Lett.* 2004, 92. 197201.

[118] Bayrakci, S. P.; Mirebeau, I.; Bourges, P.; Sidis, Y.; Enderle, M.; Mesot, J.; Chen, D. P.; Lin, C. T.; Keimer, B. *Phys. Rev. Lett.* 2005, 94. 157205.

[119] Johannes, M. D.; Mazn, I. I.; Singh, D. J. *Phys. Rev. B* 2005, 71. 214410.

[120] Sakurai, H.; Takenouchi, S.; Tsujii, N.; Takayama-Muroamchi, E. *J. Phys. Soc. Jpn.* 2004, 73. 2081-2084.

[121] Miyoshi, K.; Morikuni, E.; Fujiwara, K.; Takeuchi, J.; Hamasaki, T. *Phys. Rev. B* 2004, 69. 132412.

[122] Sakurai, H.; Tsujii, N.; Takayama-Muromachi, E. *J. Phys. Soc. Jpn.* 2004, 73. 2393-2396.

[123] Motohashi, T.; Ueda, R.; Naujalis, E.; Tojo, T.; Terasaki, I.; Atake, T.; Karppinen, M.; Yamauchi, H. *Phys. Rev. B* 2003, 67. 064406.

[124] Sugiyama, J.; Itahara, H.; Brewer, J. H.; Ansaldo, E. J. *Phys. Rev. Lett.* 2003, 67. 214420.

[125] Bayrakci, S. P.; Bernhard, C.; Chen, D. P.; Keimer, B.; Kremer, R. K.; Lemmens, P.; Lin, C. T.; Niedermayer, C.; Strempfer, J. *Phys. Rev. B* 2004, 69. 100410.

[126] Mendels, P.; Bono, D.; Bobroff, J.; Collin, G.; Colson, D.; Blanchard, N.; Alloul, H.; Mukhamedshin, I.; Bert, F.; Amato, A.; Hillier, A. D. *Phys. Rev. Lett.* 2005, 94. 136403.

[127] Rivadulla, F.; Bañobre-López, M.; García-Hernández, M.; López-Quintela, M. A.; Rivas, J. *cond-mat/0503010.*

[128] Li, S. Y.; Taillefer, L.; Hawthorn, D. G.; Tanatar, M. A.; Paglione, J.; Sutherland, M.; Hill, R. W.; Wang, C. H.; Chen, X. H. *Phys. Rev. Lett.* 2004, 93. 056401.

[129] Sakurai, H.; Tsujii, N.; Takayama-Muromachi, E. *Proceedings of SCES '05*, to be published in Physica B.

[130] Ihara, Y.; Ishida, K.; Michioka, C.; Kato, M.; Yoshimura, K.; Sakurai, H.; Takayama-Muromachi, E. *J. Phys. Soc. Jpn.* 2004, 73. 2963-2966.

[131] Ray, R.; Ghoshray, A.; Ghoshray K.; Nakamura, S. *Phys. Rev. B* 1999, 59. 9454-9461.

[132] Gavilano, J. L.; Rau, D.; Pedrini, B.; Hinderer, J.; Ott, H. R.; Kazakov, S. M.; Karpinski, J. *Phys. Rev. B* 2004, 69. 100404.

[133] Carretta, P.; Mariani, M.; Azzoni, C. B.; Mozzati, M. C.; Bradarić, I.; Savić, I.; Feher, A.; Ševek, J. *Phys. Rev. B* 2004, 70. 024409.

[134] Sakurai, H., et al. unpublished.

[135] Iliev, M. N.; Litvinchuk, A. P.; Meng, R. L.; Sun, Y. Y.; Cmaidalka, J.; Chu, C. W. *Physica C* 2004, 402. 239-242.

[136] Uemura, Y. J.; Russo, P. L.; Savici, A. T.; Wiebe, C. R.; MacDougall, G. J.; Luke, G. M.; Mochizuki, M.; Yanase, Y.; Ogata, M.; Foo, M. L.; Cava, R. J. *cond-mat/0403031*.
[137] Huang, Q.; Foo, M. L.; Lynn, J. W.; Zandbergen, H. W.; Lawes, G.; Wang, Y.; Toby, B. H.; Ramirez, A. P.; Ong, N. P.; Cava, R. J. *J. Phys.: Condens. Matter* 2004, 16. 5803-5814.
[138] Lee, K.-W.; Kuneš, J.; Novak, P.; Pickett, W. E. *Phys. Rev. Lett.* 2005, 94. 026403.
[139] Chen, X. H.; Wang, C. H.; Zhang, H. T.; Lu, X. X.; Wu, G. *cond-mat/0501181*.
[140] Terasaki, I.; Sasago, Y.; Uchinokura, K. *Phys. Rev. B* 1997, 56. R12685-R12687.
[141] Terasaki, I.; Tsukuda, I.; Iguchi, Y. *Phys. Rev. B* 2002, 65. 195106.
[142] Mukhamedshin, I. R.; Allol, H.; Collin, G.; Blanchard, N. *Phys. Rev. Lett.* 2004, 93. 167601.

In: New Research on Superconductivity and Magnetism
Editor: Lannie K. Tran, pp. 175-208
ISBN: 1-60021-540-8


*Chapter 6*

# FABRICATION AND TRANSPORT PROPERTY OF DIRECTIONALLY SOLIDIFIED RE123

***Yuichi Nakamura, Ryoji Inada and Akio Oota***
Toyohashi University of Technology

## Abstract

$REBa_2Cu_3O_y$ (RE123: RE=Y, Sm, Gd etc.) superconductive oxides have potential for many applications since they exhibit a high critical current density ($J_c$) under high magnetic fields at 77 K. In this paper, a directional solidification of RE123 (RE=Y and Sm) samples was carried out. The effect of fabrication conditions on the microstructure and transport $J_c$ properties of Y123 fibers were reported, and the continuous growth condition of Sm123 are also investigated. The transport $J_c$ property was measured by the standard d.c. 4 probe method. The continuous growth of the 0.5 mmϕ Y123 fiber was achieved in the samples grown at high pulling rates up to 10 mm/h while the usual Y123 rods of 2 mmϕ could grow continuously up to several mm/h. The size of Y211 particles trapped in Y123 was found to decrease with increasing pulling rate. The Sm123 can continuously grow at higher pulling rates than Y123 for both fibers and rods. The transport $J_c$ properties of Y123 fibers showed the maximum in the sample pulled at 6 mm/h; the best $I_c$ and corresponding $J_c$ values achieved were 134 A and $1 \times 10^5$ A/cm$^2$ at 77K self-field, respectively. This $J_c$ value is the best transport $J_c$ of the melt grown samples to date and is comparable to the best $J_c$ value of small Y123 bulks. This high $J_c$ value may attributed to the well-controlled microstructure and crystal orientation of the sample fabricated at high pulling rate. That is, the preferential crystal orientation as well as fine Y211 particles is thought to result this high transport $J_c$ of the fiber sample. On the other hand, the enlargement of continuous growth condition for the fiber samples could be fundamentally explained by the difference in effective nucleation rate, which was defined as the product of nucleation frequency and undercooled volume in the liquid. Because of the difference in the cross sectional area, the small nucleation volume of thin fiber leads the reduction of the nucleation rate, and may result in the increase of the continuous growth rate. This difference in nucleation rate is, however, found not to be enough to explain the destruction of continuous growth, and the additional factor such as compositional shift at the interface is needed. To think of a reason causing the compositional shift, a non-steady state growth model is proposed, and the effect of process parameters and growth conditions on the compositional shift is discussed.

# 1 Introduction

Since the discovery of high $T_c$ superconductor, many efforts have been made to utilize them in practical applications. Among the high $T_c$ superconductors, $REBa_2Cu_3O_{7-x}$ (RE123; RE=Y, Gd, Sm, etc.) superconductive oxides are expected for high magnetic field applications since they show the high critical current density ($J_c$) under high magnetic field at 77 K. Many applications such as superconducting bulk magnets and magnetic suspensions are proposed applying the high $J_c$ and high flux pinning properties using large melt-grown bulk RE123 superconductors [1-2]. The superconducting current leads for low temperature superconducting devices such as superconducting magnets, SMES and superconducting transformers cooled with liquid helium or cryocoolers are also a favorable application since the use of oxide superconductor that has a low thermal conductivity and a lack of resistivity can markedly reduce the heat load [3-14]. Although Bi based superconducting wires are also used for the current leads, the use of Ag-alloy sheath increases the thermal conductivity compared to that of superconductive oxide. Instead of them, the use of RE123 current leads results the further reduction of thermal conductivity and the compactness of the current leads, and such the leads are especially favorable to the relatively compact size superconducting magnet system cooled with cryocooler.

To achieve the high $J_c$ values for practical applications of RE123, the melt and growth process is widely used for attaining large and highly oriented textures, which reduce a weak link problem at the grain boundaries. It is also effective to disperse fine high temperature stable phase ($RE_2BaCuO_5$: RE211) particles into the RE123 phase matrix for introducing strong magnetic flux pinning sites through a peritectic solidification of the RE123 crystals. For applying the melt textured RE123 materials to superconducting current lead, we have to fabricate a continuously grown RE123 crystal in an entire length to achieve the high critical current density. A directional solidification method is suitable to fabricate the RE123 rods with such the single domain structure, and the high $I_c$ values over 1000 A were reported using the directionally solidified samples with 3-5 mm in diameter [5,6].

The better understanding of the growth mechanism of the RE123 crystal is required to control the microstructure of the sample during the melt processing. The peritectic solidification of RE123, RE211 + liquid = RE123, is widely recognized to proceed by the solute diffusion between the RE211 particles dispersed in the liquid and the RE123 growing interface through the liquid phase [15-19]. The liquid composition close to the interface is an important factor to control the growth of RE123 crystal because the solute diffusion for RE123 growth is caused by the compositional difference between the RE211/liquid and the RE123/liquid interfaces. In addition to the solute diffusion, some degree of supersaturation at the RE123 interface is also required to grow the RE123 crystal [19]. Because of the difference in the interface kinetics of the *ac*- and *ab*-planes of Y123, the temperature dependence of the growth rates of the planes is different from each other [20,21]. To attain the high growth rate of RE123 crystal for high production rate, the large undercooling is required to achieve the large driving force for solute diffusion as well as supersaturation. However, the excessive undercooling and corresponding supersaturation lead the destruction of the continuous growth of the RE123 single domain crystal due to the undesired nucleation of RE123 crystal in liquid ahead of the growth interface, and cellular or blocky structures are easily formed. Such the polycrystalline microstructures include grain boundaries that act as weak links and can not use for the high current applications. Consequently, the growth rate to attain the continuous growth of Y123 rod with a few mm in diameter is usually limited less than a few mm/h [15, 16, 18].

On the other hand, the RE123 crystals in which the usual Y element is substituted by light rare earth elements show better properties compared with Y123 such as the higher critical temperature, higher $J_c$ properties under high magnetic field including peak effect [22, 23], and lower thermal conductivity [24, 25]. Such the properties superior to the Y123 are also useful for current lead applications.

In this report, the effects of fabrication conditions on the microstructure and the transport Jc property of directionally solidified RE123 crystal are investigated. At first, the fundamentals of solidification growth mechanism of RE123 crystal are briefly introduced in the next section. In the third section, the details of the experimental procedure about fabrication and evaluation of the samples are shown. In the fourth section, the continuous growth conditions of Y123 and Sm123 crystals concerning to the sample diameter and the transport Jc properties of the Y123 fibers are reported. The effect of pulling rate on the size distribution of RE211 particles dispersed in Y123 crystals is also mentioned. Based on these experimental results, the mechanisms controlling the microstructure of directionally solidified RE123 and the effect of them on the transport properties are discussed in detail, in the fifth section. Concerning to the breakdown of continuous growth, a non-steady state solidification model is also proposed. Finally, the results and discussion are summarized in the last section.

## 2 Fundamentals of Growth Mechanism of Directionally Solidified RE123 Crystal

As mentioned briefly in previous section, the RE123 superconductive oxides are solidified through the peritectic solidification of RE123, RE211 + liquid = RE123, by the solute diffusion between the RE211 particles dispersed in the liquid and the RE123 growing interface through the liquid phase [5-9]. This solute diffusion is crucial to determine the growth of RE123 crystal, and the interface undercooling is an important factor to determine this solute diffusion and the microstructure of the RE123 crystal. A simple image of the compositional difference as a driving force for RE123 growth is schematically shown in Figs. 1 and 2 where the bulk liquid is assumed to be in equilibrium with the RE211 phase. In this image, $\Delta C_r$ means the compositional difference due to the curvature of RE211 particle, $\Delta C_{\Delta T}$ means that due to the interface undercooling, and $\Delta C_G$ means that due to the temperature gradient at the interface, respectively. These three factors are the driving force for solute diffusion. On the other hand, the growth interface of RE123 crystal shows clear faceted interfaces. This means that the growth of RE123 is also controlled by the interface kinetics as well as solute diffusion in liquid. Hence, some degree of interface supersaturation is required for the growth of RE123 and is shown as $\Delta C_k$ in these figures.

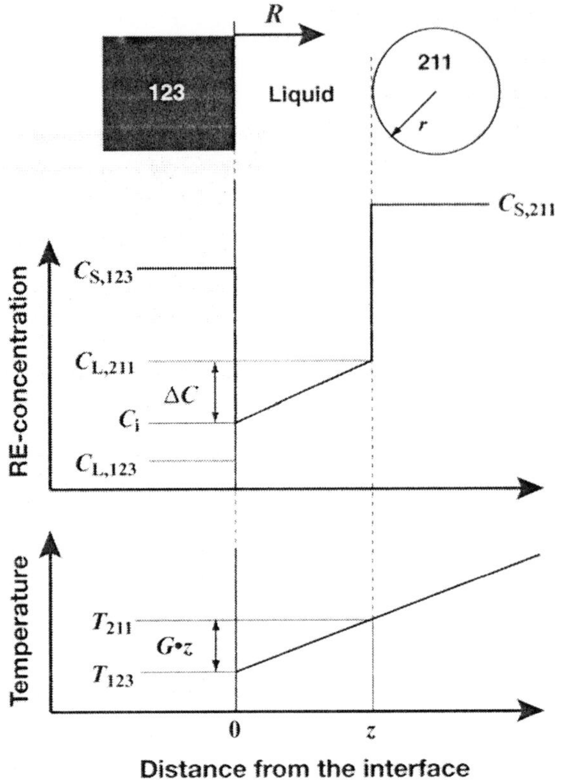


Fig. 1 Schematic image of the solute diffusion for peritectic solidification of RE123 showing composition and temperature profiles between the growth interface and RE211 particle in liquid.

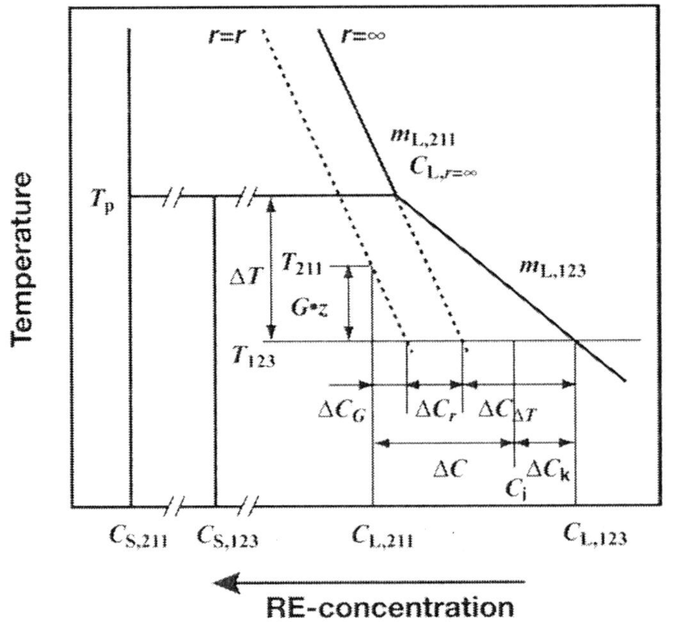


Fig. 2 Schematic quasi-binary phase diagram showing the compositional differences as a driving force for the peritectic solidification of RE123.

Based on this image, the solute diffusion flux from the RE211 to the RE123 growth interface, $J_{211}$, is described as:

$$J_{211} = -(1 - V_{L,211})D_L \frac{\partial C}{\partial z} = -(1 - V_{L,211})D_L \frac{\Delta C}{\Delta z}$$
$$= -(1 - V_{L,211})D_L \frac{(\Delta C_r + \Delta C_{\Delta T} + \Delta C_G - \Delta C_k)}{\delta} \quad (1)$$

where $D_L$ is the diffusivity in liquid and δ is the thickness of diffusion boundary layer, and $V_{L,211}$ is the volume fraction of RE211 particles in bulk liquid, respectively. On the other hand, the necessary flux for RE123 growth, $J_{123}$, is described as:

$$J_{123} = (1 - V_{S,123})(C_i - C_{S,123})R_{123} \quad (2)$$

where $V_{S,123}$ is the volume fraction of RE211 in RE123, $C_i$ is the composition in liquid at the RE123 interface, $C_{S,123}$ is the composition in RE123, and $R_{123}$ is the growth rate of RE123 crystal. In the steady state growth condition, these fluxes should be same.

$$J_{211} = J_{123} \quad (3)$$

In addition to this mass balance condition, the growth rate is described as a function of supersaturation at the growing interface as followings:

$$R_{123} = f(\sigma) = f((C_i - C_{L,123})/C_{L,123}) \quad (4)$$

where $\sigma$ is the supersaturation, and $C_{L,123}$ is the equilibrium composition in liquid at RE123 interface, respectively. The function $f$(s) is determined from the mechanism of interface kinetics and could be generally approximated as a power law function:

$$f(\sigma) = k\sigma^n \ (1 \leq n \leq 2) \quad (4)'$$

where $k$ is a kinetic coefficient. Accordingly, the growth rate of the RE123 crystal is determined as the balance of solute diffusion and interface kinetics by solving equations (1) through (4) simultaneously.

Another important factor to understand the transport current and mechanical properties by controlling growth feature of RE123 especially in directional solidification is the growth anisotropy of the RE123 crystal between the *ab*- and *ac*-planes since it determines the preferential growth direction of RE123 during directional solidification [26]. The crystal orientation of RE123 along the longitudinal direction, in which direction transport current flows, is important because it affects the critical current and mechanical properties through the electrical and mechanical anisotropy of RE123. Due to the faceted growth of RE123, the maximum growth rate and its direction of RE123 crystal are determined by composing the growth rates of the *ab*- and *ac*-planes. This feature is schematically shown in Fig. 3. As shown in this figure, the maximum growth rate $R_{max}$ is determined as the moving rate of the corner of the faceted crystal and given by the relation of:

$$R_{max} = \frac{R_a}{\cos\theta_a} = \frac{R_c}{\cos\theta_c} \tag{5}$$

where q$_a$ is the angle from the direction with maximum growth rate to the *a*-axis, q$_a$ is that to the *c*-axis, and $R_a$ and $R_c$ are the growth rates of the *ab*- and *ac*-planes, respectively. In the directional solidification process, the precursor sample has a polycrystalline structure and no seed crystal is used. In this case, the crystal orientation along the pulling direction is initially random in precursor while the crystal grains which have the highest growth rate along the pulling direction grow preferentially among the grains as shown in Fig. 4. Consequently, the preferential growth orientation along the pulling direction is determined in the initial growth stage as the result of grain selection since the crystal with the highest growth rate along the pulling direction will be survived to determine the crystal orientation.

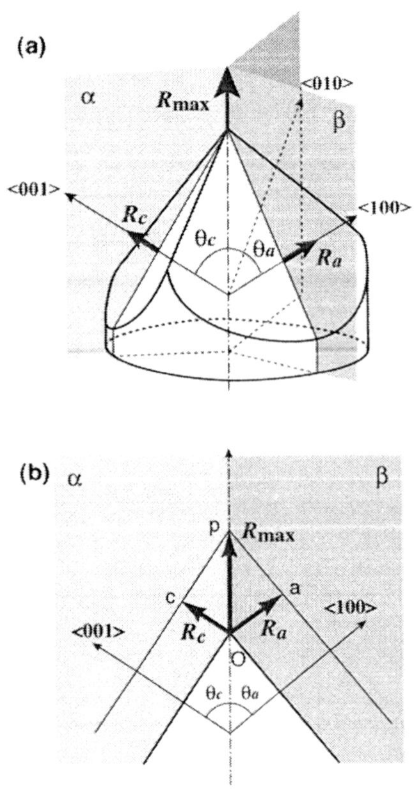


Fig. 3 Schematic illustration of directionally solidified RE123 growth interface showing the relation between the maximum growth rate of the three-dimensional crystal and the growth rate of the faces forming the interface. (a) three dimensional illustration of growth interface (b) illustration of two spread sections shown in (a); section a including $R_{max}$ and $R_c$ and section b including $R_{max}$ and $R_a$. The growth interface composed of the {100} and {001} faces, and the maximum growth rate of the RE123 crystal is determined by the combination of the growth rate of each face.

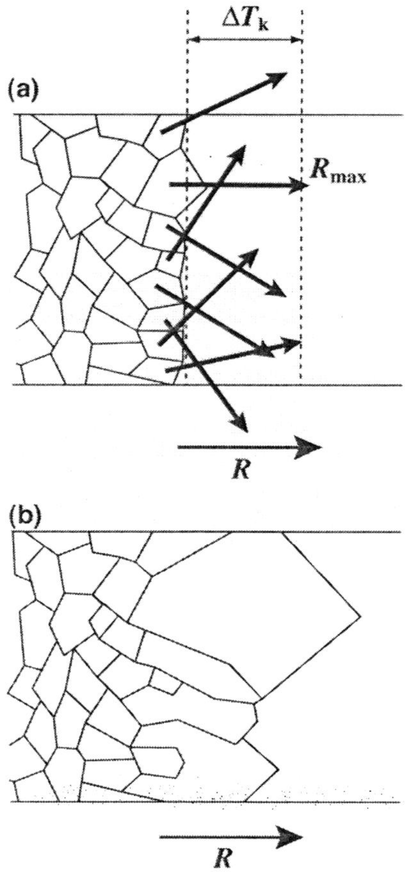


Fig. 4 Schematic illustration at the onset of solidification: (a) before growing where growth directions of each crystal with the maximum growth rate are at random due to the random crystal orientation of precursor. (b) the crystal with the direction of maximum growth rate parallel to the pulling direction can grow preferentially.

# 3 Experimental Procedures

Sample fibers with a diameter of 0.5 mm and rods with 2 mm in diameter were directionally solidified by zone melting method. Raw materials of $RE_2O_3$ (RE=Y and Sm), $BaCO_3$ and CuO were mixed in the ratio of RE:Ba:Cu = 1.4:2.2:3.2 and calcined at 900 °C for 45 hours in air with two intermediate grindings to prepare precursor powders. The precursor powder was extruded with organic binder to form thin fibers with a diameter of 0.5 mm. On the other hand, the precursor powder was packed into a rubber tube to form a 2mm rod by cold isostatic pressing (CIP). The formed precursor fibers and rods were sintered at 900 °C before zone melting. Figure 5 shows a schematic illustration of the zone melting set up used in this experiment. The maximum temperature in the furnace was controlled to be in the range between 1060 and 1100 °C The sample fiber or rod was lowered into the preheated furnace rapidly and then pulled upward at predetermined rates ranging from 1 to 20 mm/h more than 70 mm for transport $J_c$ measurement. To observe the morphology of the growing interface and the microstructure, some of the samples were quenched in air after about 30 mm growth.

The microstructure of the solidified samples was observed with a polarized optical microscope. To evaluate $I_c$ and $J_c$ properties of the fibers, electrodes for current leads and potential taps were formed with silver paint. The painted samples were sintered at 900 °C for 1 hour and subsequently annealed at 450 °C for about 100 hours in flowing oxygen condition before measuring $J_c$. The transport $J_c$ was measured by the standard d.c. four-probe method in liquid nitrogen using a criterion of 1 μV/cm

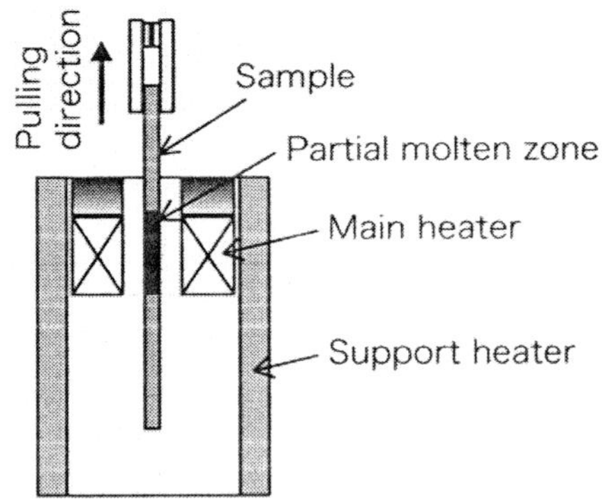


Fig. 5 Schematic illustration of the zone melting set-up.

# 4 Results

## 4.1 Growth Morphologies of Y123 Fibers

Figure 6 shows the longitudinal cross sections around the growth interfaces of the directionally solidified Y123 fibers grown with several pulling rates under the temperature gradient of 26 °C/mm. A light gray region in the left side is the grown Y123, and a dark region on the right side of Y123 is the liquid phase before quenching, respectively. Y211 particles were finely dispersed in both regions. A continuously grown single faceted interface was observed in the samples grown at the pulling rates lower than 10 mm/h, while a cellular array structure was observed in the samples grown at higher rates than 15 mm/h and no continuously grown structure was observed. The relation among the microstructures and the process parameters of temperature gradient, $G$, and pulling rate, $R$, are summarized in Fig. 7, where the reported microstructures for the Y123 rods of about 2 mm in diameter [18] were also shown. In this figure, open symbols indicate the 0.5 mmϕ fibers, and filled symbols indicate the 2 mmϕ rods, respectively. As clearly shown in this figure, although the pulling rate to achieve continuous growth is as high as 2 mm/h for the samples with about 2 mm in diameter, the continuously grown structure could be obtained at the higher growth rates in thin fibers under the similar temperature gradient condition. The larger $G/R$ condition is preferable to attain the continuously grown structure in both samples.

Figure 8 shows the microstructures in the Y123 single domain crystals grown with several pulling rates. In this figure, dark particles are Y211 phase and the matrix is the grown Y123 crystal. We can see in this figure that the increase in pulling rate leads the reduction of Y211 particles trapped in Y123 crystal. Many Y211 particles with diameters larger than 10 μm exist in the sample pulled at 2 mm/h (Fig. 8 (a)). Comparing to this sample, the number of such the large Y211 particles decreases as increasing the pulling rate, and most of the Y211 particles have diameters of a few μm or less in the sample pulled at the rates higher than 6 mm/h. Figure 9 shows the size distribution of the Y211 particles trapped in the Y123 crystals shown in Fig. 8. The average particle sizes (*m*) and standard deviations (s) are also shown in this figure. The average particle size and standard deviation of 211 particles slightly decreases at higher pulling rates. Especially, the fraction of Y211 particles larger than 2 μm decreases and that less than 1.5 μm increases as increasing the pulling rate, simultaneously. This decrease in the size of Y211 trapped in Y123 at high pulling rates will be helpful to achieve a high $J_c$ property through the better flux pinning properties.

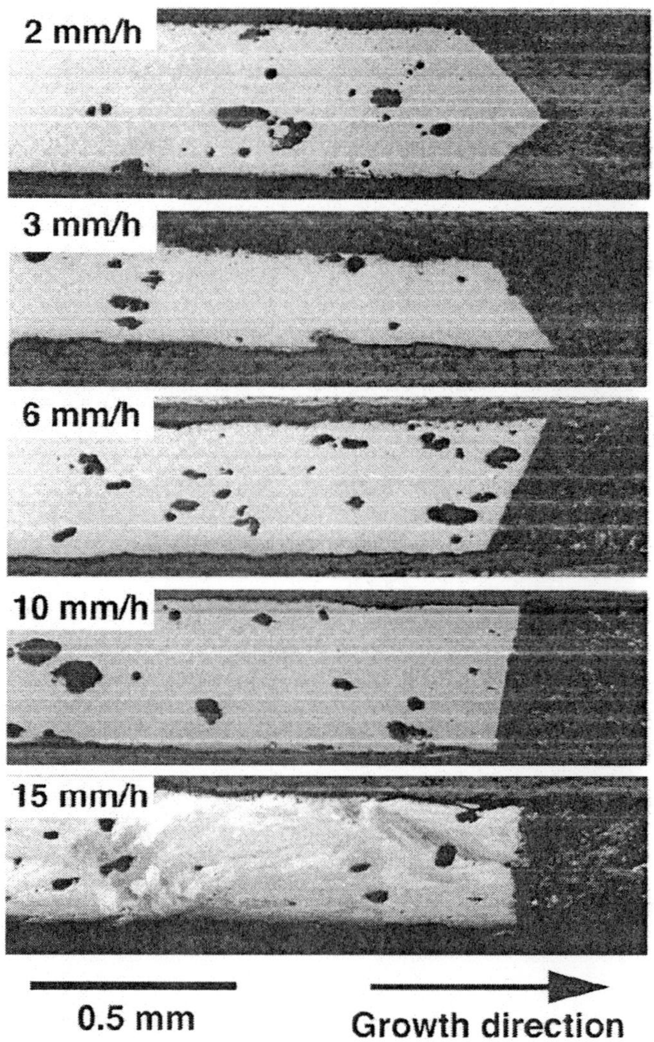


Fig. 6 Longitudinal cross sections around the growth interface of the directionally solidified Y123 fibers with the diameter of 0.5 mm grown at several pulling rates. The light gray region in the left side is the grown Y123 crystal, and the dark gray region in the right side is the liquid phase with Y211 particles before quenching, respectively. The black spots in Y123 are pores.

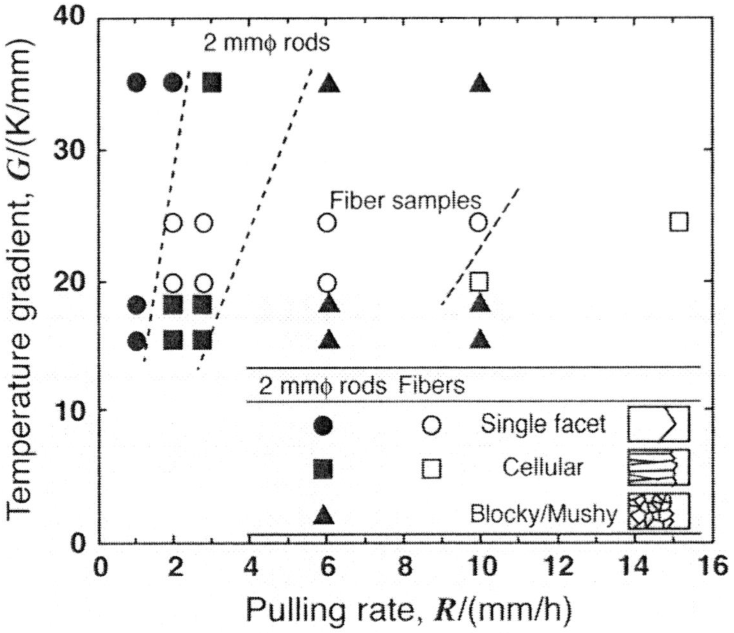


Fig. 7 The relation between the microstructures of directionally solidified Y123 and the process parameters of temperature gradient and growth rate for 0.5 mmϕ fibers and 2 mmϕ rods. Open symbols indicate the thin fiber and filled symbols indicate the fat rods. The continuous growth structure in fiber samples could be obtained in high growth rate conditions compared to those for the 2 mm rods.

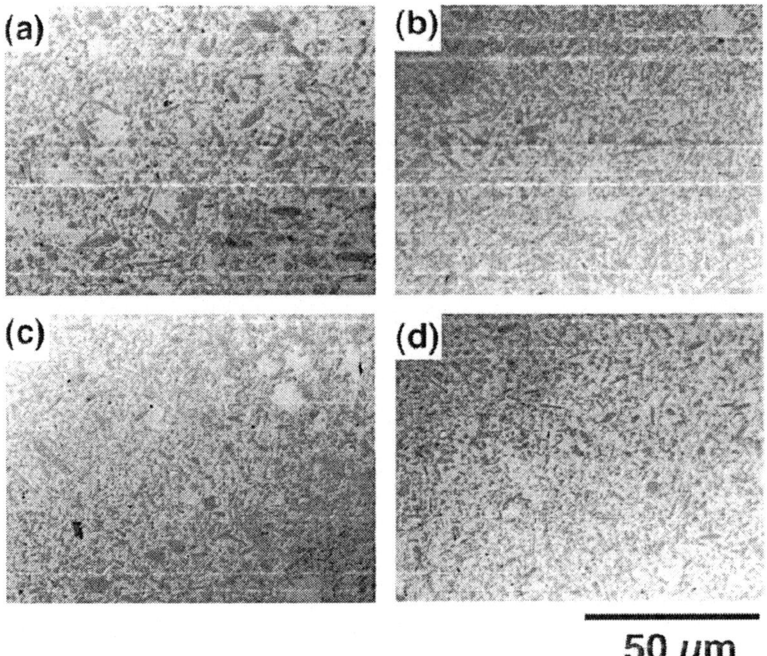


Fig. 8 The microstructures in the Y123 fibers grown at several pulling rates; (a) 2 mm/h, (b) 3 mm/h, (c) 6 mm/h and (d) 10 mm/h. The dark particles are Y211, and the matrix is Y123 crystal. The number of large 211particles decreases with increasing pulling rate.

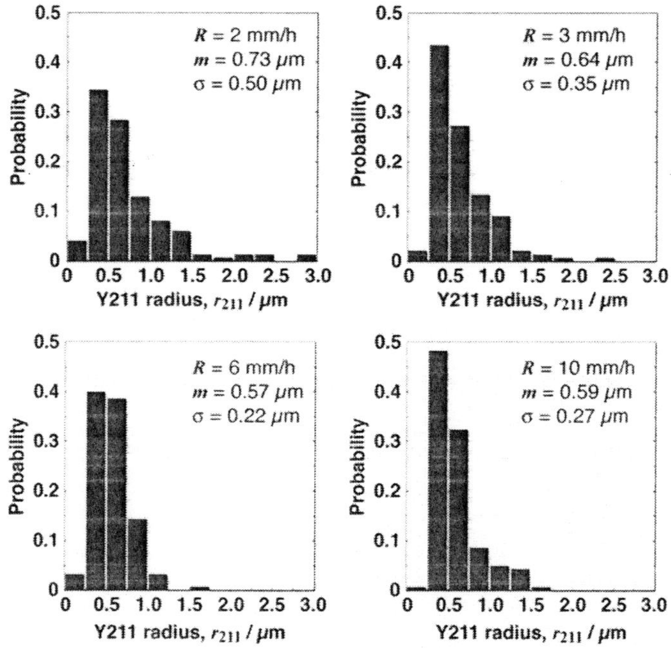


Fig. 9 Size distributions of 211 particles trapped in the Y123 crystals grown at several pulling rates. The average particle size ($m$) and standard deviation ($\sigma$) of each condition are also shown as insets.

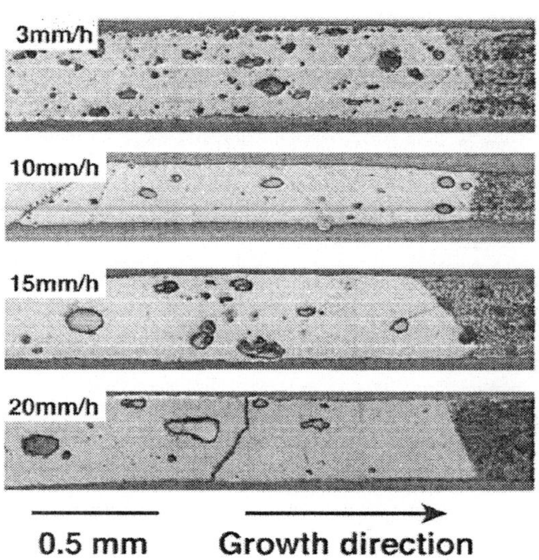


Fig. 10 Longitudinal cross sections around the growth interface of the directionally solidified Sm123 fibers with the diameter of 0.5 mm grown at several pulling rates.

## 4.2 Growth Morphologies of Sm123 Rods and Fibers

Figure 10 shows the longitudinal cross sections around the growth interface of the directionally solidified Sm123 fiber with the diameter of 0.5 mm. The light gray region in the left side is the grown Sm123 and the dark gray region in the right side is the liquid with

Sm211 particles before quenching, respectively. A continuously grown single faceted structure was obtained in the samples grown at the pulling rates up to 15 mm/h while a polycrystalline structure was observed at 20 mm/h. Similarly, Fig. 11 shows the microstructures around the growth interface of Sm123 rods with 2 mm in diameter grown under the temperature gradient of 12.5 K/mm. The continuously grown single faceted structure is achieved at the pulling rates up to 12 mm/h, which is very close to that of Sm123 0.5 mmϕ fiber, and the difference in maximum pulling rate for continuous growth between fiber and rod is small compared to the case of Y123. Figure 12 summarizes the continuous growth condition of Sm123 of 0.5 mm□ fibers and 2 mmϕ rods. Comparing with the continuous growth condition of Y123 shown in Fig. 7, the maximum pulling rate to achieve the continuous growth of Sm123 is larger than that of Y123 both for fibers and rods. This can be fundamentally attributed to the large compositional difference for solute diffusion through the difference in liquidus slope. Figure 13 (a) shows the liquidus lines of Y123 and Sm123 systems reported by Krauns et al [28], and Fig. 13 (b) and (c) show the schematic images of compositional differences at the same undercooling for different RE123 systems. As clearly shown in this figure, the compositional difference between Sm123 and Sm211 liquidus is larger than that between Y123 and Y211 liquidus at the same undercooling condition because of the relatively gentle slope of Sm123 liquidus line. In other words, the driving force for solute diffusion in the Sm123 system can easily attain larger value than that in the Y123 system. This leads the high potential for achieving high growth rates of Sm123 crystal.

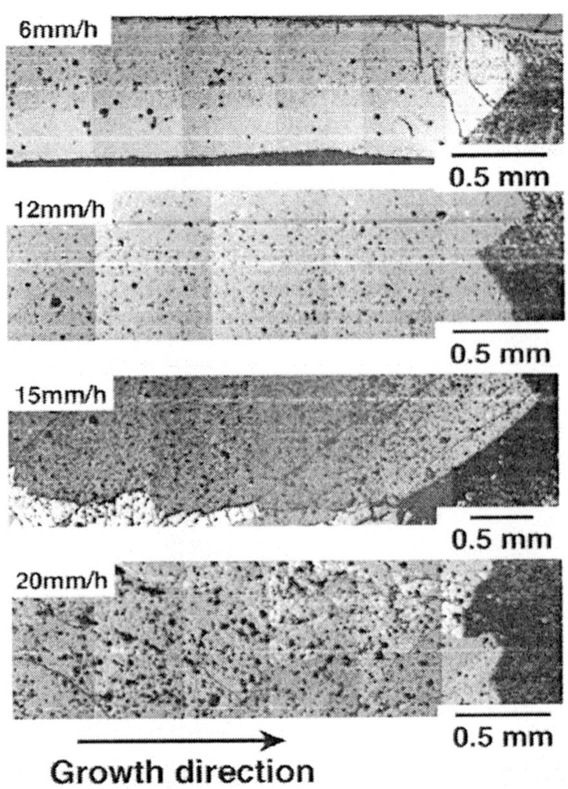


Fig. 11 Longitudinal cross sections around the growth interface of the directionally solidified Sm123 rods with the diameter of 2 mm grown at several pulling rates.

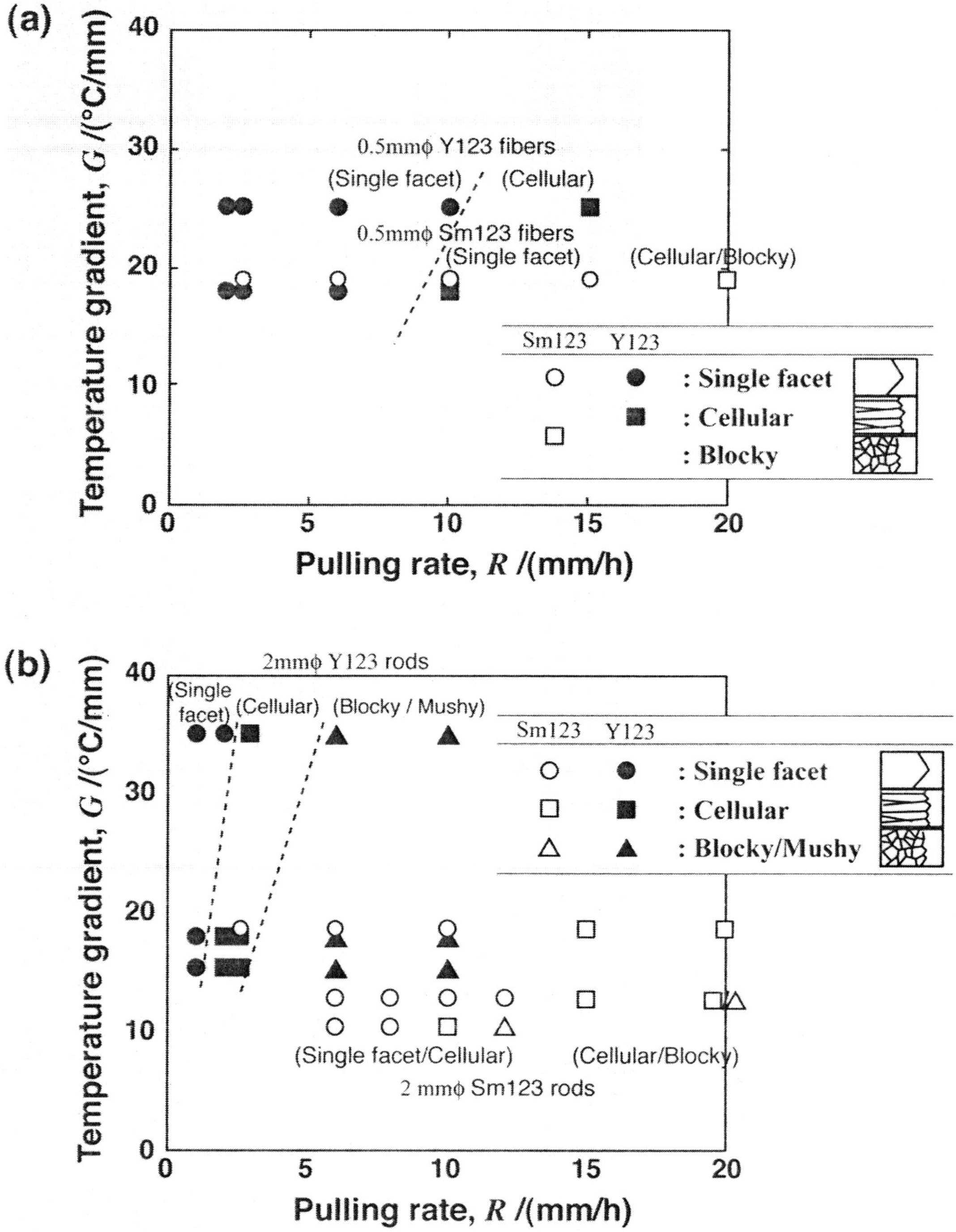


Fig. 12 The relation between the microstructures of directionally solidified Sm123 and the process parameters of temperature gradient and growth rate for (a) 0.5 mmϕ fiber and (b) 2 mmϕ rod. The relation about Y123 is also shown for comparison.

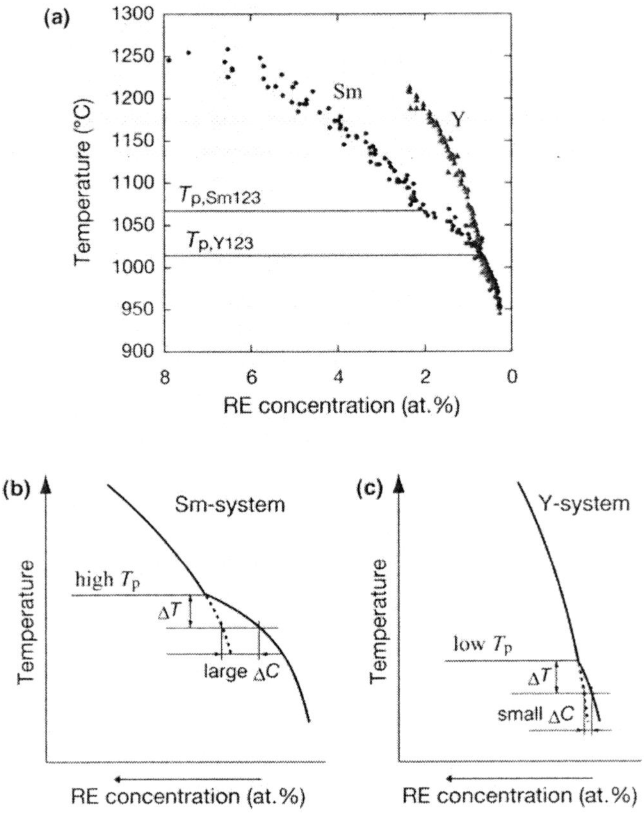


Fig. 13 (a) The liquidus lines of Y123 and Sm123 systems reported by Krauns et al [28], and the schematic images of compositional differences as the driving force for RE123 growth at the same undercooling in (b) Sm123 and (c) Y123 systems.

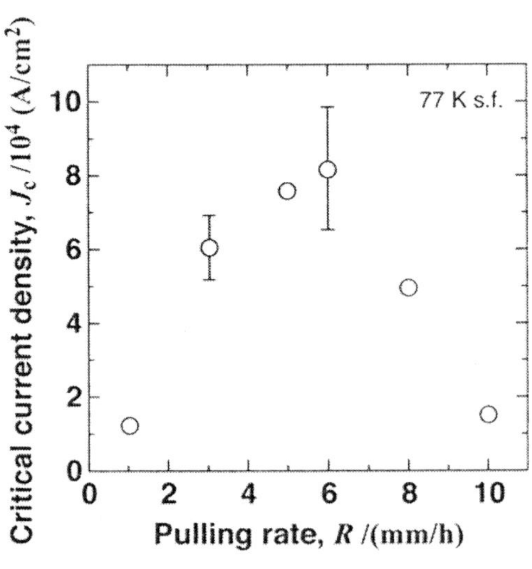


Fig. 14 The critical current densities of the Y123 fiber samples grown at several pulling rates at 77 K under self field. The plots show the average $J_c$ values and the bars mean the standard deviations, respectively.

### 4.3 Critical Current Density of Y123 Fibers

Figure 14 shows the transport $J_c$ values of the Y123 fibers grown at several pulling rates. As shown in this figure, the $J_c$ value increases with increasing the pulling rate up to 6 mm/h, while it decreases with further increase of the pulling rate larger than 8 mm/h. The best $I_c$ and corresponding $J_c$ values of 134 A and $1\times10^5$ A/cm$^2$ are obtained in the sample grown at 6 mm/h. This high transport $J_c$ value would be the best one measured for the melt grown samples to date, and is comparable to the best $J_c$ of Y123 measured by magnetization method.

## 5 Discussion

### 5.1 Effect of Pulling Rate on the Size of Y211 Particles

As shown in Figs 8 and 9, the size of Y211 particles trapped in Y123 crystal decreased with increasing the pulling rate. This reduction of the Y211 particle size could be explained by mainly two mechanisms; one is the suppression of coarsening in the molten zone and the other is the change of dissolution rate of Y211 particles ahead of the growth interface for Y123 growth due to the relative change of driving force. Since the length of partially molten zone of the sample is almost same in the range of pulling rates used in this condition, the increase of the pulling rate leads the decrease of the staying time of Y211 particles in liquid region.

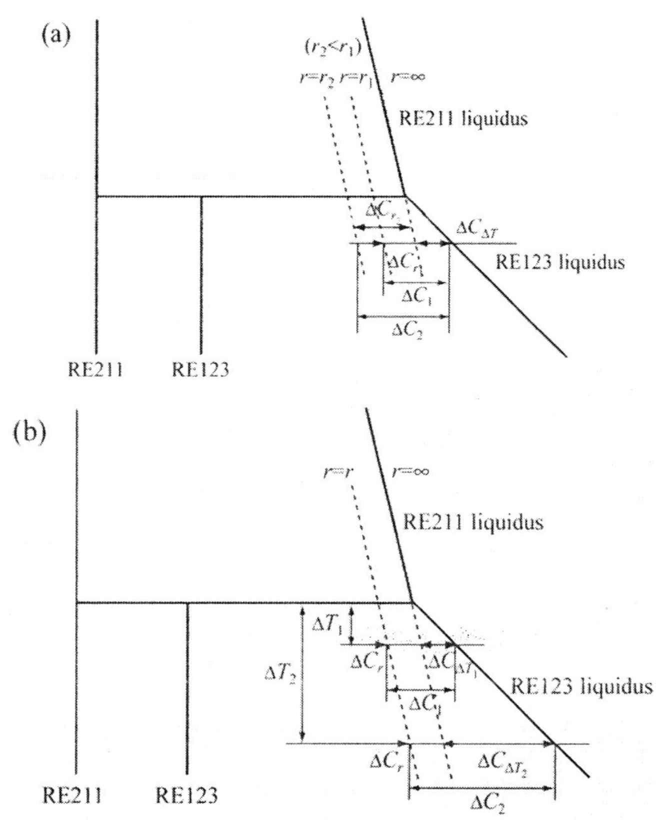

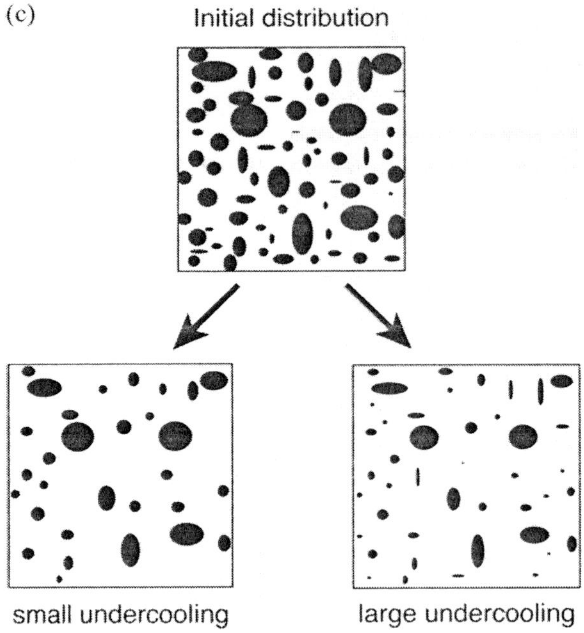


Fig. 15 Schematic phase diagram showing the effect of curvature and undercooling on the compositional difference for Y123 growth. (a) The effect of curvature is relatively large and small Y211 particles have large compositional difference. (b) The effect of curvature becomes relatively small under the large undercooling. (c) Schematic image showing the effect of growth condition on the size of trapped Y211 particles (black particles) in Y123 crystal. In the case of small undercooling or low growth rate, small Y211 particles dissolve preferentially, and the trapped particle size is relatively large. In the case of large undercooling or high growth rate, all Y211 particles may dissolve in the similar rate and the trapped size tends to decrease.

This shortening of the duration time in liquid suppresses the coarsening of Y211 particles and results in the decrease of the size of Y211 particles in liquid ahead of the Y123 growing interface. In addition, since the Y211 particles in liquid are dissolved to supply the Y-solute for the growth of Y123 crystal, the final Y211 size in Y123 is also affected by the dissolution rate of each Y211 particle. The driving force for the Y123 growth and the corresponding Y211 dissolution rate are determined by the effects of curvature of Y211 particles ($\Delta x_r$) [15,17-19] and the interface undercooling ($\Delta x_{\Delta T}$) [16-19] as mentioned in section 2; the former is relatively large in the slow growth rate and the small undercooling condition, while the latter becomes dominant in the high growth rate and large undercooling condition as schematically shown in Fig. 15. Hence, the smaller Y211 particles dissolve preferentially in the small growth rate (small undercooling) condition, while the larger Y211 particles may also dissolve as fast as small ones. As the result of this difference in driving force, the size of trapped Y211 particles become small in the Y123 crystals grown in the high pulling rate condition [17]. This fine Y211 particle size in Y123 may be a reason for the high $J_c$ value of the fiber sample grown at high pulling rate.

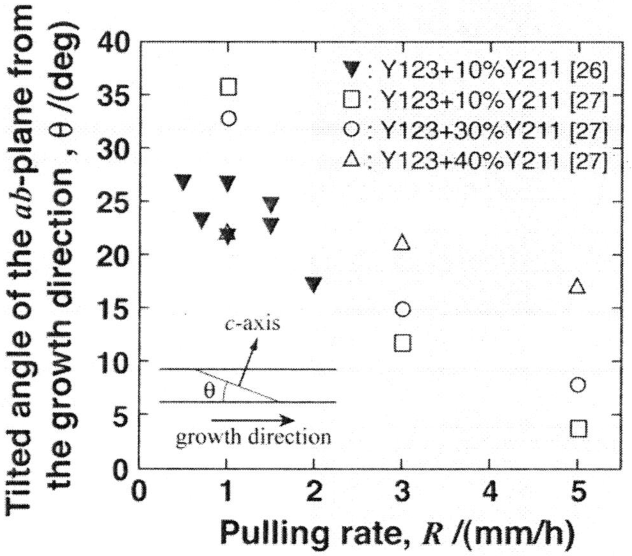


Fig. 16 The reported relation between the pulling rate and the tilted angle of the *ab*-plane from the growth direction of the directionally solidified Y123 samples.

## 5.2 Preferential Growth Direction and $J_c$

In addition to the effect of the RE211 size reduction, the increase of $J_c$ with increasing pulling rate up to 6 mm/h will be explained from the point of crystal orientation of the fibers. As mentioned in section 2, the preferential crystal orientation of the directionally solidified sample is determined by the relation between the growth rates of the faceted faces of Y123 crystal [26]. Figure 16 summarizes the reported relation between the pulling rate and the tilted angle of the *ab*-plane of Y123 crystal from the growth direction of the sample [26,27]. Since the ratio of the growth rates of the *ac*-planes to the *ab*-plane of Y123 crystal increases as increasing the undercooling [20,21], the preferential growth direction of directionally solidified Y123 changes as increasing the pulling rate which leads the increase of the interface undercooling. As the result of this growth anisotropy, the *ab*-plane of Y123 crystal would become almost in parallel to the growth direction of the sample at high pulling rate. When the *ab*-plane, in which plane the superconducting current flows, is almost parallel to the longitudinal axis at high pulling rate, such the desirable crystal orientation as well as the fine Y211 particles could result in the high $J_c$ value up to $10^5$ A/cm$^2$. According to the above discussion, the $J_c$ should be kept high values under the high pulling rate conditions when the continuous growth is maintained. However, the measured $J_c$ values decreased in the samples grown at the pulling rates higher than 8 mm/h while they still show the $J_c$ values over $10^4$ A/cm$^2$. The detailed observation of the microstructure with polarized microscopy revealed the existence of slight difference in color meaning the sub-domain structure in those samples. This suggests that the low angle grain boundaries caused by sub-domain structure may be easily formed although the continuous growth of Y123 crystal is possible under the high pulling rates about 10 mm/h, and these low angle grain boundaries would limit the $J_c$ value.

## 5.3 Continuous Growth Condition

### 5.3.1 Effect of Nucleation on Continuous Growth Condition

The Y123 thin fibers with the diameter of 0.5 mm show the single faceted interface at the pulling rates up to 10 mm/h, while such the single domain structure could be obtained at the pulling rates only below 2 mm/h for the samples with about 2 mm in diameter. A similar tendency was also observed in the Sm123 system although the difference due to the sample diameter was small. Thinking of a general reason for the break down of the continuous growth, a nucleation of the RE123 crystals must be occurred ahead of the growth interface. This means that a reason for this enlargement of continuous growth condition by thinning sample diameter is the reduction of the nucleation of the RE123 crystal in the undercooled liquid. In other words, the continuous growth structure is achieved when the nucleation of the RE123 crystal in the liquid is suppressed during the solidification.

During directional solidification processing, the partial molten zone ahead of the growth interface is undercooled below the peritectic temperature of RE123 to give the driving force for the growth of RE123 crystal. Since the (RE)-Ba-Cu-O liquid would have a quasi-equilibrium composition with the RE211 phase, the liquid composition in front of the RE123 growth interface is supersaturated to the RE123 phase; that is, the driving force for the continuous growth of RE123 simultaneously acts as the driving force for the nucleation of RE123 crystal. The RE123 crystal is, therefore, possible to nucleate anywhere in the undercooled liquid ahead of the growth interface corresponding to the nucleation frequency and the volume of undercooled region. This nucleation of RE123 crystal disturbs the continuous growth and forms a polycrystalline structures. The interface energy is an important factor to think of the nucleation frequency of RE123. Since the RE211 particles are reported to be pushed by the RE123 crystal [29], the RE123 crystal is not wet on the RE211 phase. Hence, the nucleation of RE123 in liquid would be the homogeneous nucleation or the nucleation on the surface of the sample. In the case of homogeneous nucleation, the nucleation frequency per unit volume per time, $I$, is the function of undercooling or supersaturation in the liquid phase and is generally expressed as:

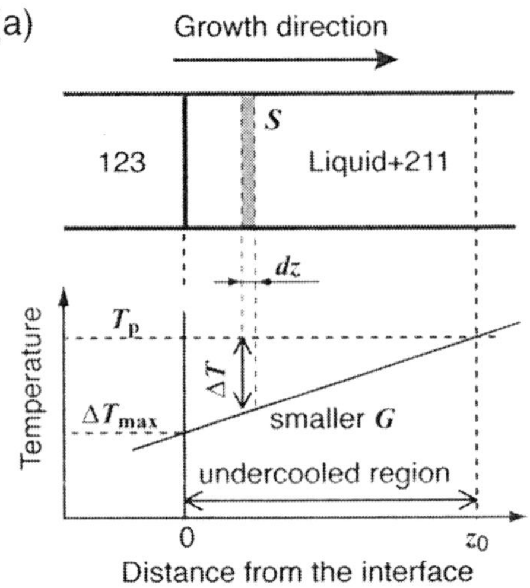

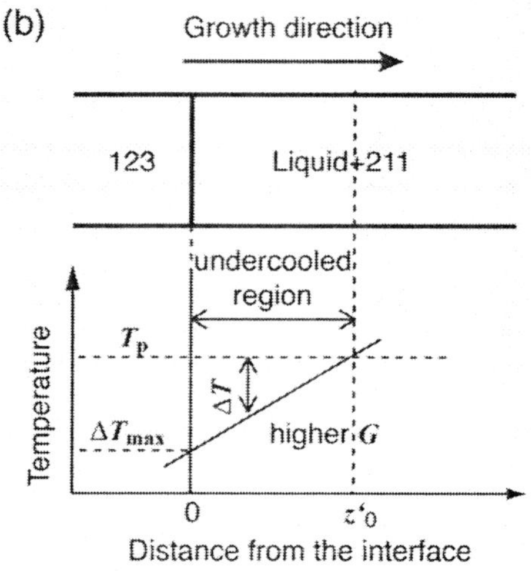


Fig. 17 Schematic image of the undercooled region ahead of the growth interface where the nucleation of RE123 crystals can occur for (a) small temperature gradient and (b) high temperature gradient cases. The length of undercooled region under high temperature gradient is shorter than that under small temperature gradient.

$$I = I_0 \exp\left(-\frac{\Delta G^*}{k_B T}\right) \tag{7}$$

by the classical steady-state homogeneous nucleation theory in condensed system [30], where $I_0$ is the constant concerning to the self-diffusion, $k_B$ is the Bolzman constant and $\Delta G^*$ is the activation energy of nucleation concerning to interface energy, undercooling, shape of nuclei, and so on. The activation energy, $\Delta G^*$, of an isotropic sphere is in inverse proportion to the square of the undercooling as $\Delta G^* = A / \Delta T^2$, where $A$ is the constant concerning to the surface energy between nuclei and liquid [30]. We presume that this nucleation theory can apply to the nucleation of the RE123 crystal in liquid during directional solidification process to understand the effect of nucleation qualitatively. Figure 17 shows the schematic image of undercooled region during directional solidification under steady state condition with the same growth rate condition. Based on equation (6), when the temperature difference along the radial direction is negligible, the nucleation rate of RE123 per time in a slice of the undercooled liquid region, $dV=Sdz$, with a cross sectional area, $S$, and a thickness, $dz$, at a temperature, $T$, is expressed as:

$$IdV = I_0 \exp\left(-\frac{\Delta G^*}{k_B T}\right) Sdz \tag{7}$$

The nucleation rate of RE123, , in the whole undercooled region per time is calculated by integrating equation (7) over the undercooled region:

$$\Psi = \int IdV = I_0 S \int_0^{z_0} \exp\left(-\frac{\Delta G^*}{k_B T}\right) dz \tag{8}$$

where $z_0$ is the length of undercooled region. Thinking of the directionally solidified RE123 crystal, the RE123 crystals grown at the same growth rate with the flat interface have the same interface undercooling, $\Delta T_{max}$. In this case, the integral term of the equation (8) has the same value for all samples independent of the cross sectional area under the same growth condition. Since the nucleation rate in the undercooled liquid is proportional to the cross sectional area of the sample from eq. (8), the nucleation rate of 0.5 mmϕ fibers is one order smaller than that of 2 mmϕ rods for the same pulling rate. This reduction of nucleation rate in thin fiber may expand the limit of pulling rate for continuous growth of RE123 fiber. Accordingly, the thin RE123 fibers may grow continuously with high pulling rate condition, and this high productivity may be an advantage to fabricate directionally solidified thin current leads.

In the similar way, the effect of temperature gradient is discussed. Considering the samples pulled with the same condition except for the temperature gradient, the length of undercooled region is shorter in the sample under higher temperature gradient as shown in Fig. 17 (b). This means that the integral term in equation (8) becomes small by increasing the temperature gradient due to the decreased length of undercooling region, $z'_0$. Consequently the nucleation rate in the undercooled region could be suppressed by increasing the temperature gradient, and the limit for pulling rate for continuous growth, therefore, expand under high temperature condition.

### 5.3.2 Condition for the Destruction of Continuous Growth

As mentioned previously, since the (RE-)Ba-Cu-O liquid ahead of the growth interface is undercooled with the composition in quasi-equilibrium with RE211, there is a driving force for the nucleation of RE123 crystal as well as that for the continuous growth of RE123. Although this nucleation of RE123 ahead of the growth interface is essential to break down the continuous growth of the directionally solidified samples, it is not enough to disturb the continuous growth. The overall growth rate of the interface is determined by the undercooling at the bottom of facet interface, where the interface undercooling is the largest under the temperature gradient, and the nucleation may occur near the bottom of the facet. Fig. 18 (a) shows the schematic growth feature of the continuously grown pre-existing crystal and the crystals nucleated a little ahead of the pre-existing one, where every crystal has almost the same growth rate. In addition, the pre-existing crystal is assumed to have the direction with the maximum growth rate along the pulling direction through the preferential growth in the initial growth stage [26]. Under this condition, the pre-existing crystal is not overgrown by other crystals even though the maximum growth direction of the nuclei is along the pulling direction as shown in Fig. 18 (a).

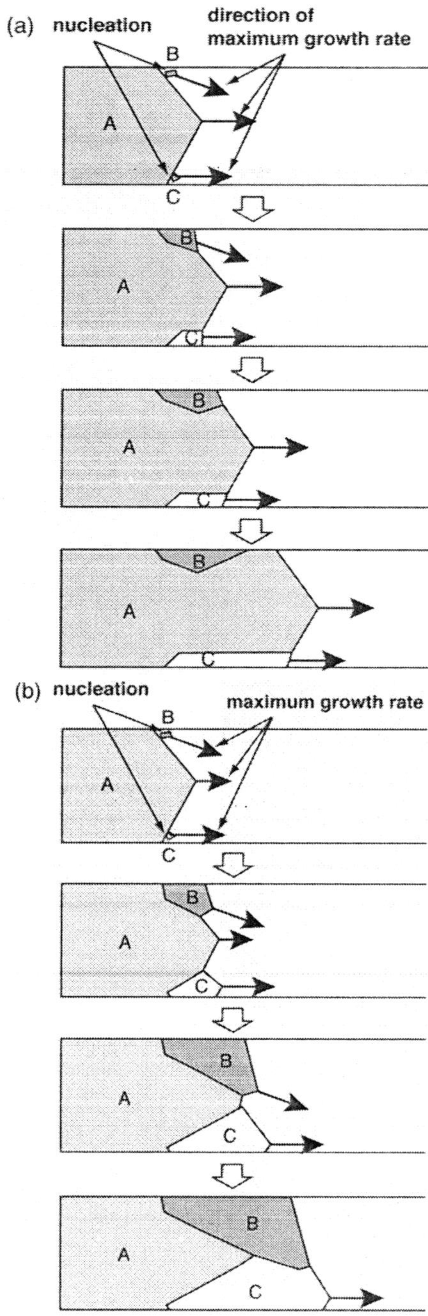


Fig. 18 Schematic image of the development of growth morphology after the nucleation ahead of the growth interface. The arrow shows the direction and speed of maximum growth rate of each crystal. Two crystals "B" and "C" nucleate ahead of the pre-existing crystal "A"; where (a) the direction with the maximum growth rate of crystals "A" and "C" is parallel to the pulling direction, and that of crystal "B" is different. As a result of grain selection, crystals "A" and "C" survives while the crystal "C" cannot overgrow the pre-existing crystal "A". (b) the directions with the maximum growth rate are same as the case of (a), but the growth rate of crystal "A" is smaller than other nucleated crystals. As a result of grain selection, the pre-existing crystal "A" is overgrown by crystals "B" and "C". Finally, the crystal "C" will overgrow the crystal "B".

In order to overgrow the pre-existing crystal to break down the continuous growth, the nucleation should occur near the top of the facet, or the growth rate of the pre-existing crystal should be lower than those of newly nucleated crystals. Since the nucleation frequency increases exponentially as decreasing activation energy through the increase of undercooling from the eq. (6), the nucleation near the top of the facet is less probable than that near the bottom. On the other hand, a possible reason for the decrease of the growth rate is proposed as the compositional shift in the liquid phase [31-33]. Since the RE-Ba-Cu-O system is not binary, the liquid compositions at the RE123/liquid and RE211/liquid interfaces have a freedom to move on the RE123 and RE211 liquidus faces under a constant temperature. This image is shown in Fig. 19 as a schematic isothermal cross section of the ternary phase diagram, where we think of the systems such as the Y-Ba-Cu-O system to neglect the effect of solid solution. When the liquid compositions $C^{jo}_{L,211}$ and $C^{jo}_{L,123}$ ($j$=RE, Ba and Cu, filled circles) are just on the line connecting compositions of RE211, RE123 and 035, the liquid compositions do not shift anywhere as described in the quasi-binary phase diagram. However, if either of the compositions is not on the tie-line (open circles) by some reasons such as a slight deviation of the initial average composition, those compositions may shift toward the corresponding side of the RE123-RE211-liquid three phase equilibrium point A or A'. This compositional shift may lead the decrease in compositional difference between $C^{j}_{L,211}$ and $C^{j}_{L,123}$, and will results the decrease in the growth rate. In usual, even if the small deviation of the liquid composition from the tie-line exists, the compositional shift will occur as a final transient and has less effect on the steady growth condition. However, the volume fraction of RE211 particles is enough large to decrease the macroscopic effective diffusivity in the bulk due to the reduction of cross sectional area through liquid, such the small effective diffusivity may enhance the compositional shift locally within the peritectic diffusion boundary layer [31-32]. Under the iso-thermal growth conditions, the compositional shift near the growth interface results in the decrease in the growth rate of RE123 crystal because of the decrease in compositional difference between the RE123/liquid and RE211/liquid interfaces. In the case of directional solidification under the temperature gradient, if the growth rate of solidified interface decreases due to the compositional shift, the interface temperature is decreased by pulling the growth interface to the lower temperature region faster than the growth rate of the interface as schematically shown in Fig. 20 (a). Since the compositional difference between RE123 and RE211 liquidus becomes large at lower temperature, the compositional difference between them needed for the growth could be maintained as the result of this decrease in the interface temperature as shown schematically in Fig 20 (b). Hence, the growth rate of the zone melting sample could be maintained almost constant while the interface composition and the temperature are continuously changing during growth. On the other hand, the liquid composition away from the interface is kept the initial average composition $C^{j,*}_{L,211}$ shown as the filled triangle in Figs 20(a) and 21(a). Accordingly, the large supersaturation for RE123 nucleation and growth would be achieved in such the bulk liquid region away from the interface. Under this condition, if the nucleation occurred in the bulk liquid region, the nucleated crystal has the growth rate higher than that of the pre-existing RE123 crystal and can overgrow to form the microstructure with the break down of continuous growth as shown in Fig. 18 (b).

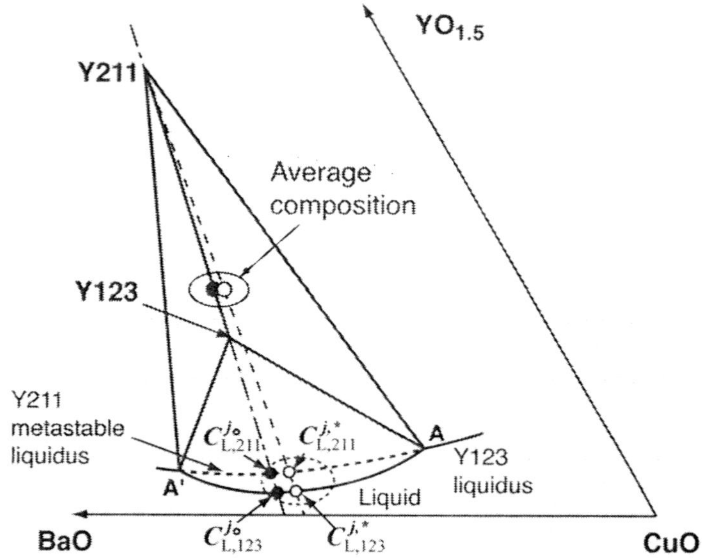


Fig. 19 Schematic illustration of the ternary phase diagram in the Y-Ba-Cu-O system showing the relation between the average composition and the compositions at Y123/liquid and Y211/liquid interfaces. The filled symbols show the case when the average composition is on the Y211-Y123 tie-line, and the open symbols show the case when the average composition is slightly shift to the CuO rich side.

## 5.4 Non-steady Growth Model

### 5.4.1 Basic Idea

Based on the fundamental idea mentioned previously, a growth model is considered to simulate the effect of process parameters on the compositional shift near the growth interface. The basic idea of this model is similar to the model for constant temperature [31] except for the temperature condition in the liquid. Here, the Y123 system is selected as the model system to exclude the effect of solid solution of RE123 for simplicity. The assumptions used in this model are as follows, (i) the temperature gradient near the interface is constant, (ii) the dissolution of Y211 for peritectic reaction occurs at the end of the peritectic boundary layer ($z=\delta$), (iii) the liquid composition is locally in equilibrium with the Y123 crystal at the growth interface ($z=0$) and with the Y211 in the bulk liquid region ($z>\delta$), (iv) no curvature effect of Y211 particles is considered, (v) the growth interface of Y123 is flat, and (vi) physical and thermodynamic properties are constant with temperature and composition. The equilibrium liquid composition in bulk liquid region is continuously changed at each position under the temperature gradient.

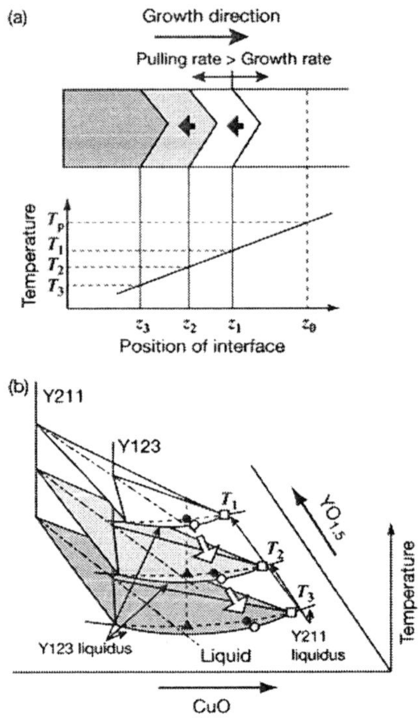


Fig. 20 Schematic illustrations of the compositional shift and changes of interface temperature during directional solidification. (a) the relation between the position of interface and temperature during the growth. When the growth rate of the crystal is smaller than the pulling rate due to the compositional shift, the growth interface moves to the low temperature region. (b) three dimensional illustration showing the relation between the compositional shifts near the interface and the interface undercooling in the schematic 3D ternary phase diagram. The open circles are the composition equilibrium with Y123 and the filled circles are those with Y211. Even when the interface composition shifts to the Y123/Y211/liquid three phase equilibrium point (open square), the compositional difference between Y123/liquid and Y211/liquid interfaces and growth rate of the crystal could be maintained due to the decrease in the interface temperature.

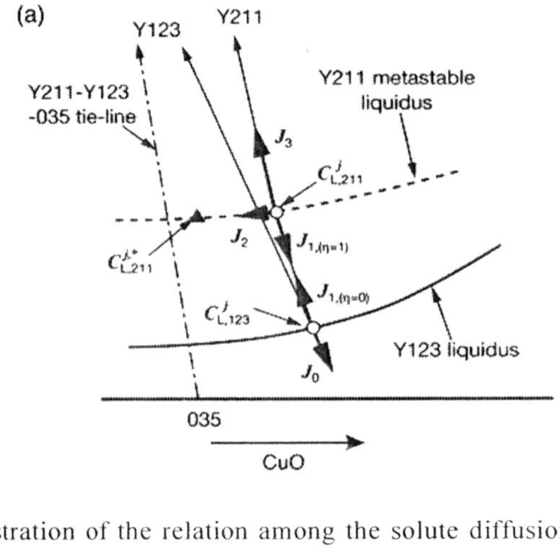


Fig. 21 (a) Schematic illustration of the relation among the solute diffusion flux vectors for the Y123 growth shown on the liquidus lines of the ternary phase diagram in the Y-Ba-Cu-O system when the compositional shift occurs.

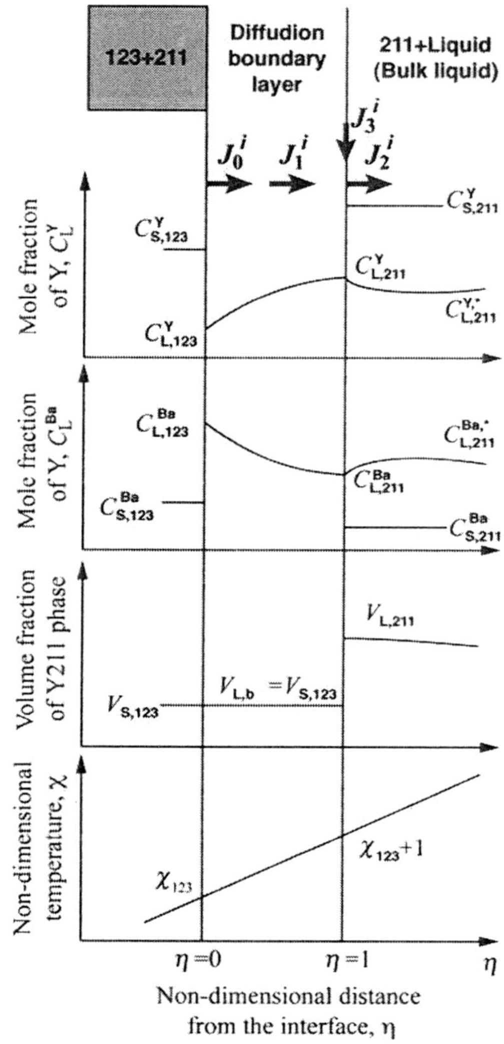


Fig. 21 (b) Schematic illustration of the compositional profiles, volume fraction of RE211, and temperature profile around the growth interface considered in the non-steady state directional solidification model.

Based on these assumptions, the fundamental images of compositional shift and the solute flux vectors during Y123 growth are schematically shown in Fig. 21 (a) as the schematic ternary phase diagram, where we consider a slight compositional deviation to CuO rich side. Fig. 21 (b) shows the corresponding compositional profiles, volume fraction of Y211, and temperature profile near the growth interface, respectively.

The mass fluxes of $j$ ($j$=Y, Ba and Cu) solute element concerning to the growth are defined as the following manner. The diffusion flux in peritectic diffusion boundary layer ($0<z<\delta$) is described as the following equation:

$$J_1^j = -\left(1 - V_{L,211}\right) D_L \frac{\partial C_L^j}{\partial z} = \frac{D_L}{\delta}\left\{-\left(1 - V_{L,211}\right)\frac{\partial C_L^j}{\partial \eta}\right\} = \frac{D_L}{\delta} J'^j_1 \quad (0 \leq z \leq \delta)$$

$$J'^{j}_{1} = -\left(1 - V_{L,211}\right)\frac{\partial C^{j}_{L}}{\partial \eta} \quad (0 \leq \eta \leq 1) \tag{9}$$

where, $J'^{j}_{1}$ is the non-dimensional flux, and $\eta = z/\delta$ is the non-dimensional distance from the interface, respectively. The diffusion flux in bulk liquid ($\delta < z$) is similarly described as follows considering the difference in effective diffusivity:

$$J^{j}_{2} = -\left(1 - V_{L,211}\right)D_{L,e}\frac{\partial C^{j}_{L,211}}{\partial z} = \frac{D_L}{\delta}\left\{-\left(1 - V_{L,211}\right)\kappa\frac{\partial C^{j}_{L,211}}{\partial \eta}\right\} = \frac{D_L}{\delta}J'^{j}_{2} \quad (\delta \leq z)$$

$$J'^{j}_{2} = -\left(1 - V_{L,211}\right)\kappa\frac{\partial C^{j}_{L,211}}{\partial \eta} \quad (1 \leq \eta) \tag{10}$$

where $D_{L,e}$ is the effective diffusivity in bulk liquid, and $\kappa = D_{L,e}/D_L$ is the reduction ratio of effective diffusivity. The fluxes at boundary conditions are described as follows; the flux of ejected $j$ solute element, $J^{j}_{0}$, due to the RE123 growth at the growth interface ($z=0$) is described as:

$$J^{j}_{0} = \left(1 - V_{S,123}\right)\left(C^{j}_{L,123} - C^{j}_{S,123}\right)R_{123} = \frac{D_L}{\delta}\left(1 - V_{S,123}\right)\left(C^{j}_{L,123} - C^{j}_{S,123}\right)\nu = \frac{D_L}{\delta}J'^{j}_{0} \quad (z=0)$$

$$J'^{j}_{0} = \left(1 - V_{S,123}\right)\left(C^{j}_{L,123} - C^{j}_{S,123}\right)\nu \quad (\eta=0) \tag{11}$$

where $V_{S,123}$ is the volume fraction of RE211 in RE123 crystal, and $\nu = R_{123}(t)\delta/D_L$ is the non-dimensional growth rate. The mass supply due to the dissolution of 211 phase particles at the outer end of peritectic diffusion boundary layer ($z=\delta$), $J^{j}_{3}$, is described as:

$$J^{j}_{3} = \left(V_{L,211} - V_{L,b}\right)\left(C^{j}_{S,211} - C^{j}_{L,211}\right)R = \frac{D_L}{\delta}\kappa\left(V_{L,211} - V_{L,b}\right)\left(C^{j}_{S,211} - C^{j}_{L,211}\right)\nu_0 = \frac{D_L}{\delta}J'^{j}_{3}$$

$$J'^{j}_{3} = \kappa\left(V_{L,211} - V_{L,b}\right)\left(C^{j}_{S,211} - C^{j}_{L,211}\right)\nu_0 \quad (\eta=1) \tag{12}$$

where $\nu_0 = R\delta/D_L$ is the non-dimensional pulling rate, and $V_{L,b}$ is the volume fraction of RE211 in peritectic diffusion boundary layer assuming to be same as $V_{S,123}$. In these equations, one should give attention that the compositions in bulk liquid, $C^{j}_{L,211}$, and the volume fraction of RE211 phase, $V_{L,211}$ is the function of temperature at each position, which is defined as:

$$T = T_{123} + Gz$$

and:

$$\chi = \chi_{123} + \eta \tag{13}$$

in the non-dimensional form, where $\chi = T/G\delta$ is the non-dimensional temperature. The position of growth interface and corresponding interface temperature may change during growth due to the difference between pulling rate and growth rate of the crystal. The relations are described as follows in non-dimensional form:

$$(R_{123} - R)dt = dz \quad \therefore (v - v_0)d\tau = d\eta \tag{14}$$

and from (13) and (14):

$$d\chi = (v - v_0)d\tau \tag{15}$$

where $\tau = Fo = D_L t/\delta^2$ is the non-dimensional time generally called Fourier Number. The symbols and non-dimensional parameters are summarized in Table 1.

Table 1 nomenclatures and non-dimensional parameters

| | |
|---|---|
| $t$ | : time |
| $z$ | : distance from the growth interface |
| $\delta$ | : length of diffusion boundary layer |
| $C^j_L$ | : mole fraction of $j$ ($j$ = Y, Ba and Cu) element in liquid |
| $C^j_{S,211}$ | : mole fraction of $j$ element in RE211 |
| $C^j_{S,123}$ | : mole fraction of $j$ element in RE123 crystal. |
| $R_{123}$ | : growth rate of RE123 crystal |
| $R$ | : pulling rate |
| $V_{L,211}$ | : volume fraction of RE211 phase in bulk liquid. |
| $V_{S,123}$ | : volume fraction of RE211 phase in RE123 crystal. |
| $V_{L,b}$ | : volume fraction of RE211 phase in diffusion boundary layer. |
| $G$ | : temperature gradient |
| $D_L$ | : diffusion coefficient in diffusion boundary layer |
| $D_{L,e}$ | : effective diffusion coefficient in bulk liquid |
| $\tau = Fo = \frac{D_L t}{\delta^2}$ | : non-dimensional time (Fourier number) |
| $\eta = \frac{z}{\delta}$ | : non-dimensional distance from the growth interface |
| $v = \frac{R_{123}\delta}{D_L}$ | : non-dimensional growth rate of RE123 crystal |
| $v_0 = \frac{R\delta}{D_{L,e}}$ | : non-dimensional pulling rate |
| $\chi = \frac{T}{G\delta}$ | : non-dimensional temperature |
| $\kappa = \frac{D_{L,e}}{D_L}$ | : ratio of effective diffusivity to diffusivity in liquid |

### 5.4.2 Mass Balance Condition

Under these conditions, we can consider the non-steady solute diffusion ahead of the growth interface in the non-dimensional form. The basic equation based on mass balance condition in liquid is described as following non-dimensional formula:

$$\frac{\partial C_{\mathrm{L}}^{j}}{\partial \tau} = \frac{\partial J'^{j}_{i}}{\partial \eta} + v\frac{\partial C_{\mathrm{L}}^{j}}{\partial \eta} + \left(C_{\mathrm{L}}^{j} - C_{\mathrm{S,211}}^{j}\right)\frac{\partial V_{\mathrm{L,211}}}{\partial \tau} \quad (0<\eta<1 \text{ for } i=1,\ 1<\eta \text{ for } i=2) \qquad (16)$$

where the first term in right hand means the difference in mass fluxes, the second term means moving coordinates, and the third term is the solute supply due to the dissolution or precipitation of the 211 particles. As a boundary condition at the 123/liquid interface, the degree of compositional shift is described as:

$$\frac{\partial C_{\mathrm{L,123}}^{j}}{\partial \tau} = \frac{J'^{j}_{0} - J'^{j}_{1,(\eta=+0)}}{\Delta\eta} + v\frac{\partial C_{\mathrm{L}}^{j}}{\partial \eta} + \left(C_{\mathrm{L}}^{j} - C_{\mathrm{S,211}}^{j}\right)\frac{\partial V_{\mathrm{L,211}}}{\partial \tau} \quad (\eta=0) \qquad (17)$$

where $\Delta\eta$ is the minimal thickness, and $J'^{j}_{1,(\eta=+0)}$ is the diffusion flux to the growth interface defined as:

$$J'^{j}_{1} = -\left(1 - V_{\mathrm{L,b}}\right)\left.\frac{\partial C_{\mathrm{L}}^{j}}{\partial \eta}\right|_{\eta=1+0} \qquad (18)$$

Similarly, the boundary condition at the outer end of the peritectic boundary layer is described as:

$$\frac{\partial C_{\mathrm{L,211}}^{j}}{\partial \tau} = \frac{J'^{j}_{1,(\eta=1-0)} - J'^{j}_{2,(\eta=1+0)} + J'^{j}_{3}}{\Delta\eta} + v\frac{\partial C_{\mathrm{L}}^{j}}{\partial \eta} + \left(C_{\mathrm{L}}^{j} - C_{S,211}^{j}\right)\frac{\partial V_{\mathrm{L,211}}}{\partial \tau} \quad (\eta=1) \qquad (19)$$

where $J'^{j}_{1,(\eta=1-0)}$ and $J'^{j}_{2,(\eta=1+0)}$ are defined as:

$$J'^{j}_{1} = -\left(1 - V_{\mathrm{L,b}}\right)\left.\frac{\partial C_{\mathrm{L}}^{j}}{\partial \eta}\right|_{\eta=1-0} \qquad (20)$$

$$J'^{j}_{2} = -\left(1 - V_{\mathrm{L,211}}\right)\kappa\left.\frac{\partial C_{\mathrm{L}}^{j}}{\partial \eta}\right|_{\eta=1+0} \qquad (21)$$

Under the local equilibrium condition, the compositions at the growth interface is on the Y123 liquidus face, and that in the bulk liquid is on the Y211 liquidus face, respectively. We can get the relation among the growth conditions, compositions and temperature from these equations. In this paper, the liquidus faces of Y123 and Y211 are assumed as flat faces for the first approximation with the forms of:

$$C^{\mathrm{Y}}_{\mathrm{L},123} + \alpha_{123} C^{\mathrm{Ba}}_{\mathrm{L},123} + \beta_{123} T + \gamma_{123} = 0 \tag{22}$$

$$C^{\mathrm{Y}}_{\mathrm{L},211} + \alpha_{211} C^{\mathrm{Ba}}_{\mathrm{L},211} + \beta_{211} T + \gamma_{211} = 0 \tag{23}$$

where $\alpha_k$, $\beta_k$ and $\gamma_k$ ($k$=123 and 211) are the constants estimated from the phase diagrams [28, 34]. The values of the constants and other parameters used in calculation are summarized in Table 2. In this paper, we only think the case of the deviations to the Cu-rich side. The deviations to the Ba-rich side can be treated in the same manner using the corresponding parameters describing the liquidus for equations (22) and (23).

Table 2 Properties used in calculation

| | | |
|---|---|---|
| Temperature gradient: | $G$ | |
| Boundary layer thickness: | $\delta$ | |
| Coefficients for 123-liquidus : | $\alpha_{123}$ | 0.0551432 |
| | $\beta_{123}$ | $-1.48´10^{-6}$ |
| | $\gamma_{123}$ | 0.0728144 |
| Coefficients for 211-meta stable liquidus : | $\alpha_{211}$ | 0.0425532 |
| | $\beta_{211}$ | $-5.94´10^{-7}$ |
| | $\gamma_{211}$ | 0.0178787 |

As shown in Fig. 20 (a), when the liquid composition $C^{j}_{\mathrm{L},211}$ and $C^{j}_{\mathrm{L},123}$ are not on the Y211-Y123-035 tie-line but deviate toward the CuO rich side, the interface composition, $C^{j}_{\mathrm{L},123}$ will shift to CuO rich side along the liquidus caused by the combination of the flux vectors $\boldsymbol{J}_0$ and $\boldsymbol{J}_{1,(\eta=0)}$. Correspondingly, there is the driving force to shift the composition of the bulk liquid, $C^{j}_{\mathrm{L},211}$, to the same direction as the result of combination of the flux vectors $\boldsymbol{J}_{1,(\eta=1)}$ and $\boldsymbol{J}_3$. However as described in equation (17), the driving force for this compositional shift is partially canceled by the flux, $\boldsymbol{J}_2$ which is caused by the compositional difference between $C^{j}_{\mathrm{L},211}$ and the initial composition, $C^{j,*}_{\mathrm{L},211}$. As the results the balance of each flux vector determines the degrees of these compositional shifts.

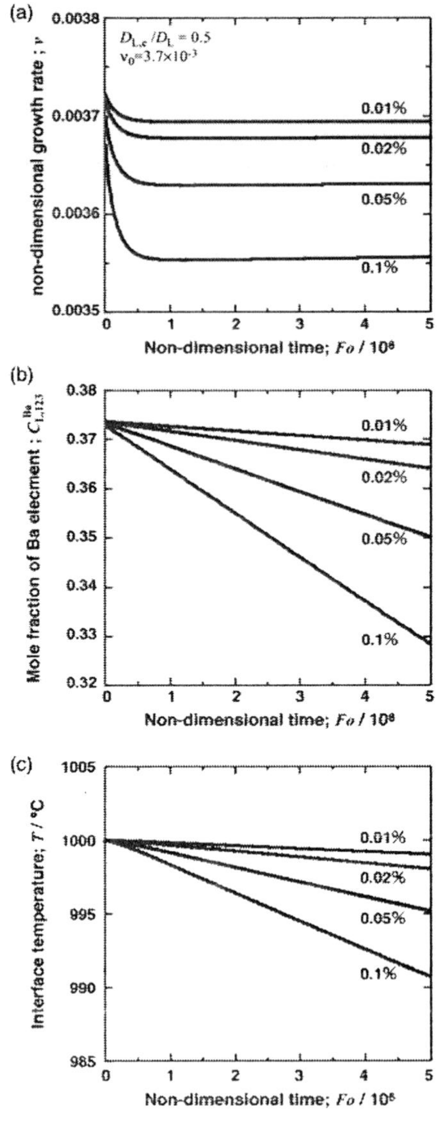


Fig. 22 Effect of the deviation of initial composition from the Y123-Y211 tie-line on the changes of (a) non-dimensional growth rate of Y123 crystal, (b) interface composition and (c) interface temperature during solidification calculated by non-steady state directional solidification model.

### 5.4.3 Calculation Results and the Effect of Growth Parameters

The relation among the growth conditions, compositions and interface temperature could be obtained numerically by calculating these equations (9) thorough (23) simultaneously. Figure 22 shows the effect of the deviation of initial average composition from the Y123-Y211 tie-line defined as $\left|C^{\mathrm{Ba}}_{\mathrm{L,211}} - C^{\mathrm{Ba,o}}_{\mathrm{L,211}}\right| / C^{\mathrm{Ba,o}}_{\mathrm{L,211}}$ on the growth rate of the crystal, the mole fraction of Ba element at the growth interface and the interface temperature under the condition of $\kappa = 0.5$ and $v_0 = 0.0037$. As shown in Fig. 22 (a), the growth rates of the crystal initially decrease and reach steady growth rates depending on the initial deviations. This steady growth rate is slower for the larger initial deviation condition. The compositions at the interface continuously shift to the Ba-poor side and the interface temperatures also decrease correspondingly. The rate of compositional shift and corresponding decrease in interface temperature becomes large as increasing the initial deviation of average composition. This corresponds to the large difference between the pulling rate and the actual growth rate of crystal under the large initial deviation. These calculation results suggest that the effect of compositional shift may not be negligible for even a small deviation about 0.1 % and that it is difficult to achieve the quasi-steady state growth during directional solidification. In this calculation, the effective diffusivity was assumed about the value of average volume fraction. If the reduction of effective diffusivity is the order of the minimum cross sectional area in the liquid, the κ value reduces to about 0.25 in the case of volume fraction of $V_{L,211} = 0.5$. Figure 23 shows the effect of the reduction of effective diffusivity on the change in interface temperature for several κ values under the condition of $v_0 = 0.011$. As shown in this figure, the reduction of effective diffusivity strongly enhances the decrease of interface temperature caused by the compositional shift. Figure 24 shows the effect of pulling rate on the change of interface temperature. This clearly shows that the composition at the growth interface shifts faster away form the initial value by increasing the pulling rate, and that the decrease in interface temperature is also enhanced correspondingly.

These results are explained by considering the flux balance near the interface shown in Fig. 21 (a). In the case of small effective diffusivity, the reduction of effective diffusivity directly results the decrease in diffusion flux, $\boldsymbol{J}_2$ from eq. (10), which is the flux away from the peritectic diffusion boundary. Since the driving force for the compositional shift due to the difference between $\boldsymbol{J}_1$ and $\boldsymbol{J}_3$ could be partially canceled by the flux, $\boldsymbol{J}_2$ shown in Fig. 21 (a), the reduction of flux directly leads the enhancement of compositional shift and corresponding decrease in interface temperature. In addition, the effect of pulling rate is considered similarly. In the case of high pulling rate, the solute flux necessary for Y123 growth, $\boldsymbol{J}_0$, should be large. Such the large value of necessary flux requires also the large value of flux $\boldsymbol{J}_1$ to maintain the growth rate of the Y123 crystal. Since these large fluxes act as the large driving force for compositional shift, the compositional shift and corresponding decrease in interface temperature will easily occur under the higher growth rate conditions. The higher growth rate and large deviation of initial average composition would cause the large extra undercooling in addition to the fundamental undercooling for the growth.

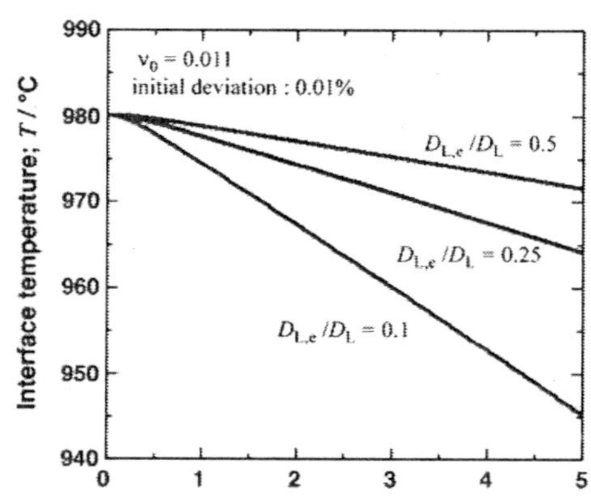


Fig. 23 Effect of reduction of effective diffusivity in bulk liquid on the change of interface temperature.

Although the precise information about the liquidus compositions of RE123 and RE211 as the ternary system is required to evaluate accurately the effect of the compositional shift during the solidification, the calculation results in this work suggest that the effect of compositional shift may not be negligible during solidification even under small compositional deviations from the Y211-Y123-035 tie-line. Consequently, the breakdown of continuous growth caused by the compositional shift will easily occur under the conditions of large deviation from the tie-line, large volume fraction of RE211 in liquid and high pulling rate.

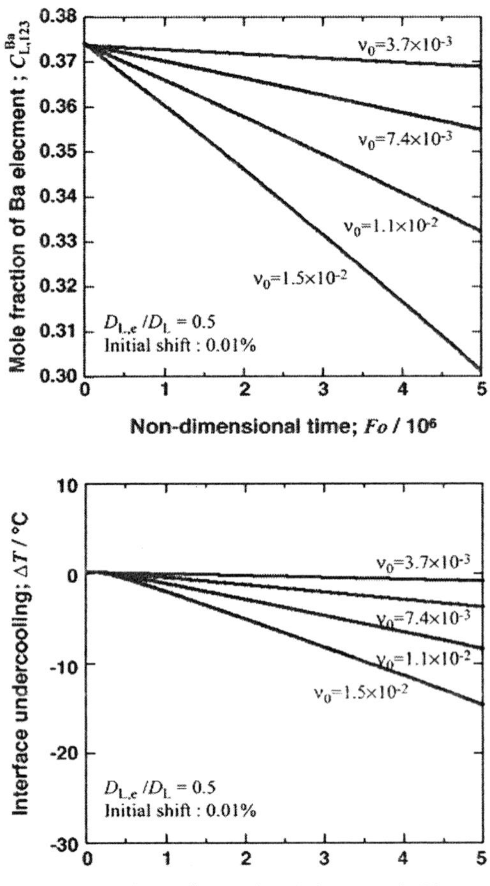


Fig. 24 Effect of non-dimensional pulling rate on the changes of (a) interface composition and (b) interface temperature during solidification.

# 6 Conclusion

The RE123 (RE=Y and Sm) samples with about 0.5 and 2 mm in diameter were directionally solidified by a zone melting method with various conditions, and the effect of fabrication conditions on the microstructure and transport $J_c$ properties of RE123 were investigated. The continuous growth structure of the 0.5 mmϕ Y123 fiber was achieved in the samples grown at high pulling rates up to 10 mm/h while the usual Y123 2 mmϕ rods could grow continuously up to a few mm/h. The size of Y211 particles trapped in Y123 was found to decrease with increasing the pulling rate. On the other hand, the Sm123 showed the continuous growth structure up to 15 mm/h for fiber samples and up to 10 mm/h for 2 mmϕ rods. The transport $J_c$ properties were evaluated for the continuously grown Y123 fibers and showed the maximum in the sample pulled at 6 mm/h; the best $I_c$ and corresponding $J_c$ values achieved were 134 A and $1\times10^5$ A/cm$^2$ at 77K self-field, respectively. This $J_c$ value is the best transport $J_c$ of the melt grown samples to date and is comparable to the best $J_c$ value of small Y123 bulks measured by the magnetization method. This high $J_c$ value may attribute to the preferential crystal orientation as well as fine Y211 particles of the sample fabricated at high pulling rate. That is, since the *ab*-plane of the Y123 crystal tends to align parallel with the longitudinal direction of the sample grown at high pulling rate, the superconducting current can flow in the ab-plane along the sample grown at high pulling rate. On the other hand, the enlargement of continuous growth condition for the fiber samples comparing to the 2 mmϕ rods could be fundamentally explained by the difference in effective nucleation rate determined by integrating the product of nucleation frequency and undercooled volume in the undercooled liquid region. In the directionally solidified sample, the nucleation frequency is almost the same under the same growth condition while the undercooled volume of the thin fibers is smaller than that of the rods. This small nucleation volume of thin fiber may reduce the nucleation rate and is an important factor to increase of the limit of continuous growth rate. This difference in nucleation rate is, however, found not to be enough to explain the destruction of continuous growth, and the additional factor such as compositional shift at the interface is needed. To think of a reason causing the compositional shift, a non-steady state growth model is proposed, and the effect of process parameters and growth conditions on the compositional shift is discussed. The large initial deviation of the average composition and high pulling rate condition may easily cause the compositional shift at the interface and corresponding decrease in interface temperature. This suggests that the breakdown of continuous growth will easily occur under such the conditions.

## Acknowledgements

This work was partially supported in The 21st Century COE Program "Intelligent Human Sensing" from the ministry of Education, Culture, Sports, Science and Technology, in part from Research Foundation for the Electrotechnology of Chubu (No. R-15114), and by a Grant-in-Aid for Scientific Research (No. 16560647) from JSPS

# References

[1] H. Ikuta; A. Mase; Y. Yanagi; M. Yoshikawa; Y. Itoh; T. Oka; U. Mizutani *Supercond. Sci. Technol* 1998 11 1345-1347.

[2] T. Oka; Y. Itoh; Y. Yanagi; M. Yoshikawa; H. Ikuta; U. Mizutani *Physica C* 2000 335 101-106.

[3] R. Wesche; A. M. Fuchs *Cryogenics* 1994 34 145-154.

[4] S. Marinel; D. Bourgault; O. Belmont; A. Sotelo; G. Desgardin *Physica C* 1999 315 205-214.

[5] L. Porcar; D. Bourgault; J. M. Barbut; M. Barrault; P. Germi; T. Tournier *Physica C* 1997 275 293-298.
[6] Y.R. Cai; T.C. Wang; B.Q. Fu; Y. Feng; P.X. Zhang; L. Zhou *Physica C* 2000 337 339-341.
[7] L. Martini; F. Barberis; R. Berti; G. Volpini; L. Bigoni; F. Curcio *Physica C* 2000 341-348 2513-2516.
[8] M. Putti; E. Bellingeri; C. Gerdeghini; W. Goldacker; P. Vase *Supercond. Sci. Technol.* 2001 14 L1-L3.
[9] R. Endoh; H. Kato; T. Izumi; Y. Shiohara *Physica C* 2003 392-396 1167-1170.
[10] M. Tomita; K. Nagashima; T. Herai; M. Murakami *Physica C* 2002 372-376 1216-1220.
[11] H. Teshima; M. Sawamura; H. Hirano *Physica C* 2002 378-381 827-832.
[12] T. Isono; K. Kawano; K. Hamada; K. Matsui; Y. Nunoya; E. Hara; T. Kato; T. Ando; K. Okuno; T. Bohno; A. Tomioka; Y. Sanuki; K. Sakaki; M. Konno; T. Uede *Physica C* 2003 392-396 1219-1224.
[13] Koichi Ohsemochi; Michitaka Ono; Shunji Nomura; Toru Kuriyama; Hirofumi Kasahara; Shirabe Akita; Seiichi Koso *Cryogenics* 2003 43 643-648.
[14] Haigun Lee; Ho Min Kim; Yukikazu Iwasa; Keeman Kim; Paul Arakawa; Greg Laughon *Cryogenics* 2004 44 7-14.
[15] T. Izumi; Y. Nakamura;Y. Shiohara *J. Mater. Res.* 1992 **7** 1621-1628.
[16] M. J. Cima; M. C. Flemings; A. M. Figuredo; M. Nakade; H. Ishii; H. D. Brody; J. S. Haggerty *J. Appl. Phys.* 1992 72 179-190.
[17] Yuichi Nakamura; Teruo Izumi; Yuh Shiohara; Shoji Tanaka J. *Japan Inst. Metals* 1992 56 810-818 (in Japanese).
[18] T. Izumi; Y. Nakamura; Y. Shiohara *J. Cryst. Growth* 1993 128 757-761.
[19] Y. Nakamura; Y. Shiohara *J. Mater. Res.* 1996 11 2450-2468.
[20] Y. Nakamura; A. Endo; Y. Shiohara *J. Mater. Res.* 1996 11 1094-1100..
[21] Endo; H. S. Chauhan; Y. Nakamura; Y. Shiohara *J. Mater. Res.* 1996 11 1114-1119;.
[22] M. Murakami; S. I. Yoo; T. Higuchi; N. Sakai; J. Weltz; N. Koshizuka; S. Tanaka *Jpn. J. Appl. Phys.* 1994 33 L715-L717.
[23] S. I. Yoo; M. Murakami; N. Sakai; T. Higuchi; S. Tanaka *Jpn. J. Appl. Phys.* 1994 33 L1000-L1003.
[24] M. Ikebe; H. Fujishiro; T. Naito; K. Noto; S. Kohayashi; S. Yoshizawa *Cryogenics* 1994 34 57-61.
[25] H. Fujishiro; S. Kohayashi *IEEE Trans. Appl. Supercond.* 2002 12 1124-1127.
[26] Y. Nakamura; K. Furuya; T. Izumi; Y. Shiohara *J. Mater. Res.* 1994 9 1350-1356.
[27] Y. Imagawa; Y. Shiohara *Physica C* 1996 262 243-248.
[28] Ch. Krauns; M. Sumida; M. Tagami; Y. Yamada; Y. Shiohara *Z. Phys. B* 1994 96 207-212.
[29] Endo; H. S. Chauhan; T. Egi; Y. Shiohara *J. Mater. Res.* 1996 **11** 795-803.
[30] D. Turnbull; J. C. Fisher *J. Chem. Phys.* 1949 **17** 71-73.
[31] Y. Nakamura; M. Kambara; T. Izumi; Y. Shiohara *Physica C* 2000 335 56-60.
[32] Y. Nakamura; M. Kambara; T. Izumi; Y. Shiohara *Physica C* 2000 341-348 2417-2420.
[33] Yuichi Nakamura; Tsuyoshi Kawase; Teruo Izumi; Kiyoshi Murata; Yuh Shiohara *J. Cryst. Growth* 2001 229 369-373.
[34] J. Lee; D. N. Lee *J. Am. Ceram. Soc.* 1991 74 78-84.

# INDEX

## D



## E



## F

## T